Quantum Physics, Retrocausation, PreCognition, Entanglement, Consciousness, Mental Time Travel

By
Deepak Chopra, Giovanni Berlucchi, Brandon Carter, Menas C. Kafatos, Subhash Kak, Chris King, John Smythies, Andrea Nani, Andrea E. Cavanna, Shaun Gallagher, Juan C. González, Steven Bodovitz, Franz Klaus Jansen, Michael C. Corballis, Liliann Manning, John T. Furey, Vincent J. Fortunato, Francois Martin, Federico Carminati, Giuliana Carminati, Lan Tao, R. G. Joseph

Edited by Deepak Chopra and Menas C. Kafatos

Quantum Physics, Retrocausation, PreCognition, Entanglement, Consciousness, Mental Time Travel

By
Deepak Chopra, Giovanni Berlucchi, Brandon Carter, Menas C. Kafatos, Subhash Kak, Chris King, John Smythies, Andrea Nani, Andrea E. Cavanna, Shaun Gallagher, Juan C. González, Steven Bodovitz, Franz Klaus Jansen, Michael C. Corballis, Liliann Manning, John T. Furey, Vincent J. Fortunato, Francois Martin, Federico Carminati, Giuliana Carminati, Lan Tao, R. G. Joseph

From an Edition Edited by Deepak Chopra and Menas C. Kafatos

Cosmology Science Publishers, Cambridge, 2015
Copyright © 2015, 2017

Published by: Cosmology Science Publishers, Cambridge, MA
Cover Design: R. Joseph, Ph.D.

All rights reserved. This book is protected by copyright. No part of this book may be reproduced in any form or by any means, including photocopying, or utilized in any information storage and retrieval system without permission of the copyright owner.

The publisher has sought to obtain permission from the copyright owners of all materials reproduced. If any copyright owner has been overlooked please contact: Cosmology Science Publishers at Editor@Cosmology.com, so that permission can be formally obtained.

Key Words: Quantum Physics, Relativity, Consciousness, Neuroscience, Brain, Mind, Time Travel, Mental Time Travel,

ISBN: 978-1-938024-48-1

Contents

I. Quantum Physics, Relativity, Retrocausation, Precognition, Multiple Dimensions, Entanglement, and Consciousness

1. Past Present Future Exist Simultaneously. Entanglement, Dream Time, PreCognition, Retrocausation, Deja Vu, and Premonitions   -7

2. Synchronicity, Entanglement, Quantum Information and the Psyche   -69

II. Uncertainty Principle, Multiple Worlds, Wave Functions, Entanglement, Violations of Causality, and Paradoxes of Time Travel

3. Classical Anthropic Everett Model: Indeterminacy in a Preordained Multiverse   -79

4. Quantum Paradoxes: The Uncertainty Principle, Wave Function, Probability, Entanglement, and Multiple Worlds   -89

III. Time Travel Through Black Holes

5. Time Travel Through Black Holes Holes in the Fabric of Space-Time   -119

IV: Time, Reality, Temporal Non-Locality Vendanta, Upanishads and Quantum Mechanics

6. The Nature of Reality, the Self, Time, Space and Experience — -151

7. Perceived Reality, Quantum Mechanics, and Consciousness — -158

V: Brain, Mind, Cosmology, Causality, Mental Time Travel

8. Space, Time and Consciousness — -175

9. Many Mansions: Special Relativity, Higher-Dimensional Space, Neuroscience, Consciousness and Time — -204

10. Brain, Consciousness, and Causality — -271

11. Time, Altered States of Consciousness, And Neuroscience — -230

12. Consciousness of Continuity of Now — -249

13. The Observer's Now, Past and Future in Physics from a Psycho-Biological Perspective — -256

14. Mental Time Travel: How The Mind Escapes From The Present — -274

15. How The Mind Escapes From The Present — -287

16. Mental Time Travel And The Self-Concept   -295

17. Continuity In Hippocampal Function As A Constraint On The Convergent Evolution Of Episodic-Like Cognition   -308

18. The Theory of MindTime   -313

I. Quantum Physics, Relativity, Retrocausation, Precognition, Multiple Dimensions, Entanglement, and Conscious Time

1. The Past Present Future Exist Simultaneously. Entanglement, Quantum Time, Dream Time, PreCognition, Retrocausation, Deja Vu, and Premonitions

R Gabriel Joseph

Cosmology.com / BrainMind.com

Abstract:

There is no "universal now." The distinctions between past present and future are illusions. As predicted by Einstein's field equations space-time may be a circle such that the future leads to the present and then the past which leads to the future, thereby creating multiple futures and pasts and which allows information from the future to effect the present. Causes may cause themselves. Coupled with evidence from entanglement where choices made in the future effect measurements made in the present and theoretical tachyons which travel at superluminal speeds from the future to the present and then the past, this may account for precognition, deja vu, and premonitions. In quantum mechanics, where reality and the quantum continuum are a unity, time is also a unity such that the future present past are a continuum which are linked and the same could be said of consciousness which exists in the future and in the present and past. If considered as a "world line" and in space-like instead of time-like intervals, then consciousness from birth to death would be linked as a basic unity extending not in time but in space and the same could be said of time. Time-space and consciousness are also linked and interact via the wave function and as demonstrated by entanglement and the Uncertainty Principle. Evidence from space-time contraction, atomic clocks and the twin paradox as functions of gravity and acceleration also demonstrate that the future already exists before it is experienced by consciousness in the present. Likewise, under conditions of accelerated consciousness (such as in reaction to terror) and dream states where various brain structures are in a heightened state of activity, space-time may also contract, such that time may slow down and consciousness may be given glimpses of the future in advance of other conscious minds thereby providing again for experiences such as precognition, premonitions, and deja vu. Closed time curves, conscious time, relative time, dream time, and quantum time are also discussed.

Keywords: Consciousness, retrocausation, Time Travel, Space-Time, Relativity, Quantum Physics, deja vu, premonitions, precognition, Length

Contraction, twin paradox, atomic clocks, dreaming, tachyons, entanglement, Uncertainty Principle, Wave function, Everett Multiple Worlds, rotating universe, closed time curves, violations of causality.

Relativity: The Future is the Past. The Past is the Future

Time is relative to the observer (Einstein 1961). Since there are innumerable observers, due to gravity, velocity and other variables, there is no universal "past, present, future" (Einstein 1905a,b, 1906, 1915a, 1961) all of which overlap and are infinite in number and yet interconnected and entangled in the basic oneness of the spacetime quantum continuum. It is the unity and relativity of time which makes time travel possible as well as some of the unique features of consciousness such as "premonition" and deja vu during which an observer experiences or is effected by the future before it becomes the present.

Because time is relative, and due to entanglement (Lee et al. 2011; Matson 2012; Olaf et al. 2003), the future can effect the past and may be experienced in the present depending on the observer's frame of reference. Consider the moon, the sun, and the stars up above. From a vantage point on Earth, the moon we see is actually the moon from 1 second ago; the moon we see in the present is from the past. The sun we observe is a sun from 8 seconds in the past. Upon gazing at the nearest star, Alpha Centauri, the star we see is from years ago since it takes 4.3 light years for its light image to reach Earth. If you stand in front of a full length mirror, just 3 feet away, and since light travels at 1 foot per nanosecond, you see yourself as you looked 6 nanoseconds ago. You are staring into the past. You are always younger in a mirror because what you see is the "you" from moments before. Mirrors are gateways to the past even though the past image you see is experienced in the "present." However, although from the past, the images are in the future until they arrive in the present relative to an observer. Before the image from the past arrives it is still in the future, relative to the observer. A stream of photons which just left the surface of the sun will not arrive on Earth until the future, 8 seconds from now. Until the splash of light arrives, it is in the future, relative to those on Earth but in the past relative to the sun.

If an alien observer living on a planet in Alpha Centauri was gazing at Earth, then the present on Alpha Centauri overlaps with the past on Earth. The alien sees an Earth from 4.3 light years ago. The reverse is true for an observer on Earth gazing at this distant star. However, until those images from the past reach the observer, they are in the future relative to that observer and overlap and exist simultaneously. The past can be the future and both may exist before they arrive in the present only to again become the past thereby creating a circle of time. Innumerable futures, presents, and pasts exist simultaneously albeit in different locations within space-time all of which are in motion. Observers located in New York, Shanghai, Tokyo, Paris, Mexico City, and in distant galaxies, are also in motion, as planets spin and orbit the sun, the sun orbits the galaxy, and galaxies move about in the universe. Observers, regardless of what planet, solar system, or in what galaxy they reside, are continually moving though space-time, often at different velocities and effected by varying degrees of gravity, and all are continually coming into contact with different times which are effect by velocity as well as the consciousness and emotional state of the observer.

Contraction of Time: The Future and Present Come Closer Together
As predicted by relativity and quantum mechanics, the experience of time and the distinctions between the past, future and the "present" are shaped and affected by distance, gravity, acceleration, consciousness, and our emotions, contracting and speeding up under conditions of pleasure and slowing down and sometimes splitting apart or even running backwards under conditions of fear and terror (Joseph 1996, 2010a). Acceleration contracts space-time thereby decreasing the distance between the future and the now, albeit depending on the location and frame of reference of the observers (Einstein 1905c, 1961; Einstein et al. 1923; Lorentz 1892, 1905). When time-space contracts, more time is squeezed into a smaller space such that it may take one consciousness less time to reach the future (vs the consciousness of a second observer) if that consciousness experiences time contraction. Two observers, with two different inertial frames of reference, may experience time as slowing down or speeding up. Consciousness can also accelerate and contract the space-time continuum, particularly during dream states, or under conditions of terror in which case, time may speed up or slow down and there may be a splitting of consciousness (Joseph 1996, 2010a). Likewise, emotions such as pleasure can also speed up the experience of time; a phenomenon observed by Einstein nearly 100 years ago:

"Put your hand on a hot stove for a minute, and it seems like an hour. Sit with a pretty girl for an hour, and it seems like a minute. THAT'S relativity." -Einstein

Time is a dimension, not in Euclidian space, but in "Minkowski space" (Minkowski 1909). Euclidian space consists of 4 spatial dimensions which include movement and geometric space; but none of which encompass time. By contrast, in "Minkowski space" which is incorporated within Einstein's special relativity, time is the 4th dimension (Einstein 1961). More specifically, 3 of the Euclidian dimensions of space are combined with a dimension of time thereby creating a four-dimensional manifold known as "space-time." Space-time, however, is effected by gravity and acceleration, and can shrink and contract as gravity and velocity increase and in response to alterations in cerebral activity and thus, consciousness. As demonstrated in quantum mechanics, consciousness and the act of perceptual (or mechanical) registration directly impacts the quantum continuum through interactions via the wave function and through entanglement.

The relationship between time dilation and the contraction of the length of space-time can be determined by a formula devised by Hendrik Lorentz in 1895. As specified by the Lorentz factor, γ (gamma) is given by the equation $\gamma = $, such that the dilation-contraction effect increases exponentially as the time traveler's velocity (v) approaches the speed of light c.

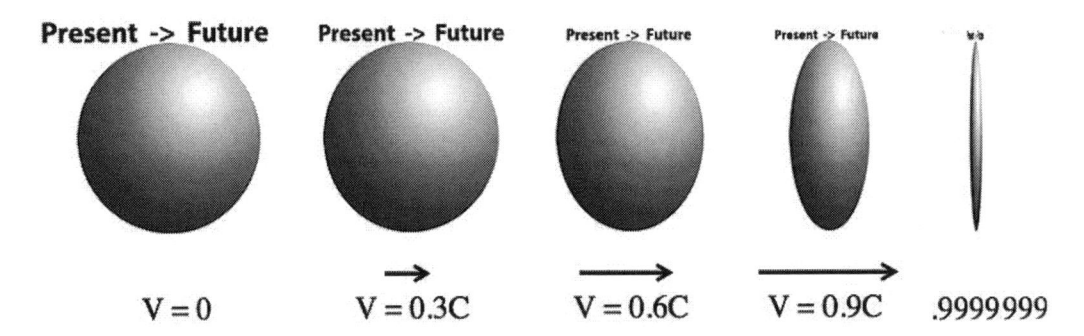

Present -> Future **Present -> Future** Present -> Future Present -> Future

V = 0   V = 0.3C   V = 0.6C   V = 0.9C   .9999999

When time-space contracts more time is compacted into smaller spaces and it takes less time to reach the future which is squeezed closer to the present, whereas from the perspective of a dissociated consciousness, time may appear to slow down, thus paralleling some of the paradoxes of time travel: the time traveling consciousness experiences the future more quickly, and in less time, thereby providing the foundations of deja vu, premonitions, anticipation, and the ability to make accurate predictions and to plan for the future.

Not just space-time, but the time machine, the time traveler, and any ticking clocks inside also shrink with increased gravity and velocity such that the passage of time inside the time machine shrinks relative to time and any observer located outside and is looking inside the time machine. For those inside the time machine, time appears to pass at the same rate inside, and this is because everything inside the time machine has shrunk to the same degree. According to Einstein (1961), an observer inside the moving object or traveling alongside at the same speed would not notice this contraction. It is only apparent to an outside observer with a separate frame of reference; and the same appears to be true of dissociated consciousness under conditions of terror (Joseph 1996, REF). Time is relative and it is only an outside observer at a safe distance from the time machine (or the accelerated consciousness) who will perceive the contraction of time-space surrounding the time traveler and that the time traveler's clock has slowed. By contrast, if the time traveler were to look outside the time machine it would seem that the outside observer's clock is ticking faster. Thus as predicted by Einstein, clocks run more slowly (time contraction) as velocity and acceleration increase and a time traveler in a time machine would appear to be slowing down from the perspective of an outside observer, whereas from the perspective of the time traveler the outside observer would be speeding up.

This concept is brilliantly anticipated by H.G. Wells in "The Time Machine." Well's Time Traveler and his time machine began the voyage through time in his laboratory and looking outside the time machine he could see celestial events, people, and even a snail whiz by: "The laboratory got hazy and went dark. Mrs.

Watchett came in and walked, apparently without seeing me, towards the garden door. I suppose it took her a minute or so to traverse the place, but to me she seemed to shoot across the room like a rocket." Well's Time Traveler also kept his eye on the laboratory clock and noted "a moment before it had stood at a minute or so past ten; now it was nearly half-past three!" A clock outside the time machine was therefore ticking away rapidly (relative to his clock inside the time machine), whereas his clock inside the time machine, from the perspective of Mrs. Watchett would run very slowly; exactly as predicted by Einstein's theories of relativity (Einstein 1914, 1915a,b, 1961). Although the laboratory clock sped up and Mrs. Watchet, the housekeeper, from the perspective of the Time Traveler, seemed to race across the room, from Mrs. Watchet's perspective the Time Traveler would appear to be frozen in time or moving exceedingly slowly. Likewise, victims subject to extreme terror may experience accelerated consciousness and from the perspective of an outside observer they may appear catatonic and frozen in time (Joseph 1996).

Twin Paradox

The shrinkage of space-time under accelerated conditions has given rise to the famous "twin paradox," a thought experiment based on special relativity (Langevin 1911; von Laue 1913. Because clocks inside the time machine run more slowly whereas the distance between the now and the future decreases, a time traveling twin will arrive more quickly in the future (since velocity does not shrink) and will age more slowly than her twin back on Earth. Consider for example, 30 feet of space which contracts to 10 feet. Those inside the time machine need only walk 10 feet whereas those outside the time machine must walk 30 feet. Likewise because the time traveler's clock runs more slowly, and since more time is contracted into a smaller space, it might take him 10 minutes to get 30 minutes into the future. Thus, the Earth-bound twin will be much older (as more time has passed) and may have already turned to dust if the time traveler arrived hundreds of years into the future. As summed up by Einstein (1911; see also Langevin 1911):

> "If we placed a living organism in a box ... one could arrange that the organism, after any arbitrary lengthy flight, could be returned to its original spot in a scarcely altered condition, while corresponding organisms which had remained in their original positions had already long since given way to new generations. For the moving organism, the lengthy time of the journey was a mere instant, provided the motion took place with approximately the speed of light."

If one twin leaves Earth and accelerates toward light speed, that twin will arrive in the future in less time than the twin left behind on Earth (since more time passed for that twin whose clock ran faster). By contrast, because it took less time for the time traveling twin she does not age as much (since her clock ticked slower) whereas the twin left on Earth ages at the normal rate. Hence, the time traveling twin will be younger: it took her less time (clock ticks slower) whereas the twin on Earth took more time (clock ticks faster) to reach the same destination in the future. The time traveling twin arrives in the future more quickly--and the same can be said of accelerated and dissociated consciousness thereby providing the foundations of deja vu and premonitions.

Because of time dilation and the contraction of space, once the time traveler lands on Earth, and depending on how fast and far into the future she is propelled, all her friends and relatives back on Earth may have died and a completely new generation of Earthlings may greet the time traveler upon her return. By contrast, since time slows down and time-space become squeezed together, the time traveler who arrives in the future may not have aged appreciably.

For example, say the time traveler is born in the year 2100, had a life expectancy of 80 years and would have died in the year 2180 if she had never left on her journey into the future. If she began her journey at age 20 in the year 2120, achieved 0.999999999999999 light speed and arrived in the future date of 2180, she would still have a life expectancy of 60 years (minus the 20 she already lived and time spent in the time machine). Upon arriving in the year 2180, she would still be 20 years old instead of 80 and could now expect to live another 60 years until the year 2240 (vs the year 2180 if she had never left home).

This premise is based on achieving near light speeds almost instantaneously and is supported by experiments with non-living, ultra-short-lived particles. For example, the muon particle is given a new lease on life when accelerated to a velocity of 99.92% light speed and its life span is nearly 25 times longer (Houellebecq 2001; Knecht, 2003). The muon particle not only lives longer but travels 25 times further due to its expanded life span. Particles, including phi mesons, which have been accelerated to velocities of 99.9% light speed also achieve significant life span extensions with a γ factor of around 5,000 (Houellebecq 2001). Presumably particles live longer because they arrived in the future more quickly vs their counterparts traveling at their normal, slower speeds. Therefore, it could be predicted that a time traveler who journeys at near light speeds will live longer compared to friends left back on Earth; and this is because the contraction of time space enabled them to reach the future before those back on Earth. Premonitions and deja vu, work on the same principles.

A time traveling consciousness will also experience time as speeding up or slowing down depending not just on acceleration and gravity, but emotion and neural activity. The distance between the "present" and the future decrease because of the shrinkage of space. In consequence the time traveling consciousness gains access to information in the future more quickly than other observers.

These same principles can be applied when traveling great distances across space to other stars and planets. If the journey takes place at near lights speeds, the space-time traveler may visit a distant star and then return home still fresh and young whereas her relatives and friends will have grown old and infirm and may have already died.

For example, if Gaia stays on Earth and her twin, Aurora travels at 80% the speed of light to Proxima Centauri which is 4.2 light years away, then Aurora's trip will take 5.25 Earth years (4.2/0.8 = 5.25). One day in the time machine at 80% light speed is equal to 1.67 days on Earth. Thus 1,916.25 days in the time machine (5.25 years) is equal to 3,200 days on Earth (8.76 years). Hence, Aurora's clock will tick 0.599% more slowly than Gaia's clock on Earth (5.25/8.76) and Aurora will age only 3.15 years during the journey (0.599 x 5.25 = 3.146) whereas Gaia will age 5.25 years. If Aurora immediately returns to Earth at 80% light speed she will be 6.3 years older and her twin will be 10.5 years older.

Atomic Clocks

Alterations in consciousness, gravity, and velocity can shrink or stretch space-time. Therefore, time-space is also warped, shrunk, stretched, and may even curl up and fold upon and over itself depending on local conditions. Time is asymmetric. Like the weather, time is not the same everywhere, even when measured by atomic clocks.

Atomic clocks tick off time as measured by the vibrations of light waves emitted by atoms of the element cesium and with accuracies of billionths of a second (Essen & Parry, 1955). However, these clocks are also effected by their surroundings and run slower under conditions of increased gravity or acceleration (Ashby 2003; Hafele & Keating 1972a,b). In 1971 Joe Hafele and Richard Keating placed atomic clocks on airplanes traveling in the same direction of Earth's rotation thereby combining the velocity of Earth with the velocity of the planes (Hafele & Keating 1972a,b). All clocks slowed on average by 59 nanoseconds compared to atomic clocks on Earth. These clocks arrived in the future in less than than their counterparts.

It has also been demonstrated that atomic clocks at differing altitudes will show different times; a function of gravitational effects on time. The lower the altitude the slower the clock, whereas clocks speed up as altitude increases; albeit the differences consisting of increases of only a few nanoseconds (Chou et al. 2010; Hafele & Keating, 1972; Vessot et al. 1980). "For example, if two identical clocks are separated vertically by 1 km above the surface of Earth, the higher clock gains the equivalent of 3 extra seconds for each million years (Chou et al., 2010). The speeding up of atomic clocks at increasingly higher altitudes has been attributed to a reduction in gravitational potential which contributes to differential gravitational time dilation.

Accelerated Consciousness: Dissociative Mind And The Slowing Of Time

Evidence from relativity, quantum mechanics, atomic clocks, and space-tome contraction, demonstrates that the future, or at least, "a" future must have existed so this future could contract closer to the present experienced by the time traveler;. This is also proved by experiments in entanglement (REF), and Einstein's field equations which demonstrate that time is a circle (REF) where the present leads to a future which already exists and that this future leads to the present and then the past. That is, instead of the present moving toward a future which does not yet exist, the future already exists and streams toward consciousness, becoming the present, and then continues into the past relative to that observing consciousness. However, consciousness may also accelerate as reflected by increased brain activity, and this too would contract space-time. Buried deep within the brain are a series of structures referred to collectively as the limbic system, and which includes the amygdala, hippocampus, and hypothalamus. The limbic system governs all aspects of sexual behavior and emotion, including emotional and non-emotional memory as well as anxiety, fear, and the ability to visualize one's self (Joseph, 1992, 2011). Limbic system structures, such as the amygdala are able to receive sensory information from multiple modalities at the same time and excessive activity in these areas, as reflected by increased EEG activity, are associated with the the reception of information which is normally inhibited and filtered, including the experience of deju vu and other precognitive phenomenon (Daly, 1958; Halgren 1990, Gloor, 1990; Joseph, 1996, 2011; Penfield, 1952; Penfield & Perot 1963; Williams, 1956), such as a splitting of consciousness where time slow down and the dissociated consciousness can observe itself as if up in the air looking down.

> "I had a clear image of myself... as though watching it on a television screen." "The next thing I knew...I was looking down from 50 to 100 feet in the air...I had a sensation of floating. It was almost like stepping out of reality. I seemed to step out of this world."

Electrode stimulation, or other forms of heightened activity within limbic system structures such as the amygdala, hippocampus and overlying temporal lobe can also cause time to speed up or slow down (Joseph, 1998, 1999b, 2001). Likewise, in response to extreme trauma, stress and fear, the amygdala, hippocampus and temporal lobe become hyper activated resulting not only in a "splitting of consciousness" but the sensation that time has slowed down while the dissociated consciousness seems to speed up (Courtois, 2009; Grinker & Spiegel, 1945; Noyes & Kletti, 1977; van der Kolk 1987).

One individual, after losing control of his Mustang convertible while during over 100 miles per hour on a rain soaked freeway, reported that:

> "time seemed to slow down and then... part of my mind was a few feet outside the car zooming above it and then beside it and behind it and in front of it, looking at and analyzing the respective positions of my spinning Mustang and the cars surrounding me. Simultaneously I was inside trying to steer and control it in accordance with the multiple perspectives I was given by that part of my mind that was outside. It was like my mind split and one

consciousness was inside the car, while the other was zooming all around outside and giving me visual feedback that enabled me to avoid hitting anyone or destroying my Mustang."

"Tiffany" describes her experience as follows:

"I was a passenger in my boyfriend's sports car and we were laughing and racing along highway 17, going 80, 90 mph. I remember he reached his cigarettes when we were going around a corner and then the car began to slide sideways toward the embankment and all the trees. Everything just suddenly slowed down, like in slow motion, and I could see the car sliding very slowly toward the trees, and I turned and looked at my boyfriend and he had this look of fear and determination on his face. He was gritting his teeth which were very white. I remember looking at his hands tightly gripped on the steering wheel, and I could see the ring I gave him. And outside the car there were other cars and they were also moving in slow motion. We were still sliding, and I turned my head and I could see we were going to slide right into this big tree, and everything was still so slow, and I could see the trunk and bark of the tree, the tree limbs, coming closer and I could see this bird flying out of the tree flapping its wings real slow, then we hit the tree with the back of the car which made the car spin around the tree, but it was all in slow motion, and all this glass blew out the side and back window and I could see little pieces of glass going everywhere moving very slowly through the air. I was wearing my seat belt and was spun toward my boyfriend but he wasn't there. Instead, the driver's side door was open and the car was turning upside down and I could see my purse falling upward and my wallet and phone and eyeliner and lipstick and a pencil were all falling out but going upward very slowly, like floating right in front of me, and I remember thinking that I hoped that pencil did not stick me in the eye, and then the air bags popped out and it was also going in slow motion billowing out toward me and I could see the trees down below because we were falling over and down the embankment and everything was upside down and going sideways and then the airbag hit me in the chest and time suddenly sped up and the car landed upside down and slid down the embankment and hit some trees."

The slowing down of time and the splitting of consciousness, creating twin consciousnesses, are not uncommon under conditions of terror. Terror can accelerate the mind and brain by releasing a cascade of "fight or flight" neurochemicals such as norepinephrine (Joseph 1992, 1994, 1996, 20011). When the brain and mind are accelerated under these conditions one aspect of consciousness may split off and observe itself and the body which houses it. Under certain accelerated conditions, time will also slow down for the dissociated consciousness which has split off and is observing (Joseph 1996, 2000); exactly as predicted by relativity and the twin paradox: acceleration slows time for one observer and speeds it up for the other (Einstein et al. 1927; Einstein 1961).

In fact, time slows down under conditions of terror and accelerated, hyper-brain activity to such a degree that seconds may last minutes, and minutes hours (Joseph 1996). Individuals may become completely motionless, almost catatonic from the perspective of outside observers and may fail to make any effort to save their lives or to respond to assistance such as attempting to evacuate a burning plane or sinking ship even though they have been uninjured (Courtois, 1995; Galliano et al., 1993; Miller, 1951; Nijenhuis et al., 1998). From the perspective of outside observers, those so afflicted appear to be frozen in time; which is exactly how a time traveler accelerating toward light speed would appear to those outside the time machine. Likewise, from the perspective of the dissociated consciousness time also appear to slow down and to contract which gives that dissociated consciousness more time in less space to observe its surroundings; and this too is predicted by relativity and the Lorentz transformations of length and space-time contraction. Under accelerated conditions space-time shrinks and more time is compacted into less space.

> "I began moving at a tremendous speed... and I was aware of trees rushing below me. I just thought of home and knew I was going there... I saw my husband sitting in his favorite armchair reading the newspaper. I saw my children running up and down the stairs... I was drawn back to the hospital, but I don't remember the trip; it seemed to happen instantaneously" (Eadie & Taylor 1992).

Thus time slows down for the consciousness attached to the body, but may speed up for the dissociated consciousness. However, this multiplicity of mind, although dissociated, is still entangled, and as such, time may speed up and slow down simultaneously.

In fact, under conditions of extreme fear and terror and accelerated consciousness, time slows to such a degree and corresponding movements becomes slowed to such a degree that to outside observers the person may appear to be dead: "Far down below I could see houses and towns and green land and streams... I was very happy now. I kept on going very fast... then I started back, going very fast...Then I was lying on my back in bed and the girl and her father and a doctor were looking at me in a queer way...I had been dead three days (they told him)...and they were getting ready to buy my coffin" (Neihardt 1989).

Consider the case of Lisa, a wild blonde 22 year old beauty, a passenger in a sport's car with the top down that struck a telephone pole:

> "It felt like a movie in slow motion and everything slowed down just before we crashed. I had a seat belt and my arms and hands stretched out in front on the dashboard... and the windshield shattered and I could see all these cracks forming in the glass in slow motion. Everything was slowed down and the windshield just broken in half and it fell toward me all in slow motion and cut off my arm....I could see it cutting the skin and droplets of blood and then

all this blood and my arm falling slowly to the floor of the car. Everything was so slow... and I got out of the car and my arm was spraying blood everywhere. I walked only a few feet and in slow motion fell down.... Then I was in the air, watching everything. Part of the time I was on the ground looking up and at the same time I was in the air looking down at me. I could see people getting out of their cars. They were all around me and I could see them while I was on the ground and at the same time I could see them like I was 50 feet in the air looking down at them. Then the ambulance came and they put a tourniquet on my arm and put me inside... and then everything started going real fast. I was outside the ambulance, like I was inside sticking out and my mind was racing up and down the streets, like I was running very fast alongside.... then I was in the hospital. But I was no longer part of my body. I was racing along, tripping out, bobbing up and down the halls, just checking everything out and everything was going very fast. Then I saw all these doctors and nurses working on this body of a girl. I peaked over their shoulders and then I realized it was me, that girl was me and I could see that my hair was all bloody and this bothered me. It needed to be washed. But I wasn't moving. I looked dead.... The doctors also thought I was dead.... and that's when I fell back into my body with this thump and I started moving and that's when the doctors and nurses realized I was still alive."

Of course, time can also slow down under conditions of extreme boredom. However, boredom does not induce a splitting of the mind or dissociated states of consciousness. Instead of more time in less space, there is less time in more space. The clock ticks more slowly for the consciousness of the observer and more slowly external to that observer. Conversely, under conditions of pleasure, time speeds up and passes more quickly as so eloquently summed up by Einstein.

Time is entangled with and relative to consciousness and the multiplicity of mind.

The Event Horizon of Consciousness: The Eternal Now

As a thought experiment Einstein imagined that if he flew away from a big clock in the town square precisely at 12 noon, and traveled at the speed of light, the clock would appear to stop and would remain 12 noon forever--and this is because Einstein would be traveling at the same rate of speed as the light coming from the clock, in tandem and in parallel with it. Time would also essentially stop for Einstein, for if he were looking at the light beams on either side of him, they would look like stationary waves of electromagnetic activity consisting of crests and valleys--and this is because he would be moving in tandem and relative to these light beams; like two trains traveling at exactly the same speed, side by side and the only view is of the other train. At light speed Einstein would be captured in an "eternal now" with the future on one side and the past on the other.

All observers in uniform motion (like two trains traveling side by side) view themselves as at rest (so long at they can only see the two trains). If traveling at the speed of light, a light from a flashlight held in that time traveler's hand will never escape from the flashlight. The light from the flashlight will be frozen in place, in an eternal now.

If a star, astronaut, or space-time machine were to approach a supermassive black hole at the center of this galaxy, they would accelerate toward the "event horizon" at light speed (Dieter 2012; McClintock, 2004)--the "event horizon" being the point of no-return, the vortex forming the mouth of the hole. The Time Traveler's clock would tick increasingly slower and light trailing behind would become redder (red shifted) as the event horizon is approached. However, for the time traveler, time continues as before.

Once caught by the gravitational grip of the vortex spinning round the event horizon, the star, astronaut, or space-time machine would have a velocity of light speed (Dieter 2012). Time stops. They would be captured and held in the grip of what could best be described as an "eternal now." Light could not escape, and the outside of the hole would appear black, whereas the event horizon would be blazing brightly illuminated with light.

Just as a star will accelerate toward light speed as it approaches the event horizon of a supermassive black hole in the fabric of space-time (Bethe et al. 2003; Dieter 2012; McClintock, 2004), the multiple futures flowing toward the event horizon of consciousness may also accelerate toward light speed. Once captured by the event horizon of consciousness these futures have a velocity of light speed becoming the "eternal now." Consciousness of the "present" could be likened to an event horizon illuminated with light. The present, the "eternal now" is the illumination of the event horizon of consciousness at light speed. On one side of the event horizon of "now" would be the future, and on the other, the past.

Predicting A Future Which Exists Before It Is Experienced

Relativity and quantum physics both predict the future exists before it is experienced. However, due to the fact that time is entangled in the frenzied activity of the quantum continuum, the future, or rather "a" future may continually change until the moment it is perceived by consciousness.

Since futures and pasts overlaps and as time-space is coextensive, then time, including local time relative to a single observer, is entangled. The past may effect the future, the future can effect the past, time effects consciousness and alterations in consciousness effect the passage of time.

As a "future" flows toward Earth it can also be effected by whatever it encounters on the way to the consciousness of "now," relative to an observer on Earth--exactly as befalls light. All futures are also entangled with space-time, the quantum continuum, and subject to the Uncertainty Principle. Therefore, future time may be continually altered until perhaps just moments before these futures are experienced by observers who are also entangled with what they experience.

Hence, although one may anticipate and predict the future, just like they may predict the weather, the ability to accurately anticipate and predict the future, like predicting future weather, may increase the closer that future is to the present. Planning skills, goal formation, strategy, long term investments, concern for consequences, and even the most basic of calendars, all rest upon the ability to make predictions about the future.

The future is like the weather, with the ability to forecast the weather decreasing in accuracy as time and distance from the present increases. In other words, and because of entanglement and classic concepts governing "cause and effect", the future is not already determined but is in flux and subject to continual alteration. The act of observing and other forces related to cause and affect alter the quantum continuum and continually change the future as it approaches. The future may not become fixed until the moment it is perceived by an observer relative to that observer, at which point it is in the present. Hence, predictions about the future will seldom be completely accurate, and become less accurate regarding increasingly distant events in the future, but more accurate but not completely accurate regarding events in the immediate future; a consequence of entanglement and the Uncertainty Principle.

Since the past is also relative and can exist in the future for some observers and in the present for others, and as the past is entangled with the quantum continuum, then the past is also subject to change after it has been experienced and before it is experienced by another observer at a downstream location in space-time. Two historians writing about history interpret and experience the past differently. A husband and wife discussing what happened at a party the night before, disagree. Eye-witness accounts differ among eye-witnesses (REF). A peasant living in a small village in western China in 1963 may have never heard of the assassination of president John F. Kennedy. The past is relative. There is no universal "past."

Time is entangled and is affected by consciousness and relative to and effected by the act of observation and measurement--as predicted by quantum mechanics (Bohr, 1958, 1963; Dirac, 1966a,b; Planck 1931, 1932, Heisenberg 1927, 1958; Neumann 1937, 1955).

Causes and Effects Are Relative To Consciousness

Every particle, person, planet, star, galaxy, has a wave function. The brain and consciousness have a wave function (Penrose & Hameroff 2011). Reality, including the reality of time, is a manifestation of wave functions and alterations in patterns of activity within the quantum continuum which are perceived as discontinuous (Bohr, 1958, 1963; Planck 1931; Heisenberg, 1958). This also gives rise to the perception of temporal order and what comes first, second, third, and what is in the present and in the past. The perception of temporal order, and structural units of information are not just perceived, but inserted into the quantum state which causes the reduction of the wave-packet and collapse of the wave function.

The brain and mind of a time traveler also has a wave function. As predicted by Einstein's field equations, consciousness can be accelerated into the future, and from the future, into the past. The Time traveler, upon observing his surroundings causes a collapse of the wave function, as predicted by the Copenhagen school of quantum physics: "The discontinuous change in the probability function takes place with the act of registration...in the mind of the observer" (Heisenberg, 1958).

The loss of coherence, the creation of discontinuous states in the quantum continuum is the result of entangled interactions within the environment which results in an exchange of energy and information: quantum entanglements. These entanglements, or blemishes in the quantum continuum, may be observed as shape, form, cause, effect, past, present, future, first, second, last, an so on, all of which are the result of a decoupling of quanta from the quantum (coherent) continuum which leaks out and then couples together in a form of knot which is observed as a wave form collapse. Every moment in time, is a wave form collapse of space-time at the moment of observation (Bohr, 1958, 1963; Heisenberg 1958; Von Neumann 1932, 1937).

However, in the Copenhagen model, the observer is external to the quantum state and is not part of the collapse function but a witness to it (Bohr, 1958, 1963; Heisenberg 1958). The observer is not the creator of reality but registers the transition of the possible to the actual: "The introduction of the observer must not be misunderstood to imply that some kind of subjective features are to be brought into the description of nature. The observer has, rather, only the function of registering decisions, i.e., processes in space and time, and it does not matter whether the observer is an apparatus or a human being; but the registration, i.e., the transition from the possible to the actual, is absolutely necessary here and cannot be omitted from the interpretation of quantum theory" (Heisenberg 1958).

As summed up by Von Neumann (1932), the "experiential increments in a person's knowledge" and "reductions of the quantum mechanical state of that person's brain" corresponds to the elimination of all those perceptual functions that are not necessary or irrelevant to the knowing of the event. Consciousness, therefore, could be viewed as a filter, which selectively attends to fragments of the quantum continuum which are perceived as real: the transition from the possible to the actual.

Therefore, according to the Copenhagen interpretation, the observing consciousness is external and separate from (albeit entangled with) what is observed and external to the ensuing collapse of the wave function which is collapsed by being measured and observed and this includes the observation and experience of time. However, consciousness can also be conscious of consciousness and thus consciousness can be subject to wave form collapse when observed by consciousness (Joseph 2011).

Consciousness and Mind

In a quantum universe all of existence, including consciousness, consists of a frenzy of subatomic activity which can be characterized as possessing pure potentiality and all of which are linked and entangled as a basic oneness which extends in all directions and encompasses all dimensions including time (Bohr, 1958, 1963; Dirac, 1966a,b; Planck 1931, 1932, Heisenberg 1955, 1958; von Neumann 1937, 1955). Hence, consciousness and the act of observation be it visual, auditory, tactile, mechanical, digital, are entangled with the quantum continuum and creates a static and series of impressions of just a fragment of that quantum frenzy that is registered in the mind of the observer as length, width, height, seconds, minutes, hours, days, weeks, months, first, second, third, and so on; like taking a series of pictures of continual motion and transformation and then believing it consists of temporal sequences when in fact, the conscious mind imposes temporal order (Joseph 1982, 1996, 2010). Just as, according to the quantum physics, the observing mind interacts with the quantum continuum and makes it possible to perceive shape and form, the conscious mind (and the dreaming mind) can perceive temporal sequences where there is none (Joseph 1982, 1986, 2010a); and those sequences include the illusion of future and past. That is, the act of sensory registration, be it a function of a single cell, or the conscious mind of a woman or man, selects a fragment of the infinite quantum possibilities and experiences it as real, and it is real but only to that mind or that cell at the moment of registration (Heisenberg 1955, 1958). Hence, "past present future" are a manifestation of consciousness which is entangled with time-space and the quantum continuum.

"I regard consciousness as fundamental. I regard matter as derivative from consciousness" (Max Planck, 1931).

As demonstrated by quantum mechanics and formalized by the Uncertainty Principle (Heisenberg 1925, 1927), what is known, is imprecise (Bohr, 1958, 1963; Dirac, 1966a,b; Planck 1931, 1932, Heisenberg 1955, 1958; Neumann 1937, 1955) and this includes time. To know something in its totality, would require a multi-dimensional all encompassing infinite "god's eye" view.

It could be said that consciousness is consciousness of something other than consciousness (Joseph 1982, 2011). Consciousness and knowledge of an object, such as a chair, are also distinct and separate. Consciousness is not the chair. The chair is not consciousness. The chair is an object of consciousness, and thus become discontinuous from the quantum state and entangled with consciousness.

Consciousness is consciousness of something and consciousness can be conscious of not being that object that it is conscious of. By knowing what it isn't, consciousness may know what is not, which helps define what is. This consciousness of not being the object can be considered the "collapse function" which results in discontinuity within the continuum: consciousness of consciousness being conscious.

Moreover, as demonstrated by neuroscience, the mind is not a singularity, but a multiplicity with different aspects of consciousness and awareness directly associated with specific regions of the brain (Joseph, 1992, 1996, 2011). These different mental realms and brain areas can perceive time and the quantum continuum differently. Time may be perceived by one brain region as lacking temporal order but as a continuum or gestalt. The mind is a multiplicity, which can become a duality, and which is often experienced as a singularity referred to as consciousness.

Further, it could be said that consciousness of consciousness, that is, self-consciousness, also imparts a duality, a separation, into the fabric of the quantum continuum. Hence, this consciousness that is the object of consciousness, becomes an abstraction, and may create a collapse function in the quantum continuum (Heisenberg, 1958; Joseph 2011; von Neurmann 1955, 2001). Consciousness may cause itself. That is, continuum which is consciousness, and which exists as entangled in the quantum continuum which includes time, may cause itself to experience the "eternal now" which is simply a collapse of the wave function of time.

Entanglement: The Future Causes the Past

Time, the fragmentation of time into temporal sequential units and where "causes" precede "effects," are also a "derivative of consciousness." If time is a feature of the quantum continuum, and if considered independent of "consciousness" then causes and effects may be one and the same, a unity and simultaneity, such that causes may cause themselves, or effects may be responsible for the causes.

Sometimes the association between and the classification of one event as coming first or second, or as a "cause" and the other an "effect" are also little more than illusion, as demonstrated by quantum entanglement and "spooky action at a distance" (Francis 2012; Lee et al. 2011; Matson 2012; Olaf et al. 2003; Plenio 2007; Juan et al. 2013). If time is a circle and due to entanglement, there are "effects" without any apparent "cause;" a possible consequence of the future effecting the present and the consciousness mind which experiences a premonition (Aharonov et al. 1988; Bem 2011; Radin 2006' Cho 2011). If time has a wave function and is an integral aspect of the quantum continuum which extends in all direction, not only would the future be linked to the past as a unity, but the conscious mind (with its own wave function) would be linked to the future and the past, thereby accounting for premonitions as well as anticipation of what is going to take place; and this is because what will take place has already taken place. The future is entangled with and causes the premonition. And just as likely, the premonition may cause the future due to entanglement.

However, typically, these "effects" (or premonitions) are written off as "mistakes" or due to "coincidence." Nevertheless, in an entangled universe the wave function of time representing the future can be predicted to interact with the wave function of the present (and vice versa) thereby inducing a causality-violating reduction of the wave form as perceived by the conscious mind.

If the future already exists, and if superluminal particles or information can arrive in the present from the future before they are perceived, this, coupled with entanglement (Plenio 2007; Juan et al. 2013; Francis 2012), may result in causes becoming confused with effects, whereas it is the future which is causing and effecting the present (Bem 2011; Radin 2006).

Consciousness is entangled with the space-time continuum which includes the future. Conscious observers can also engage in "mental time travel" (Suddendorf & Corballis 2007). Upon anticipating or looking into the future the observing consciousness can then engage in behaviors that are shaped and directed by that future. What constitutes a cause and what constitutes an effect, are relative and not uncommonly it is the anticipation of the future which causes the cause in the present.

A man buys a beautiful woman flowers, candy, jewelry, and an expensive dinner at a five star restaurant. He doesn't lavish these gifts upon the lucky maiden because he loves her, but because he is hoping she will reciprocate, after the date, by giving him sex. The expectation of sex in the future, and thus an event in the future, is the cause of his behavior in the present. The future is the cause which effects and causes his behavior in the present.

Before he bought her these gifts the man may have fantasized about the date, how he would take her to his home, what he would say, what he would do, how she would respond. This could be described as "mental time travel; rehearsing and practicing for a future event before it occurs. As demonstrated by Bem (2011), future practice can effect performance in the present before the practice occurs.

Time is also relative. Hence, when the beautiful woman received these gifts she decided to reward him. Therefore, relative to and from the perspective of the lucky maiden, the effect (sex) is a direct consequence of the cause (his gifts). On the other hand, she also knew that she could cause him to give her gifts by giving him sex in the future. Future sex caused his behavior.

Consciousness is also part of the quantum continuum and so too is the future, present, and past. Thus, consciousness, like gravity and electromagnetic waves, is relative and can affect distant objects and events, including, perhaps, those in the future and the past (Planck 1931, 1932). Moreover, all have a wave function, and time and consciousness are entangled. However, since consciousness is also entangled, then consciousness may also perceive a future event before it occur; a phenomenon known as "precognition"

Precognition: Experimental Proof

Precognition is a form of conscious cognitive awareness which involves the acquisition of future knowledge just prior to its occurrence. Premonitions are a form of presentiment or an emotional feeling that something may happen in the near future, but without conscious knowledge of exactly what it is that is going to happen. Both can be considered forms of quantum entanglement (Radin 2006; Bem, 2011) where some near future event exerts and makes an impression on consciousness before the event occurs even when there is absolutely no way the future event could be inferred as about to happen.

Various surveys have indicated that over 50% of adults have experienced premonitions or phenomenon which could be classified as precognition (Kennedy et al., 1994; Radin 2006). Moreover, numerous rigorous, scientifically controlled experiments and meta-analyses of these experiments have demonstrated statistically significant evidence for precognition and premonitions (Honorton & Ferrari 1989; Radin 2006). For example Honorton and Ferrari (1989) performed a meta-analysis of 309 forced-choice precognition experiments involving over 50,000 subjects, and which had been published in scientific journals between 1936 and 1997. They found a consistent, statistically significant hit rate, meaning that the results could not be due to chance.

As with deja vu, increased brain activity or arousal contributes to precognitive activity (Bem, 2011; Radin 1997, 2006; Spottiswoode & May, 2003). Presentiment effect has also been directly related to increased brain activity as demonstrated in fMRI experiments (Bierman & Scholte, 2002) and with other physiological indices of participants' emotional arousal in which case they become aroused before they see the stimulus (Radin 1997). For example, when participants viewed a series emotionally neutral or emotionally arousing pictures on a computer screen, strong emotional arousal occurred a few seconds before the picture appeared, even before the computer had selected which emotional picture was to be displayed (Radin 1997, 2006).

In 2011, a well respected scientist, Daryl Bem published extensive statistically significant evidence for the effects of future events on cognition and emotion, demonstrating that the effect is in the present whereas the cause can still be in the future. For example, Bem had subjects perform a memory test which required that each subject look at a long list of words and to remember as many as possible. After completing the memory test he had the subjects type various words from that list which were randomly selected. Subjects showed statistically superior memory for the words which they were later asked to type. That is, the practice effect was retrocausal. The practice which was to take place in the future (the typing of words they had already seen) improved their memory of those words before they typed them. Thus, rehearsing a set of words makes them easier to recall even when the rehearsal occurs in the future and after subjects recall the words.

In another set of experiments Bem (2011), allowed a computer to control the entire procedure which involved showing each subject "explicit erotic images." The instructions were as follows: "on each trial of the experiment, pictures of two curtains will appear on the screen side by side. One of them has a picture behind it; the other has a blank wall behind it. Your task is to click on the curtain that you feel has the picture behind it." Statistical analysis of the results demonstrated that based on "feelings" subjects picked the location of the pornographic image at well above chance (even though they couldn't see it), whereas the location of the non-erotic neutral pictures were chosen at the rate of chance, i.e. 49.8% of the time.

Bem (2011) performed nine rigorously controlled experiments involving over 1000 subjects involving erotic stimuli, the avoidance of negative stimuli, and retroactive priming effects on memory and recall. Eight of the nine experiments yielded statistically significant results, and thus evidence for precognition and premonition.

Criticism of Precognition Experimental Results: The Baseball Analogy

A common criticism regarding the validity of research on premonitions and precognition is: if it exists, why doesn't it happen all the time? Why doesn't everyone have these experiences?

Consider major league baseball. In 2013, Miguel Cabrera had a batting average of .348 which was the best of all major league players. Although he is the best hitter in major league baseball, he hit the ball less than 50% of the time when he was at bat, and was able to get a "base hit" less than 35% of the time. Out of 750 major league players, 726 of them got a base hit less than 30% of the time in 2013 during regular season play (http://espn.go.com/mlb/stats/batting). Given that these players had up to 5 opportunities to hit the ball each time at bat, and 3 opportunities to swing, it can be said that professional baseball players actually hit the ball less than 30% of the time. Bem (2011), Raden (1996, 2006) Bonorton and Ferrari (1989) and others have shown a precognition hit rate above 50%. But unlike major league baseball players, those displaying precognition get their hits before they see what is being thrown at them.

Precognition should be treated like all other measures of ability. We should not be surprised that there is variation (Carpenter 2004, 2005; Schmeidler, 1988). Indeed, the same complaints can be made about memory and past events: If it really happened, why does everyone remember it differently. Why do some people have a great memory and others are more forgetful? Why do different eye-witnesses remember the same event differently?

Even highly arousing and emotionally significant "flashbulb memories" are subject to considerable forgetting. For example, Neisser and Harsch (1992) had subjects fill out a questionnaire regarding where they were and how they heard about the Challenger space craft explosion soon after this national tragedy occurred in 1986. When these subjects were questioned again 32-34 months later, 75% could not recall filling out the questionnaire. Many of the subjects in fact had forgotten considerable detail regarding the Challenger explosion and where they were when the heard about it. According to Neisser and Harsch (1992), "As far as we can tell, the original memories are just gone."

Memory is poor. Batting averages are dismal. Should it be any surprise that premonitions and the experience of precognition is also variable?

The Quantum Physics of Premonition and Retrocausation

The phenomenon of premonition must be considered from the perspective of quantum physics not Newtonian physics or Einstein's theories of relativity. As summarized by John Stewart Bell in his 19964 ground breaking paper ("On the Einstein Podolsky Rosen paradox") "any physical theory that incorporates local realism, favoured by Einstein cannot reproduce all the predictions of quantum mechanical theory."

In 2006, the American Association for the Advancement of Science organized an interdisciplinary conference of research scientists and physicists to discuss evidence for retrocausation as related to quantum physics, the conclusions of which were published in 2006: "it seems untenable to assert that time-reverse causation (retrocausation) cannot occur, even though it temporarily runs counter to the macro-scopic arrow of time" (Sheehan, 2006, p vii).

As demonstrated by quantum physics and entanglement, the future may effect and even direct the past or the present. Consider again entanglement between photons. In delayed choice experiments, entanglement was demonstrated among photons even before there was a decision to make a choice regarding these photons, that is, before it was decided to do a measurement (Ma et al., 2012; Peres 2000). Entanglement has also been demonstrated among photons which do not yet exist, where the choice has not even been made to create or measure future photons. Nevertheless, decisions which will be made in the future effect the measurement of photons in the present (Megidish et al 2013). The same principles can be applied to precognition. Information in the future, information which does not yet exist in the present, can effect and is entangled with the consciousness which will directly perceive that information even before it arrives in the present.

The future, past, present, and consciousness are entangled within the quantum continuum. The future exists before it arrives and some people consciously perceive a future before it becomes the present; phenomenon which can be classified as evidence of entanglement and which are variably experienced as deja vu, premonitions, and precognition and which would only be possible if the future already exists, and if time is a circle.

The Circle of Time: The Future Leads to the Past

Einstein (1915a,b, 1961) theorized that time and space can be unified in the 4th dimension. Like the unification of mass and energy, space-time are two aspects of the same quantity, such that space can be converted into time, and time into space in the 4th dimension. Space-time and time, therefore, have energy, and can be experienced and perceived, and in this respect, time also shares characteristics with light and may have a particle wave duality.

A fundamental principle of physics is that a beam of light takes the shortest path between two points which is a straight line (Fermat's least time principle). However, light bends due to the influence of gravity (Einstein 1911), which means the path is not straight, but curved. Likewise, according to Einstein (1914, 1915a, 1961), space is curved. That curvature would not be a round circle, however,, but would have different geometric characteristics depending on and due to differences in gravity in various regions of the cosmos.

Einstein's curved universe could not be a perfect circle, as galaxy distribution is asymmetry and includes great "walls" of galaxies throughout the cosmos which have clustered together. It is this clustering, and these galactic walls which contribute to the unequal distribution of gravity, which causes space-time not just to curve, but to fold and curl up and to asymmetrically effect the flow of time.

Gravity is always strongest at the center of gravity where its most concentrated. Time-space is also pulled toward the center of gravity, which is why Einstein proposed his "Cosmological constant" a repulsive force which would prevent the universe from collapsing. Einstein later rejected his "cosmological constant" calling it "the biggest blunder" of his life, when in 1929 Hubble reported the universe was expanding. Einstein believed that if not for his "cosmological constant" he could have predicted an expanding universe. Instead, the prediction was made by Alexander Friedmann in papers published in 1922 and 1924.

However, as pointed out by Godel (1949a,b), Einstein's equations do not predict an expanding universe, but a rotating universe; a conclusion that Gamov (1946) also arrived at years before based in part of his observations of rotational patterns throughout the cosmos.

Earth orbits around the sun in a curve. This solar system has curvature and its motion follows a curving path as it orbits this galaxy. Likewise, space-time is curved and light and time follow that curvature (Einstein 1915a,b, 1961; Gödel 1949a,b). All is in motion and has velocity, but because of this curvature, one may travel in a circle and arrive where they began; and the same is true of time. Time is a circle (Gödel 1949a,b). The past leads to the future and the future can lead to the past.

If the entire universe is curved, as predicted by Einstein's theories, then just as traveling in a straight line on Earth will bring the traveler full circle to his starting point, the same could be applied to a curved universe as well as to the trajectory of light and time. Time, like time-space, has curvature; and just as a journey in a "straight" line will bring a voyager full circle around the globe, the same could be said of a journey across space-time. Time may be a circle; a cosmic clock which ticks at different speeds depending on gravity and the geometry of space-time relative to an observer's velocity and frame of reference. However, what this also implies is that a journey across time will bring the voyager full circle, such that the present leads to the future, and the future leads to present and then the past.

Because gravitational influences vary throughout the cosmos, then every infinitesimal region of space-time may have its own proper time relative to observers in different locations. The present on a distant galaxy, as conveyed by images of time-light, does not arrive on Earth until the future, such that the future and the past overlap in time-space. A logical corollary is that there is no universal "now" past or present, and that "absolute time" does not exist (Gödel 1949a,b).

Beginning in 1949, Kurt Gödel, in a series of papers based on Einstein's field equations of gravity, rejected the Newtonian conception of time and the belief that the "present" consists of infinite layers of 'now' coming into existence in continual successive and immediate sequences. According to Gödel (1949a,b, 1995) if space–time is curved, then the experience of time could be considered a consequence of that curvature.

As based on Einstein's field equations, Gödel (1949a,b) discovered that a particle traveling through space would circle round from the present to the future and then continue to circle around and meet itself in the past; and from the past that particle would circle round and meet itself in the future and from the future it would again travel round and meet itself in the past; an infinitely repeating pattern. Gödel argued, since space-time is curved, then the future and past may also be curved and circle round thereby completing the circle which then continues in an endless loop. Time is a circle.

According to Godel, because of the curvature of time and space it is possible to travel through time: "By making a round trip on a rocket ship in a sufficiently wide curve, it is possible in these worlds to travel into any region of the past, present, and future and back again."

Gödel's formulations also borrowed from George Gamow's (1946) conception of a universe which, like all astral bodies in space-time, is in orbital motion. As pointed out by Gamow, and Pythagoras 2000 years before, patterns repeat themselves in nature from the subatomic to entire galaxies (Joseph 2010c). Electrons orbit the nucleus of the atom. Planets orbit the sun. The sun is just one of billions of stars located throughout the spiral arms of the Milky Way Galaxy, and the entire galaxy is rotating. Perhaps the entire universe is also rotating and thus space-time is rotating, such that time is a circle and the future leads to the present, then the past, which leads back to the future.

Closed Time Curves In A Rotating Universe: The Future Leads to the Past

Gödel (1949a,b) explained that if the universe was rotating and space-time is curved, time should also be curved and curve back upon itself, forming infinitely repeating closed time-like curves (CTCs). Just as it is possible to circle the Earth and return to where one began, if time and time-space circles back on itself in Pythagorean endless loops thereby giving rise to CTCs, it would be possible to journey in a circle back to where one began; which means, one can travel into the future and into their own past.

In a "rotating universe" time is a circle where the future leads to the past and effects precede causes; the future can effect the present, and the past. However, the time-traveler journeying along such a loop does not experience a slowing of time as there is no contraction of space-time. Time would remain the same for the time-traveler and all those on Earth, as the time traveler is merely going in a circle.

A rotating universe and closed time-like curves violate the rules of causality. If time can circle back on itself, then the future can effect the past and the temporal discontinuity between past, present, and future is abolished. Hypothetically, and based on the concept of Karma, since cause and effect are abolished, if, as a child, you do something bad in the future when you become an adult, you may be punished for that future indiscretion while you are a still a child; "karma" in reverse.

Gödel (1949a,b) developed Gamow's (1946) concept of rotating universes as a thought experiment and as a logical extension of Einstein's field equations of gravity. However, the implications were so profound, and so contrary to the predictions of Newtonian physics, Einstein's concept of relativity, and what is now referred to as the "Standard Model" that the possibility of a rotating universe has been almost universally rejected (Buser et al. 2013). Even Gödel (1949a,b, 1995) who published his observations in the 1940s and 1950s, pointed out that there was a yet no evidence of red shifts in the distant regions of the cosmos which would support a rotation model.

Gamow (1946) who first proposed a model of a rotating universe blamed the lack of evidence on the insufficient power of the telescopes available to astronomers and physicists at that time and proposed that proof of rotating universes would have to wait until advanced telescopes became available.

As based on the observation of planets, stars, and the rotation and combined gravity of mass aggregations such as entire galaxies, Gamow (1946) thought it was only logical that the entire universe must also be rotating around some axial point in space. As pointed out by Gamow, "galaxies are found in the state of more or less rapid axial rotation" contrary to the Big Bang theory and in contradiction to the belief that galaxies formed following the condensation and angular momentum of the primordial matter. Gamow posed this question: since planets, stars, and galaxies are rotating then perhaps "all matter in the visible universe is in a state of general rotation around some centre located far beyond the reach of our telescopes?" As detailed elsewhere (Joseph REF) in the 1998, observations published by two separate teams inadvertently provided that evidence, as based on the red shifts of distant stars which had undergone supernova (Perlmutter et al., 1998; Schmidt, et al., 1998); i.e. the observable "Hubble Length Universe" appears to be in orbit around a universe-in-mass black hole (Joseph 2010REF).

Gamow (1946) based his rotating universes model on the rotation and angular momentum of galaxies which appear to orbit an axial point in space. Any rotating body, be it a galaxy, a merry-go-round, or the planets orbiting the sun, shows differential speeds of acceleration and velocity depending on how far away they are from the axial center of rotation. For example, in the inner galaxy, the rotation speed rises with the radius. By contrast, in the outer galaxy the rotation speed remains constant (Petrovskaya, 1994; Teerikorpi, 1989). The point closest to the axis rotates faster than points closer to the outer rim which rotate at a similar velocity. For example, Earth and our solar system, located on an outer arm of the Milky Way galaxy, orbit the supermassive black hole at the axial center of the galaxy, at a speed of approximately 155 miles/sec (250 km/sec) (or from 965,600 km/h, to 804,672 km/h), taking around 240 million years to complete an orbit. However, those stars closest to the axial galactic center, relative to the stars on the outer rims, are moving more rapidly and display accelerating velocities as they come closer to the central axis (Ghez et al., 2005; Petrovskaya, 1994; Teerikorpi, 1989). In fact, the speeds are so high they are beyond what would be predicted based on the universal law of gravitation (Schneider, 2006); observations which also led Gamow (1946) to question the Big Bang origins model and to propose that the universe may be in rotation.

In 2010, additional evidence, based on red shifts of exploding supernovas, appears to support Gamow's predictions. The entire Hubble Length (observable) universe, appears to be rotating around a universe-in-mass black holes, with those closest to the hole rotating at a faster rate than those future away; exactly as described for stars in the Milky Way galaxy which are closer vs further away from the black hole at the center of this galaxy (Joseph, 2010a,b). If pattern repeat, the the Hubble Length (observable galaxy) is just a spec of dust in an infinitely curved universe where time-space and thus time, are a circle, with the future leading to the present and then the past.

Multiple Earth's In the Circle of Time: Patterns Repeat

A pattern, be it recurring numbers, events, or objects, repeats itself in a predictable manner down to its essential elements (Ball 2009; Novak 2002; Wille 2010). The entire field of mathematics is the "Science of Patterns" and any sequence of numbers that may be described by a mathematical function has a pattern (Wille 2010). The pattern at the elementary level is the basis, model, or template which is repeated on a larger scale to generate larger objects or series of events all of which exhibit the same or similar underlying pattern. Hence, elementary particles have orbits, planets have orbits, stars have orbits, and it can be assumed that, collectively, galaxies have orbits which would mean the "known" Hubble length universe, is also in a rotational orbit as all share similar patterns (Joseph 2010b).

In Euclidean geometry, a pattern known as a translation involves movement of every point at a constant distance in a specified direction and the same can be said of rotation (Johnson 2007). The symmetry of the cosmos is based on the repetition of patterns found throughout nature, from sea shells to spiral galaxies (Joseph 2010b). For example, snail shells, sea shells, vortices, the cochlear nucleus of the inner ear, etc., show similar repeating patterns around an axial center or "eye." The patterns intrinsic to the shell of a snail are replicated repeatedly in nature and typify the structure of whirlpools, cyclones, hurricanes, the Milky Way galaxy and every spiral galaxy so far observed all of which rotate around an axial point (a "black hole") at their center.

Since rotating patterns repeat, as pointed out by Gamov (1946) and 2000 years earlier by Pythagoras, then the universe (including an expanding universe) and time-space would be part of this pattern. The entire universe, therefore, must orbit and rotate around an axial point, as predicted by Einstein's field equations (Gödel 1949a,b). Again, however, Einstein's equations do not predict a perfectly curved universe, but a lumpy universe with waves and crests which circles round and which would be pulled inward toward the center of gravity; that is, if the cosmos is considered as a collective single entity. If correct, then curvature of space-time would continue as a repeating pattern of curvature, curving forever inward and outward; which leads to Pythagoras and the "golden ratio" (Joseph 2010b).

Because of gravity, time-space can also be bent backwards in a circle, as happens with whirlpools and eddies along river banks where water flows in a circular motion. If the implications of Einstein's field equations are correct, then the river of time is bent round in a circle and it has no ending or beginning and may include pockets or vortexes of time which pop in and out of existence like vortexes and eddies along the river banks.

Einstein's time-space curvature coupled with gravity and the principle of repeating patterns, and the concept of close-time-like curves, raises the possibility that space-time is like a spiral staircase, so that when circling round at 360 degrees one does not end up in the same space or spot where they began, but above or below it (Buser et al. 2013; Gödel 1995). From the perspective of the Time Traveler, this spiraling circle can lead to the future or the past, or to multiple futures and pasts which may coexist, in parallel, side by side.

One feature of a CTC is that it opens the possibility of a world-line which is not connected to earlier times in this past, but to multiple possible pasts, and futures, which exist in parallel, above, or below, or alongside one another-- multiple spiraling staircases of time which lead to parallel worlds of time (Buser et al. 2013; Gödel 1949a,b); and each of which may have probable existences as predicted by Everett's "many worlds" interpretation of quantum physics. Therefore, if one were to travel in a circle across space-time, they may be taken to "a" future and then to "a" past, but not to "the" future or past--rather, to multiple possible futures and pasts some of which may exist side by side or one on top of the other--parallel times existing simultaneously.

A Gödel rotating universe implies duality, if not multiplicity; and the same is true of the bending of light. Because gravity increases the curvature of space-time and can bend and split images of light, as illustrated by galactic lensing (Renn et al., 1997; van der Wel et al. 2013), then light reflected from Earth may also be split apart as it curves through space. When these light-images of Earth cross paths with stellar objects of sufficient gravitational strength, these beams of light may be curved round in an 180 degree arc with Earth as its target. That is, light-images, or time-light, may be split apart and circle around numerous gigantic galaxies, and some of these light images will be reflected back toward Earth and become mirror images of Earth's past. Therefore, as we gaze at the various stars which twinkle in the darkness of night, some of those stars may be mirror images of Earth and our solar system from the long ago. Likewise, we may exist in the past relative to other observers on Earths which exist in the future--observers which are then looking upon an mirror image of Earth which exists in our "now" but which also exists in the past of those Earths in the future.

In a Godel spinning universe where time is a circle, the mirror would also be gazing back; meaning that this Earth could also be a mirror from the past of a future Earth; a reflection that those on a future Earth can look back upon.

The same can be said of consciousness which can anticipate the future, remember the past, and reflect upon itself.

Closed Time Light Curves: The Future Causes the Past

There are two conceptions of time-like curves, open or closed (Bonor & Steadman, 2005; Buser et al. 2013; Friedman et al. 1990). Open time-like curves follow an arrow of time straight into the future, and there is no return to the past unless one can exceed the speed of light.

Time-like curves which are closed, loop back in a circle; meaning that future events could affect past ones. A closed time-like curve (CTC) is a world line in a Lorentzian manifold, such that a particle or time traveler returns to its starting point (Buser et al. 2013). That is, because light is curved, then light can loop back on itself, and it would be possible for an object to move around this loop and return to the same place and time that it started. An object in such an orbit would repeatedly return to the same point in space-time.

The circle of space-time points forwards and backwards in time. If CTCs exist then it would be possible to travel backwards in time. The question becomes: how far back or forward in time? For example, it may be possible to follow the CTC in a negative direction and revisit the day President Abraham Lincoln was assassinated, the morning when Jesus was nailed to a cross, 65 million years ago when a giant asteroid struck this planet exterminating the last of the dinosaurs, and further back still to the Cambrian Explosion 500 million years ago when all manner of complex species with bones and brains appeared almost simultaneously in every ocean of Earth, and to the time when Earth was hellishly hot and populated by only microbes some 4.2 billion years ago.

Although seemingly paradoxical, Einstein's theories of relativity (despite his posting of a cosmic speed limit) predicts that the only way to travel into the past is to travel first to the future and then exceed the speed of light. Upon accelerating toward light speed, space-time contracts and the space-time traveler is propelled into the future. For example, because of the contraction of space-time, each day in the time-space machine at 90% light speed would propel the time traveler 2.29 days into the future of Earth.

Specifically, and if we accept that time and the speed of light are related, then if the time traveler journeys at 80% speed of light, then one day (1.197 days) from the perspective of the time traveler would be the equivalent of 2 days back on Earth. If he achieves 99% light speed, then 104 days in the time machine (time contraction) would be the equivalent of about 2 years on Earth. At 99.9% the speed of light, then 1 day (26 hours) in the time machine would be the equivalent of 6 years on Earth. If the time traveler wished to experience a future 2190 years distant she would have to spend one year in the time machine traveling at 99.999% light speed. At 99.999999% the speed of light, almost two years pass for every day in the time machine. At 99.99999999999 % of c for every day on board, nearly twenty thousand years pass back on Earth. However, upon reaching light speed, time stops and the contraction of space-time comes close to a zero point, i.e. smaller than a Planck Length. At 100.0000000001% light speed, contraction continues in a negative direction, and time runs backward. The time traveler has entered the mirror universe of the past. It is only upon accelerating beyond light speed, that time runs in reverse and the contraction of space-time

continues in a negative direction. One must accelerate toward the future to reach the past. Einstein's general theory of relativity predicts that the future leads to the past. Likewise, as shown by Gödel (1949a,b), Einstein's field equations predict that time is a circle; and this violates the laws of causality (Buser et al. 2013). Therefore, upon reaching superluminal speeds, the time traveler would be headed backwards in time from the future and would pass himself journeying from the past into the future. And the time traveler voyaging toward the future and then the past, would pass himself heading back from the future. And the time traveler heading from the future to the past, could therefore, theoretically alter the past, thereby giving rise to innumerable paradoxes (Joseph 2014).

If time is a circle, then effects cannot always be traced to an earlier cause, for the cause may occur in the future; and this is because, in a closed loop, the future can come before the past and can even catch up with itself in the past so that an an event can be "simultaneous" with its cause or occur before its cause. An event may be able to cause itself.

These are not just thought experiments. There is considerable evidence of what Einstein (1955) referred to as "spooky action at a distance" and what is known in quantum physics as "entanglement" (Plenio 2007; Juan et al. 2013; Francis

2012). It is well established that causes and effects can occur simultaneously and ever faster than light speed (Lee et al. 2011; Matson 2012; Olaf et al. 2003); a consequence of the connectedness of all things in the quantum continuum including time which flows in all directions and which can circle round such that events in the future effect events in the present and in the past. The circle of time predicts that the future, present, and past exist simultaneously as a unity, and that this interconnectedness can result in effects in the future causing themselves.

Time is a circle, and this too is predicted by Einstein's theories of relativity, where accelerating toward light speed takes the voyager to the future, and upon exceeding the speed of light, the traveler heads back from the future into the past and thus effect the present and the past, as demonstrated by experiments in entanglement.

Quantum Entanglement And Causality: The Future Effects the Past

The river of time is bent round in a circle and it may have no beginning and no end. Since space-time is curved, warped and littered with vortexes surrounding black holes and effected by the gravity of innumerable stellar objects, the river of time may also be split apart and bent backwards in a circle, with circles within circles, as happens with whirlpools and eddies along river banks. Likewise, the geometry of time may flow differently in various regions of the cosmos and split off into innumerable tributaries of time each with their own unique trajectory and velocity.

Because space is "isotropic" there is nothing in the law of physics indicating that a particular direction is preferred; down, up, sideways, backwards, its all the same. Why should space-time, or time, be any different? Since the past, present and future overlap and are relative to observers and differ according to location, gravity, and speed of movement, then as Einstein stated, the distinctions between the past present and future are an illusion. If time is a circle, then time is a unity and there is no future, past, or present, except from the perspective of an observer.

The laws of electromagnetism do not make a distinction between past and future (Pollack & Stump, 2001; Slater & Frank, 2011). And yet, although light waves travel in a direction, it is assumed that these waves are traveling from the present into the future, when in fact they are traveling into the past and the future and from the future and from the past relative to different observers on different worlds and even on the same planet.

A light wave from Earth takes 4.2 light years to reach Proxima. However, since it will not be received on Proxima for 4.2 light years it will not arrive until some future date on Proxima. The light wave from Earth is in the future relative to observers on Proxima, although it is from the past relative to those on Earth. Likewise, light-images which just left Proxima are from Proxima's past but will not arrive on Earth until some day in their future. The future and the past are relative. Moreover, once light-images from Proxima arrive on Earth, they continue into the past relative to those on Earth, but not relative to those on a planet 4.2 light years in the opposite direction from Earth and 8.4 light years from Proxima, in which case although they are from the past relative to Earth and Proxima. However, these same light waves will not be received until some future date for those denizens of that more distant alien world. This conception of time is entirely consistent with Einstein's theories of relativity and Maxwell's equations of electromagnetism.

The past and future exist simultaneously in different and overlapping locations in space. Since space is isotropic, then, theoretically, there are no roadblocks to prevent a time traveler from choosing a location at will and then speeding into the future or the past; just as he may decide to go up-river or down-river.

The past, present and future, however, are like the weather, and differ in distant locations. There is no universal "now" and there are innumerable pasts, presents and futures which increasingly diverge as distance from any particular observer increases. Time is relative and the same can be said of the future and the past which only remains approximately and generally similar relative to observers sharing the same local, or personal, frames of references. Only when frames of reference are shared locally can observers agree on what took place first and last and what is in the past and what is still in the future.

Entanglement: The Future Effects the Present and the Past
Light can travel to the future and from the past relative to the observer's frame of reference. However, light and time are not the same. The speed of light, and time, be it past or future, are not synonymous, though both may be affected by gravity (Carroll 2004; Einstein 1961). Moreover, just as light has a particle-wave duality and can physically interact with various substances, time also can be perceived and therefore must have a wave function if not a particle-wave duality.

Time-space is interactional, and can contract to near nothingness and then continue to contract in a negative direction such that the time traveler can journey into the past. Therefore, time, and time-space are embedded in the quantum continuum and can effect as well as be effected by other particle-waves even at great distances; a concept referred to as "entanglement." Time and space-time are entangled.

It is well established that particles respond to and can influence and affect distant particles at speeds faster than light. This "spooky action at a distance" has been attributed to "fields," "mediator particles," gravity, and "quantum entanglement" (Bokulich & Jaeger, 2010; Juan et al. 2013; Sonner 2013).

For example, it is believed that an electric "field" may mediate "electrostatic" interactions between electromagnetic charges and currents separated by great distances across space. However, these changes can also take place at faster than light speeds. Charged particles, for example, produce an electric field around them which creates a "force" that effects other charges even at a distance without any time lapse. Maxwell's theories and equations incorporate these electrostatic physical "fields" to account for all electromagnetic interactions including action at a distance.

Since mass can become energy and energy mass, the "field" is therefore a physical entity that contains energy and has momentum which can be transmitted across space. Therefore, "action at a distance" may be both distant and local, a consequence of the interactions of these charges within the force field they create. However, the problem is: the effects can be simultaneous, even at great distances, and occur faster than the speed of light (Plenio 2007; Juan et al. 2013; Francis 2012; Schrödinger & Dirac 1936), effecting electrons, photons, atoms, molecules and even diamonds separated by great distances instantaneously (Lee et al. 2011; Matson 2012; Olaf et al. 2003; Schrödinger & Born 1935). The effect may even precede the cause since it takes place faster than light.

For example, photons are easily manipulated and preserve their coherence for long times and can be entangled by projection measurements (Kwiat et al. 1995; Weinfurter 1994). A pump photon, for example, can split light into two lower- energy photons while preserving momentum and energy, and these photons remained maximally entangled although separated spatially (Goebel et al 2008; Pan et al. 1998); the measurement of one simultaneously effects the measurement of another although separated by vast distances. It has been repeatedly demonstrated that entanglement swapping protocols can entangle two remote photons without any interaction between them and even with a significant time-like separation (Ma et al., 2012; Megidish et al. 2013; Peres 2000). Another example, two particles which are far apart have "spin" and they may spin up or down. However, an observer who measures and verifies the spin of particle A will at the same time effect the spin of particle B, as verified by a second observer. Measuring particle A, effects particle B and changes its spin. Likewise observing the spin of B determines the spin of A. There is no temporal order as the spin of one effects the spin of the other simultaneously through the simple act of measurements. Even distant objects are entangled and have a symmetrical relationship and a constant conjunction (Bokulich & Jaeger, 2010; Plenio 2007; Sonner 2013). If considered as a unity with no separations in time and space, then to effect one point in time-space is to effect all points which are entangled; and one of those entangled connections is consciousness (Joseph 2010a). And this gives rise to the uncertainty principle (Heisenberg, 1927) as the laws of cause and effect are violated. Correlation is not causation and it can't always be said with certainty which is the cause and which is the effect and this is because the cosmos is entangled.

Moreover, the decisions and measurements made in the future can effect the present. In one set of experiments entanglement was demonstrated following a delayed choice and even before there was a decision to make a choice. Specifically, four photons were created and two were measured and which became entangled such that the measurement of one effected the other simultaneously. However, if at a later time a choice was then made to measure the remaining two photons, all four became entangled before it was decided to do a second measurement, before the choice was even made in the future (Ma et al., 2012; Peres 2000).

Entanglement can occur independent of and before the act of measurement and choices made in the future can effect the present. "The time at which quantum measurements are taken and their order, has no effect on the outcome of a quantum mechanical experiment" (Megidish et al. 2013). Moreover, "two photons that exist at separate times can be entangled" (Megidish et al. 2013). As detailed by Megidish et al (2013): "In the scenario we present here, measuring the last photon affects the physical description of the first photon in the past, before it has even been measured. Thus, the "spooky action" is steering the system's past. Another point of view...is that the measurement of the first photon is immediately steering the future physical description of the last photon. In this case, the action is on the future of a part of the system that has not yet been created."

Hence, entanglement between photons has been demonstrated even before the second photon even exists; "a manifestation of the non-locality of quantum mechanics not only in space, but also in time" (Megidish et al 2013). In other words, a photon may become entangled with another photon even before that photon is created, before it even exists. Even after the first photon ceases to exist and before the second photon is created, both become entangled even though there is no overlap in time. Photons that do not exist can effect photons which do exist and photons which no longer exist and photons which will exist (Megidish et al. 2013); and presumably the same applies to all particles, atoms, molecules (Wiegner, et al 2011).

The same principles can be applied to conscious phenomenon, including the experience of deja vu and premonitions; i.e. experiencing an event before it occurs. In fact, the same could be said of feelings such as "anxiety" about what may happen before it happens, or logical thought processes of predicting what will happen before it happens--all of which may be made possible not by anticipation but by the future effecting the present. Premonitions and entanglement also prove the future exists before it becomes the present.

Deja Vu

Entanglement commonly occurs at superluminal speeds (Francis 2012; Juan et al. 2013; Plenio 2007; Lee et al. 2011; Matson 2012; Olaf et al. 2003). However, if an entangled consciousness is effected by the passage of that superluminal information from the future into the present, this can give rise to retro-cognition (Bem 2011; Radin 2006); knowing something has happened or will happen before it happens.

As illustrated by light-images from distant stars which are from the past but which will arrive on Earth in the future, various "futures" exist prior to being experienced by various observers. If time has a wave function and is entangled with space-time and the quantum continuum, and as the brain and consciousness are part of that continuum (Heisenberg 1958; Planck 1931, 1932), then under certain circumstances a future may effect consciousness prior to being experienced by consciousness due to entanglement of their wave functions. However, as discussed, consciousness too may exist in the future which is coextensive with the consciousness existing in the now. Since entanglement takes place faster than light speed, the leading edge of a future experience may be registered in various conscious minds at superluminal speeds before the future actually arrives in the now. The experience of this "time echo" is not uncommon, and has been referred to as deja vu, pre-cognition, and premonitions.

Deja vu is the conscious experience of having experienced some events just moments before the events take place. For example, a man opens the front door, step outsides, drops his keys and then a dog barks and the phone rings, and then he again experiences himself opening the door dropping his keys and then hearing a dog bark and then the ringing of his phone; like a time echo. He thus has the experience that all this has happened before or that he has done this before it happens. He may even say: "I've done this before" and then a few nanoseconds later he experiences himself saying "I've done this before."

Deja vu has been attributed to a delay in the transfer of sensory experiences from one region of the brain to another which receives that information twice, or the transmission of the same experience to the same area of the brain by two different brain areas such that the information is received twice following a brief delay (Joseph 1996). Hence, someone may experience deja vu because two or more areas of the brain are receiving or processing the same message with a slight delay between them. For example, the right and left halves of the brain are interconnected by a massive rope of nerve fibers called the corpus callosum. Each half of the brain is capable of conscious experience (Joseph 1988a,b; 2010a). Usually information is shared between the cerebral hemispheres. However, if there is a delay in transferring these signals, then one or both halves of the brain may sense it has had this experience just moments before thereby giving a sense of familiarity (Joseph 1996).

Brain areas communicate via neurons, and neurons communicate with each other by sending signals over axons which are transmitted to and received by dendrites (at the synaptic junction) belonging to other neurons which in turn may transmit message via their axons at synaptic junctions to the dendrites of other neurons. Impulses between neurons travel at various speeds, ranging from 10 to 50 m / s (Joseph 1996) whereas the speed of light is 300,000 km/sec.

The experience of deja vu has been reported under conditions of altered and heightened brain activity (Bancaud et al., 1994; Gloor 1990; Joseph 1996). Moreover, deja vu has been reported in cases involving the ingestion of anti-viral flu vaccines, such as amantadine and phenylpropanolamine (Taiminen

& Jääskeläinen 2001) which increases brain activity by acting on dopamine receptors and increasing dopamine activity.

Heightened brain activity can be likened to an accelerated state of consciousness. Accelerated states are also associated with the contraction of space-time such that future arrives more quickly.

Deja vu, is also associated with heightened and accelerated activity in the inferior temporal lobe which houses the limbic striatum and amygdala, the later of which receives multi-modal sensory information and which normally filters out most of these sensations so the brain is not overwhelmed (Joseph 1996, 2011). Deja vu has been reported by patients when these areas of the brain have been activated due to direct electrode stimulation (Halgren 1990; Gloor 1990), drug ingestion (Taiminen & Jääskeläinen 2001) or seizure activity (Joseph 1996).

Therefore, when brain activity increases and neurons fire more rapidly and process more information, one of the consequences is Deja vu. In other words, just as a Time Traveler will come closer to the future as he accelerates toward light speed, when brain activity accelerates the future may also come closer such that the leading edge of a future event is experienced by this accelerated state of consciousness just before the event happens in the present.

Consciousness and Entanglement

As demonstrated in quantum physics, the act of observation, measurement, and registration of an event, can effect that event, causing a collapse of a the wave function (Dirac 1966a,b; Heisenberg 1955), thereby registering form, length, shape which emerges like a blemish on the face of the quantum continuum. Likewise, a Time Traveler or particle/object speeding toward and then faster than light and from the future into the past will affect the quantum continuum. By traveling into the future or the past, the Time Traveler will interact with and alter every local moment within the quantum continuum and thus the future or the past.

If the past or the future are not altered, this means that these dimensions of time are hardwired as part of the quantum continuum, that these events were already woven into the fabric of time and had always happened and always will happen and cannot be altered because they already happened, albeit in different distant locations of space-time which are linked as a unity within the quantum continuum.

If the future/past are not altered by voyaging through time then this is because the Time Traveler had already journeyed into the past and future before he journeyed into it. Likewise, a person not only exists in the present, but they will exist in the future. Thus, a future self would also be entangled with a past self and that self which exists in the present. Consciousness is thus entangled in time, and consciousness is entangled with its own consciousness which exists in the future, present, or past--such that a future self can effect a past self, including what the past self thinks and feels and anticipates.

The future already exists; a concept which is intrinsic to space-time relativity and Einstein's field equations. The future and the past exist in various overlapping locations in space-time which are in motion. And the same can be said of consciousness. Therefore, just as the end of a movie already exists as one begins watching the movie, then perhaps the same may be said of the river of time and consciousness as related to the future and the past. If this premise is correct, then one's consciousness also exists in the future and in the past.

World Lines, Causality, and Entanglement

Time is entangled. Future, past, present, are relative and overlap, and what is the future in one galaxy can be the past in another; all are entangled in the fabric of space-time and the quantum continuum. To get to the past, the Time Traveler must accelerate toward light speed into the future, and then, upon achieving superluminal velocities length contraction continues in a negative direction, time runs in reverse, and the destination becomes the past. Time is a circle and the future flows to the present, and the past and this is because time is entangled in the oneness of the quantum continuum.

If time is conceived as a spatial gestalt, an interconnected continuity of length, width, height, and extent but without temporal order, then what takes place in one location of space-time can effect what takes place in another, even if the distance is measured in miles, minutes, hours, light years, or as the future vs the past.

For example, in the Great Basin, White Mountains of California, there are "bristlecone pine" trees over 5000 years old, and which stand over 50 feet high. However, if the tree is measured in space-like intervals and not time-like intervals these same trees could be viewed as having a length of 5000 years. That is, if its "world line" is visualized as a thick strand of rope moving through space, that rope would begin with the seed and extend to the top of the tree.

The "world line" of the tree, encompasses it's entire history and although the tips of some branches and roots may have only recently grown, they are connected with the entire tree from the roots to the crown, and thus to the youngest and oldest parts of the tree. And what takes place in the roots can effect the twigs, branches, and crown of the tree, and the condition of the crown can effect the roots, branches and twigs. However, if viewed from a space-like intervals, the seeds of the tree and the 50 foot tree also becomes an interconnected continuity.

Likewise, if the orbit of Earth was viewed as a strand of rope, that rope would circle around the sun; for in fact, the movement of Earth is a continuity and is not separated into intervals which take place one after another like the ticking of a clock. If the genes of the first life forms to take root on Earth were viewed as a rope, then it would extend from the present to 4.2 billion years ago and perhaps even to the DNA of life forms whose bacterial ancestors journeyed here from the stars.

Consciousness, Neuroscience, Time Travel

If time is considered from the perspective of space-like intervals and not time-like intervals, then causality can be forward, backward, or simultaneous (Bonor & Steadman, 2005; Buser et al. 2013; Carroll 2004; Gödel 1995). The future and the past are entangled as a continuity in space-time and this means information can be transmitted from the past to the future, and from the future to the past simultaneously.

Consciousness, however, would also have a world line, which extends from the birth to the death of that consciousness. Consciousness is therefore entangled with itself, and could transmit information from itself to itself, even if that consciousness exists in the illusionary present, past, or future.

Tachyons: Messengers From The Future

It is well established that various particles have a velocity close to or at the speed of light (Houellebecq 2001). Many of these high speed particles were hypothetical until their existence was verified experimentally. Photons and electromagnetic waves travel at light speed whereas some particles, such as positrons, and hypothetical "tachyons" are, or were believed to travel faster than light (Bilaniuk & Sudarshan 1969; Chodos 2002; Feinberg, 1967; Feynman 1949; Sen 2002) whereas others, such as . Superluminal tachyons, however, if they exist, may have negative energy and negative mass (Chodos 2002; Feinberg, 1967) and this may be a requirement for traveling at superluminal speeds where time flows in reverse.

Electromagnetic waves are a fundamental quality of matter and are subject to the effects of gravity as exemplified by galactic lensing (Slater & Frank, 2011; van der Wel 2013). When the electromagnetic force is stripped of its particle, it has no mass, and this is what is believed to occur when particles, or time machines enter a black hole (Everett & Roman 2012); what emerges has no mass, and it may possses negative mass and negative energy and may journey at faster than light speeds.

Wheeler and Feynman argued in 1945 that electromagnetic waves emitted by an electron proceed into the future and the past (Wheeler & Feynman 1945, 1949). When these waves collide with waves in the future they send waves back in time and further into the future due to the collision. Those sent to the past can also collide with those in the past sending them again in opposite directions, into the future and further into the past. Depending on if these waves collide crest to trough their energy levels may double, but if they collide crest to crest (or trough to trough) they cancel each other out; a phenomenon referred to, respectively, as constructive and destructive interference.

Some hypothetical particles, such as the "tachyon" are believed to be time-independent and to travel faster than light speed (Bilaniuk & Sudarshan 1969; Chodos 2002; Feinberg, 1967; Sen 2002); meaning that these particles are constantly arriving in the present from the future and continue their high speed journey into the distant past.

In contrast to slower than light particles which have "time-like four-momentum" tachyons and other hypothetical superluminal particles have "space-like four-momentum." For example, if two events have a greater separation in time than in space, they have a time-like separation which is indicated by a negative (minus) sign. If the sum is positive, the two events have a space-like separation which is greater than their separation in time. If the result is 0, then the two events have a light-like separation and are connected only by a beam of light.

Tachyons are believed to have worldlines which are space-like and not time-like such that the temporal order of events would not be the same in all inertial frames (Bilaniuk & Sudarshan 1969; Chodos 2002; Feinberg, 1967; Gibbons, 2002; Sen 2002); meaning cause and effect would be reversed or abolished. Tachyons, because they travel from the future to the past, would violate the laws of causality.

The existence of a tachyon particle was first proposed by Gerald Feinberg in 1967. According to Feinberg's theories and calculations, a tachyon could be similar to a "quanta" of a quantum field but with negatively squared mass; that is, it would have no mass or anti-mass. Its energy, nevertheless, would be real. Objects traveling toward light speed gain energy and mass, only to implode and come to consist of negative energy and negative mass (Joseph 2014). Tachyons would have negative mass and negative energy, thus avoiding any violation of the laws of thermodynamics. If the tachyon increases it superluminal speed it loses energy as it journeys faster into the past. If it accelerates to 200% light speed its negative energy diminishes to zero, time stands still and its negative mass becomes smaller than a Planck Length due to length contraction. Upon accelerating to 200.00001 light speed, what had been a negative contraction implodes and it contracts in a positive direction and gains positive energy and positive mass and turns around and travels from the past back to the future.

Therefore, whereas objects are believed to reach "infinite velocity" upon attaining light speed, the tachyon can only reach infinite velocity upon accelerating to speeds twice that of the speed of light, at which point there would be another time reversal and the tachyon would journey from the past into the future.

Einstein's field theories predict the curvature of space-time, such that the universe and time circles back upon itself (Gödel 1949a,b). The future leads to the past; a realization which greatly troubled Einstein. The existence of particles which travel from the future to the past are a logical extension of Einstein's theories and field equations (Gödel 1949a,b). Time is a circle which may be orbited by positive and negatively charged particles. If correct, then these negative and positively charged particles would also create a neutral state of equilibrium (Feynman 2011; Pollack & Stump, 2001; Slater & Frank, 2011; Wheeler & Feynman 1949); much like the positively charged nucleus of an atom counters the negative charge of an electron--the amount of positive charge determining the number of electrons.

Electrons may also circle in and out of time and changing charges as they do so, with negatively charged electrons directed toward the past and positively charged electrons, referred to as "positrons" directed toward the future. John Wheeler proposed that all electrons and positrons (the antiparticle to the electron) have identical mass but opposite charges (see also Feynman 1949). According to Wheeler (2010; Wheeler & Feynman 1949), all electrons in the universe zig zag backward and forward in time, and when zigging backward it is an electron and when zagging forward it is a positron. And when zigging and zagging they interact as an electron-positron pair, moving in and out of the past and future. Richard Feynman (2011) incorporated these ideas in his formulations for quantum electrodynamics which earned him a Nobel Prize. However, these pairs are not necessarily being created or annihilated; though annihilation could be predicted if they were to come in contact. Rather, like the positive and negative charged tachyon, they chase each other in a circle of time.

If we were building an atom-of-time, then it could be proposed that positively charged positrons and negatively charged tachyons circle toward the past, maintaining an equilibrium of charges in the past, whereas negatively charged electrons and positively charged tachyons do the same in the future; with all four circling around each other, in and out of the future and the past and without violating the laws of conservation of energy and mass.

Therefore, if a negative energy negative mass tachyon and a positron traveling from the future to the past was able to circle round and go from the past back toward the future, the tachyon would become a positive energy anti-tachyon and the positron a negatively charged electron. If time is a circle, the positron/electron and tachyon/anti-tachyon may circle from the future to the past to the future and back again; as if time was composed of particles which orbit the nucleus of "eternal now." Positrons, electrons, and tachyons would therefore provide time with the energy and atomic structure to emerge from the quantum continuum and be perceived as something real.

The existence of "tachyon" like particles has been rejected because they would violate the laws of causality and Einstein's theory of special relativity (Aharonov et al. 1969). Feinberg (1967), however, determined that special relativity did not prohibit faster than light travel so long as the object had always maintained superluminal velocities and had never had a velocity below the speed of light. According to special relativity (but not general relativity or quantum mechanics) the acceleration of matter to beyond light speed could cause the energy of this mass to becomes infinite and the Lorentz transformations would then have no meaning. However, if superluminal velocities are the norm for these faster than light particles, then there would be no need to break the cosmic speed limit except at 200% light speed. By the same token, these particle would never be able to reduce velocity to below light speed. Therefore, if these and other particles are traveling beyond the velocity of light they may have always journeyed at superluminal speeds and never had a velocity below light speed.

Others have argued that particles with negative mass cannot travel faster than light and would have negative energy and become unstable and undergo condensation (Ahraonov et al. 1969). These arguments were countered by Chodos (1985) who proposed that neutrinos can behave like tachyons and travel at superluminal speeds. By violating Lorentz invariance, neutrinos and other particles would undergo Lorentz-violating oscillations and travel faster than light while maintaining high energy levels. However, over time superluminal neutrinos would also lose energy, probably as Cherenkov radiation (Bock 1998).

Although theoretical, the implications are that information can travel from the future to the past; thus making it possible to anticipate, or see the future before it becomes the present, thereby giving rise to premonitions and related conscious experiences, including perhaps, anxiety, or conversely, an uncanny ability to correctly plan future courses of action which almost always lead to success and follow a course or result in a specific outcome exactly as predicted.

Although there have been numerous proposals and arguments to and fro it appears that objects or particles would lose mass and energy upon reaching superluminal speeds and those which travel faster than light would have negative mass and negative energy. Further, it could be said that tachyon-like particles once they accelerate to superluminal velocities, or, if they have always journeyed at faster than light speeds, may be unable to slow down to a velocity less than the speed of light. It has also been theorized that tachyons must maintain a constant speed, for if the Tachyon were to accelerate and increase velocity it loses energy which becomes zero if the speed reaches infinite velocity; i.e. 200% light speed; a velocity which would trigger a time reversal with negative contraction imploding and continuing in a positive direction back toward the future from the past.

According to some theories, if a tachyon were to slow toward light speed, the energy of a tachyon would increase and would becomes infinite as its velocity equals the speed of light; and the same would be true at 200% the speed of light. This is a mirror image of what is theorized to occur when particles or objects reach light speed as they are also supposed to gain infinite mass and infinite energy. Thus particles which always travel above and those which always travel below light speed are mirror images of each other and may have the same barriers and non-traversable event horizon, with the past on one side and the future on the other.

In other words, the transformation from positive to negative energy/mass, may be the event horizon which separates the future from the past. Thus, relative to a conscious observer, the future may consist of positive energy, and the past becomes the negative.

The Future and Past Exist Simultaneously: Circle of Time

Consciousness, Neuroscience, **Time Travel**

As implied by the Lorentz transformation (Einstein et al 1923), a tachyon would always have negative energy. The Lorentz transformation indicates that the sign of a particle's energy is the same in all inertial frames, just as the sign of the temporal order of two points on the world line remain the same. All observers will see that the particle has positive or negative energy, though they may disagree on how much energy it has. However, if the particle has positive energy according to one observer, and negative according to another, then the observers, or the particles, are occupying different inertial frames (e.g. one in the present the other in the future/past; and this implies duality.

If tachyons or other objects did not have the same energy sign in all inertial frames, that is, if they were sometimes positive and at other instances negative, then perhaps they are looping in and out of the past and future, becoming positive when below the speed of light and negative above it as predicted by superluminal Lorentz transformations (Everett & Roman 2012). Because they would have positive energy when heading toward the future and negative energy when traveling into the past they could both exist even in the same inertial frame. The future and the past would exist simultaneously.

The dichotomy between positive vs negative energy and mass implies duality; the tachyon which voyages beyond light speed is the antithesis of the tachyon or time machine at a velocity below light speed. For example, the tachyon below light speed could be considered an anti-tachyon. The antiparticle of a tachyon would be a positive energy tachyon which is traveling forward in time. The negative energy tachyon would be coming from future heading into the past. As such, they would seem to be continually circling around each other from the perspective of an observer: one coming the other going in parallel continuously.

Negative energy and positive energy are repulsive and attractive. Therefore, if the future consists of positive energy and the past negative energy, and both consist of particles with positive vs negative energy/mass respectively, then the future and the past would be continually chasing and escaping from each other, with the positive energy tachyon showing attraction and the negative repulsion and with both maintaining the same distance from each other. Positive and negative tachyons, or a positive vs negative future and past, therefore, would create a circle of time.

For example, if a negative object and a positive object of the same size came into contact, the negative would be repelled away and the positive would accelerate toward--a push pull scenario which could result in the negative and positive objects circling round and round each other as they are attracted and repelled at the same time--like a very bad romantic relationship.

An object with negative energy falls down just like an object with positive energy. However, if a negative particle swerved near a planet, the gravitational effect would be repulsive and it would be pushed away. Negative mass is repelled by positive mass and vice versa. If both were negative, they would also be repelled and this is because the two minus signs (-m and -m) cancel each other out.

Positive energy would propel the negative energy object to accelerate in the direction it is already going. If the universe and time are curved and lead back to their starting point in a circle, then the positive (heading toward the future) and the negative (head from the future to the past) would also circle round each other, with the future leading to and following into the past. If the positive particle actually caught up with and bumped into the negative particle, such as might be expected at the event horizon, the positive would force the negative to speed up in the negative direction it is already going.

Moreover, although the the positive and the negative particles might maintain the same distance from each other, they would accelerate to greater and greater speeds--and this is because both have acceleration; despite the fact that this seemingly violates the conservation laws of momentum and energy which requires that they remain constant.

For example, the positively charged anti-tachyon would accelerate toward the future coming closer and closer to light speed, and upon crossing the event horizon separating future from past, would lose positive energy and attain negative mass and then accelerate backwards into the past at superluminal values. However, they would also chase one another, such that both increase in speed; the positively charged tachyon toward the velocity of light, and negatively charged tachyon to twice the speed of light; or, in a mirror universe where all is reversed, the negative would be forced to below the speed of light. That is, as the positive speeds up, the negative, going in a negative direction, might slow down, with both exchanging energy at the event horizon of "eternal now." Alternatively, the tachyon may accelerate until reaching twice the speed of light thereby losing negative energy and gaining positive energy as the contraction of time-space implodes and collapses in a positive direction.

At this juncture, we can only theorize and hypothesize: negative energy tachyons become positive energy anti-tachyons and positrons become electrons; and the circle of time continues to circle around with the positive chasing the negative which is chasing the positive, like the hands of a clock.

Coupled with Einstein's field equations which predict time is a circle, if tachyon-like particles exist, this would mean the future could effect and alter the present and the past. If true, then the past present and future would be in continual flux and undergoing constant change--which is exactly what might be expected if time is merely a perceived aspect of the quantum continuum (a function of wave form collapse)--such that even events which already occurred in the past may or may not have occurred. Time therefore becomes uncertain and what has or will take place can only be determined imprecisely by means of a probability distribution; all of which leads to the Many Worlds Interpretation of quantum physics. If true, this may explain why premonitions of future event are not always accurate; since that future may rapidly change, and may represent just one of many futures.

The Wave Function of Conciousness of Consciousness

Quantum mechanics, in theory, governs the behavior of all systems regardless of size (Bohr, 1934, 1947, 1958, 1963; Dirac 1966a,b; Heisenberg, 1930, 1955, 1958). Central to quantum mechanics is the wave function (Bohr, 1963; Heisenberg, 1958). All of existence has a wave function, including light. Every aspect of existence can be described as sharing particle-like properties and wave-like properties. The wave function is the particle spread out over space and describes all the various possible states of the particle. According to quantum theory the probability of findings a particle in time or space is determined by the probability wave which obeys the Schrodinger equation. Everything is reduced to probabilities. Moreover, these particle/waves and these probabilities are entangled.

Reality is a manifestation of wave functions and alterations in patterns of activity within the quantum continuum which are entangled and perceived as discontinuous, and that includes the perception of time: past, present, future, and consciousness. The perception of a structural unit of information is not just perceived, but is inserted into the quantum state which causes the reduction of the wave-packet and the collapse of the wave function. It is this collapse which describes shape, form, length, width, and future and past events and locations within space-time (Bohr, 1963; Heisenberg, 1958).

Consciousness can also reflect upon and become conscious of being conscious, and in so doing, creates a collapse of the wave function which is experienced as a dissociated consciousness observing itself; conditions which are not uncommon during accelerated states of brain-mind activity typical of terror and other emotional extremes. Consider the case of U.S. Army Specialist Bayne:

"I could see me... it was like looking at a mannequin laying there... I was burnt up and there was blood all over the place... I could see the Vietcong. I could see the guy pull my boots off. I could see the rest of them picking up various things... I was like a spectator... It was about four or five in the afternoon when our own troops came. I could hear and see them approaching... I could see me... It was obvious I was burnt up. I looked dead... they put me in a bag... transferred me to a truck and then to the morgue. And from that point, it was the embalming process....I was on that table and a guy was telling a couple of jokes about those USO girls... all I had on was bloody undershorts... he placed my leg out and made a slight incision and stopped... he checked my pulse and heartbeat again and I could see that too...It was about that point I just lost track of what was taking place.... [until much later] when the chaplain was in there saying everything was going to be all right.... I was no longer outside. I was part of it at this point" (Wilson, 1987).

One woman stated: "it was though I were two persons, one watching, and the other having this happen to me." Another patient stated "it was as though the patient were attending a familiar play and was both the actor and audience."

> "I was struck from behind...That's the last thing I remember until I was above the whole scene viewing the accident. I was very detached. Everything was very quiet. This was the amazing thing about it to me... I could see my shoe which was crushed under the car and I thought: Oh no. My new dress is ruined... I don't remember hearing anything. I don't remember anybody saying anything. I was just viewing things...like I floated up there..." (Sabom, 1982; p. 90).

In instances of dissociation, consciousness is also conscious of itself as a consciousness. The dissociated consciousness creates a collapse of the wave function which includes the body, its brain, consciousness, and the surrounding space-time continuum which includes time; time which may speed up or slow down. Similar phenomenon also occur when dreaming and can be attributed to a collapse of the wave function; consciousness creating itself by dissociating itself form the quantum continuum.

In quantum physics, the wave function describes all possible states of the particle and larger objects, thereby giving rise to probabilities, and this leads to the "Many Worlds" interpretation of quantum mechanics (Dewitt, 1971; Everett 1956, 1957). That is, since there are numerous if not infinite probable outcomes, each outcome and probable outcome represents a different "world" with some worlds being more probable than others.

For example, an electron may collide with and bounce to the left of a proton on one trial, then to the right on the next, and then at a different angle on the third trial, and another angle on the fourth and so on, even though conditions are identical. This gives rise to the Uncertainty Principle and this is why the rules of quantum mechanics are indeterministic and based on probabilities. The state of a system one moment cannot determine what will happen next. Instead, we have probabilities which are based on the wave function. The wave function describes all the various possible states of the particle (Bohr, 1963; Heisenberg, 1958).

Since the universe, as a collective, must also have a wave function, then this universal wave function would describe all the possible states of the universe and thus all possible universes, which means there must be multiple universes which exist simultaneously as probabilities (Dewitt, 1971; Everett 1956, 1957). And the same would be true of time. Why shouldn't time have a wave function?

The wave function of time means there are infinite futures, presents, pasts, with some more probable than others.

Everett's Many Worlds

As theorized by Hugh Everett the universal wave function is "the fundamental entity, obeying at all times a deterministic wave equation" (Everett 1956). Thus, the wave function is real and is independent of observation or other mental postulates (Everett 1957), though it is still subject to quantum entanglement.

In Everett's formulation, a measuring apparatus MA and an object system OS form a composite system, each of which prior to measurement exists in well-defined (but time-dependent) states. Measurement is regarded as causing MA and OS to interact. After OS interacts with MA, it is no longer possible to describe either system as an independent state. According to Everett (1956, 1957), the only meaningful descriptions of each system are relative states: for example the relative state of OS given the state of MA or the relative state of MA given the state of OS. As theorized by Hugh Everett what the observer sees, and the state of the object, become correlated by the act of measurement or observation; they are entangled.

However, Everett reasoned that since the wave function appears to have collapsed when observed then there is no need to actually assume that it had collapsed. Wave function collapse is, according to Everett, redundant. Thus there is no need to incorporate wave function collapse in quantum mechanics and he removed it from his theory while maintaining the wave function, which includes the probability wave.

According to Everett (1956) a "collapsed" object state and an associated observer who has observed the same collapsed outcome have become correlated by the act of measurement or observation; that is, what the observer perceives and the state of the object become entangled. The subsequent evolution of each pair of relative subject–object states proceeds with complete indifference as to the presence or absence of the other elements, as if wave function collapse has occurred. However, instead of a wave function collapse, a choice is made among many possible choices, such that among all possible probable outcomes, the outcome that occurs becomes reality.

Everett argued that the experimental apparatus should be treated quantum mechanically, and coupled with the wave function and the probable nature of reality, this led to the "many worlds" interpretation (Dewitt, 1971). What is being measured and the measuring apparatus/observer are in two different states, i.e. different "worlds." Thus, when a measurement (observation) is made, the world branches out into a separate world for each possible outcome according to their probabilities of occurring. All probable outcomes exist regardless of how probable or improbable, and each outcome represent a "world." In each world, the measuring apparatus indicates which of the outcomes occurred, which probable world becomes reality for that observer; and this has the consequence that later observations are always consistent with the earlier observations (Dewitt, 1971; Everett 1956, 1957).

Predictions, therefore, are based on calculations of the probability that the observer will find themselves in one world or another. Once the observer enters the other world he is not aware of the other worlds which exist in parallel. Moreover, if he changes worlds, he will no longer be aware that the other world existed

(Everett 1956, 1957): all observations become consistent, and that includes even memory of the past which existed in the other world.

The "many worlds" interpretation (as formulated by Bryce DeWitt and Hugh Everett), rejects the collapse of the wave function and instead embraces a universal wave function which represents an overall objective reality which consists of all possible futures and histories all of which are real and which exist as alternate realities or in multiple universes. What separates these many worlds is quantum decoherence and not a wave form collapse. Reality, the future, and the past, are viewed as having multiple branches, an infinite number of highways leading to infinite outcomes. Thus the world is both deterministic and non-deterministic (as represented by chaos or random radioactive decay) and there are innumerable futures and pasts.

As described by DeWitt and Graham (1973; Dewitt, 1971), "This reality, which is described jointly by the dynamical variables and the state vector, is not the reality we customarily think of, but is a reality composed of many worlds. By virtue of the temporal development of the dynamical variables the state vector decomposes naturally into orthogonal vectors, reflecting a continual splitting of the universe into a multitude of mutually unobservable but equally real worlds, in each of which every good measurement has yielded a definite result and in most of which the familiar statistical quantum laws hold."

DeWitt's many-worlds interpretation of Everett's work, posits that there may be a split in the combined observer–object system, the observation causing the splitting, and each split corresponding to the different or multiple possible outcomes of an observation. Each split is a separate branch or highway. A "world" refers to a single branch and includes the complete measurement history of an observer regarding that single branch, which is a world unto itself. However, every observation and interaction can cause a splitting or branching such that the combined observer–object's wave function changes into two or more non-interacting branches which may split into many "worlds" depending on which is more probable. The splitting of worlds can continue infinitely.

Since there are innumerable observation-like events which are constantly happening, there are an enormous number of simultaneously existing states, or worlds, all of which exist in parallel but which may become entangled; and this means, they can not be independent of each other and are relative to each other. This notion is fundamental to the concept of quantum computing.

Likewise, in Everett's formulation, these branches are not completely separate but are subject to quantum interference and entanglement such that they may merge instead of splitting apart thereby creating one reality.

Many Worlds of Quantum Dream-Time

When considered as a unity within the quantum continuum, time and consciousness exist in the future, past, present, simultaneously. Consciousness which exists in the future is entangled with consciousness which exists in the present and the past--like a rope of string stretched out and extending in all directions from the birth to death of that consciousness. Therefore, just as the hypothetical tachyons can travel from the future to the past, and the circle of time circles round from the present to the future and back again, information may also be conveyed from the future to the past, perhaps along the rope of consciousness which extends in all directions and dimensions and is an aspect of the quantum continuum.

Acceleration leads to a compression of time-space, such that the future comes closer to the present. The same can be said of accelerated consciousness, thereby giving rise to phenomenon such as deja vu, premonitions, precognition, as well as anticipation and prediction about the immediate future. Dream time is a form of accelerated consciousness; i.e. dream consciousness.

Consciousness and dreaming are not synonymous. Dreams may be observed by consciousness and as such, dreaming and consciousness are entangled as dream-consciousness. However, consciousness is generally little more than a passive witness during dreaming, an audience before the stage upon which the dreams are displayed in all their mystery and majestic glory. It is rare for consciousness to become conscious that "it" is observing a dream, and when such rarities occur the dreamer may awaken or briefly take an active role in what has been described as "lucid dreaming" (LaBerge, 1990).

Unlike conscious-time and the conscious mind, the dream-kaleidescape of dream-time and dream-consciousness could best be described as manifestation of the "Many worlds" interpretation of quantum physics where all worlds are possible and past and future and time and space are juxtaposed and intermingled; time can run backward and forward simultaneously and at varying speeds, and multiple realities come and go no matter now improbable.

During dream-time the brain is in a "paradoxical" state of accelerated activity, known as paradoxical sleep, as demonstrated by rapid eye movement (REM) and electrophysiological activity (Frank, 2012, Pagel, 2014, Stickgod & Walker, 2010). As predicted by Einstein's (1961) relativity, under accelerated states time contracts and the future arrives more quickly. Dream-time represents accelerated states of brain activity and is entangled with the "many worlds" and the space-time quantum continuum of future and past, and as such, while dreaming, the dreamer may obtain a glimpse of the future before it arrives. Therefore, in dream-time and dream consciousness one may visit the future or the past during the course of the dream.

It is through dreams that we may be transported to worlds that defy the laws of physics and which obey their own laws of time, space, motion and conscious reality, where the future is juxtaposed with the past and where time runs backwards and forwards (Campbell, 1988; Freud, 1900; Jung, 1945, 1964). Throughout history it has been believed that dreams open doors to alternate realities, to the future, to the past, and the hereafter, where the spiritual world sits at the boundaries of the physical; hence the tendency to bury the dead in a sleeping position even 100,000 years ago (Joseph 2011a,b). Although but a dream, the dream is experienced during dream consciousness much as the waking world is experienced by waking consciousness. The dream is real. Thus, throughout history dreams have been taken seriously especially when they gave glimpses of the future.

Dreams are often of events from the previous day and may concern the future. It is through dreams that the dreamer may gain insight into problems which have plagued him or which he anticipates encountering in the near future. Just as one can think about the future or the past and make certain deductions and predictions, a dream may include anticipations regarding the future, and in this respect, the dream could be considered an imaginal means of preparation for various possible realities. As such, dream-time and dream-consciousness could be considered obvious manifestations of the "Many Worlds" theory of quantum physics.

Not uncommonly the dream will include so many branching and overlapping multiple realities that it makes no sense at all, except to those skilled in the art of interpreting dream symbolism (Freud 1900; Jung 1945, 1964). Indeed, it is due to the non-temporal, often gestalt nature of dreams which require that they be consciously scrutinized from multiple angles in order to discern their meaning, for the last may be first and what is missing may be just as significant as what is there.

Relativity predicts that observers with an accelerated frame of reference experience time-contraction and a shrinking of time-space such that the future and the present come closer together relative to those with a different frame of reference (Einstein et al. 1923, Einstein 1961). Thus, since dream-consciousness and dream-time are also associated with accelerated levels of brain activity, during dream-time, the dreamer may see or experience the future before that future is experienced by the awake conscious mind or the consciousness of those external observers who have a different frame of reference as regard to the contraction of time.

Abraham Lincoln Dreams Of His Death

In April of 1965, less than two weeks before he was gunned down by an assassin's bullet, President Abraham Lincoln dreamed of his own assassination (Lamon 1911). Lincoln told this dream to his wife and to several friends including Ward Hill Lamon who was Lincoln's personal friend, body guard and former law partner. According to Lincoln:

"About ten days ago, I retired very late. I had been up waiting for important dispatches from the front. I could not have been long in bed when I fell into a slumber, for I was weary. I soon began to dream. There seemed to be a death-like stillness about me. Then I heard subdued sobs, as if a number of people were weeping. I thought I left my bed and wandered downstairs. There the silence was broken by the same pitiful sobbing, but the mourners were invisible. I went from room to room; no living person was in sight, but the same mournful sounds of distress met me as I passed along. I saw light in all the rooms; every object was familiar to me; but where were all the people who were grieving as if their hearts would break? I was puzzled and alarmed. What could be the meaning of all this? Determined to find the cause of a state of things so mysterious and so shocking, I kept on until I arrived at the East Room, which I entered. There I met with a sickening surprise. Before me was a catafalque, on which rested a corpse wrapped in funeral vestments. Around it were stationed soldiers who were acting as guards; and there was a throng of people, gazing mournfully upon the corpse, whose face was covered, others weeping pitifully. 'Who is dead in the White House?' I demanded of one of the soldiers, 'The President,' was his answer; 'he was killed by an assassin.' Then came a loud burst of grief from the crowd, which woke me from my dream. I slept no more that night; and although it was only a dream, I have been strangely annoyed by it ever since."

Dream-Time and the Many Worlds of Quantum Physics

In dream-time past-present-future and the three dimensions of space may exist simultaneously as a gestalt thereby violating all the rules of causality abut not the laws of quantum physics. During dream-time events may occur in a logical or semi-logical temporal sequence, or they may be juxtaposed and make no sense at all to an external consciousness which is dependent on temporal sequences to achieve understanding. Because the future past present may exist simultaneously and as the future may be experienced in a dream during accelerated states of brain activity, then during dream consciousness the dreamer may get glimpses of future events which may occur within days, the next morning, or which may even trigger wakefulness. In other words, just as increased velocity causes a contraction of space-time thereby decreasing the distance between the present and the future (Einstein 1961, Einstein et al. 2913), accelerated dream-consciousness has the same effect.

In dream-time and dream-consciousness all worlds are possible simultaneously and in parallel. These many worlds include those of the future and the past and where time and space are juxtaposed and every probable outcome is equally likely, and where the world is continually splitting into alternate worlds. Dream-time-consciousness is a manifestation of and in many respects obeys the laws of the "Many Worlds" theory of quantum physics as first proposed by Hugh Everett (1956, 1957).

Hugh Everett's "theory of the universal wavefunction" (Many Worlds) is distinguished from the Copenhagen model, as there is no special role for an observing consciousness. Everett also removed the "wave function" collapse which he believed to be redundant, and instead insisted that what is observed must be clearly defined (thereby answering one of Einstein's criticism of quantum theory). According to Everett's theory, every action, every measurements, every behavior, every choice, even not choosing, can create a new reality, another world, generating a bifurcation between what happened and what did not happen, such that innumerable possibilities and possible worlds arise from every action, including realties which do not obey the laws of physics and cause and effect.

As conceived by Everett (1956, 1957) and Dewitt (1971), when a physicist measures an object, the universe splits into two distinct universes to accommodate each of the possible outcomes. In one universe, the physicist measures the wave form, in the other universe the physicist measures the object as a particle. Since all objects have a particle-wave duality, this also explains how an object can be measured as a particle and can be measured as a wave, but not both at the same time in the same world, and how it can be measured in more than one state, each of which exists in another world. The simple act of measurement creates two worlds both of which exist at the same time in parallel, and each separate version of the universe contains a different outcome of that event.

Instead of one continuous timeline, the universe under the many worlds interpretation looks more like a forest of trees with innumerable branches and twigs each of which represents a different possible world. According to Everett the entire universes continuously exists in a superposition and juxtaposition of multiple states. In many respects, Everett's theory defines dream-time and dream-consciousness.

According to Everett (1957), observation and measurement does not force the object under observation to take any specific form or to have any specific outcome. Instead, all outcomes are possible; much like a dream. For example, an NFL football player, a receiver, is running down the field and the quarterback throws him the ball. According to the "Many Worlds" interpretation of quantum physics, every conceivable and incomprehensible outcome is possible: The receiver catches or doesn't catch the ball. A female cheerleader runs out into the field and catches the ball. The receiver and the cheerleader ignore the ball and take off their clothes and have sex on the field. The head coach takes out a shotgun and begins shooting at the football. Spectators run onto the football field and erect circus tents and it becomes a giant carnival with rides. Some of the football players dress up as clowns and circus performers. Players and spectators lay on the grass and swim toward the goal posts. An alien space ship crashes into the football stadium and aliens emerge selling popcorn. Terrorists attack the football players and steal the football, and so on.

All outcomes are possible in Multiple Worlds, from the most probable to the least probable (Dewitt 1971). Every probable outcome is possible; trillions of outcomes including those where the defiance of physics may become the law of the land. Moreover, each of these multiple realities exist, simultaneously, side-by-side, in parallel. They exist simultaneously with the reality in which the observer resides; and whatever reality houses the observer is just one probable reality.

PreCognition in Dream-Time and Dream-Consciousness

During dream-time and during dream-consciousness the reality being dreamed is characterized by every possible outcome. Some dream worlds exist in the future, others in the past, and yet others in a world where past, present and future are juxtaposed and exist simultaneously and where every possible outcome is possible. Thus, in dream-time, the dream-consciousness can witness any number of these possible worlds including those which exist in the future.

However, these futures and possible futures which are observed by dream-consciousness are not "just a dream." According to the Many Worlds interpretation, they actually exist. In terms of space-time, these future worlds exist in the future, in a distant location. As predicted by quantum mechanics, the observer is entangled with that future. However, in dream-time the observer (dream consciousness) directly observes that future; including those futures which are improbable or most probable.

The Many Worlds of dream-consciousness provides the foundation for dream-time precognition. The dreamer may dream of the future just before it occurs. And upon waking from that dream of the future, the conscious mind may remember it and then experience it as it occurs in real time.

Dream-time access to the future is made possible because the brain is in a state of accelerated activity during the course of the dream. As predicted by Einstein (1961) an accelerated frame of reference brings the future closer to the present and makes time travel possible. Accelerated states of consciousness not only bring the future closer, but provide glimpses of those futures before they occur; a phenomenon best described as pre-cognition in dream time.

Aberfan Disaster Dream-Time Precognition

Aberfan is a small village in South Wales. Throughout late September and October 1966, heavy rain lashed down on the area and seeped into the porous sandstone of the hills which surrounded the town and against which abutted the village school (Barker 1967).

On September 27 1966, Mrs SB of London dreamed about a school on a hillside, and a horrible avalanche which killed many children.

On October 14, 1966 Mrs GE from Sidcup, dreamed about a group of screaming children being covered by an avalanche of coal.

On October 20, 1966 Mrs MH, dreamed about a group of children who were trapped in a rectangular room and the children were screaming and trying to escape.

On October 20, 1966, a 10 year old child living in Aberfan woke up screaming from a nightmare. She told her parents that in her dream she was trying to go to school when "something black had come down all over it" and there was "no school there."

Quantum Physics, **Retrocausation, PreCognition, Entanglement,**

On October 21, 1966, part of the rain soaked hills of Aberfan gave way and half a million tons of debris slid toward the village of Aberfan and slammed into the village school. The 10 year old girl who dreamed of the tragedy and 115 other schoolchildren and 28 adults lost their lives when the school was smashed and covered with mud. There were less than a dozen survivors (Baker, 1967).

Assassination of Archduke Francis Ferdinand: Dream-Time Precognition

In June of 1914, Austria was seeking to expand it's central European empire; plans which were resented by neighboring states, including Serbia, who wished to remain independent. That same month, the Archduke Francis Ferdinand, nephew of the Austrian Emperor Francis Joseph, went on a diplomatic tour accompanied by his wife, to build alliances with the leaders of these independent nations. In late June he and his wife arrived in Sarajevo, Serbia.

On the evening of June 27, 1914, Bishop Joseph Lanyi prepared for bed and upon falling asleep he began to dream. The Archduke Franz Ferdinand of Austria, heir to the throne of Austria, had been the Bishop's student and pupil, and late that night the Archduke appeared in Bishop Lanyi's dream. The dream became a nightmare and at 3:15 AM Bishop Joseph Lanyi awoke, frightened, upset and in tears. He glanced at the clock, dressed himself, and because the dream was so horrible, he wrote it down:

> "At a quarter past three on the morning of 28th June, 1914, I awoke from a terrible dream. I dreamed that I had gone to my desk early in the morning to look through the mail that had come in. On top of all the other letters there lay one with a black border, a black seal, and the arms of the Archduke. I immediately recognized the letter's handwriting, and saw at the head of the notepaper in blue colouring a picture which showed me a street and a narrow side-street. Their Highnesses sat in a car, opposite them sat a General, and an Officer next to the chauffeur. On both sides of the street there was a large crowd. Two young men sprang forward and shot at their Highnesses."

In the dream, Bishop Lanyi read the dream-letter, which had been written by the Archduke. According to the Bishop's account, which he wrote down in the early predawn hours of June 28, the dream letter from the Archduke was as follows: "Dear Dr Lanyi: Your Excellency. I wish to inform you that my wife and I were the victims of a political assassination. We recommend ourselves to your prayers. Cordial greetings from your Archduke Franz. Sarajevo, 28th June, 3.15 a.m." Bishop Joseph Lanyi was convinced that the Archduke had been assassinated, and called his parishioners and household staff to tell them of the terrible news. Later that morning of June 28, 1914, the Bishop held a mass for the Archduke and his wife. But, the Archduke were still alive and would not be shot dead for another 2 hours.

On June 28, 1914, at 11 a.m., as the Archduke and his wife were leaving a ceremony at Sarajevo, a Serbian nationalist leaped from the crowd and killed them both. It was the Archduke's assassination which triggered World War One.

Death of Mark Twain's Brother: Dream-Time Precognition

In May of 1858, Mark Twin had a dream about his younger brother Henry who was working on a riverboat as a "mud clerk." As related by Mark Twain:

> "The dream was so vivid, so like reality, that it deceived me, and I thought it was real. In the dream I had seen Henry a corpse. He lay in a metallic [burial case]. He was dressed in a suit of my clothing, and on his breast lay a great bouquet of flowers, mainly white roses, with a red rose in the [centre]. The casket stood upon a couple of chairs...it suddenly flashed upon me that there was nothing real about this--it was only a dream. I can still feel something of the grateful upheaval of joy of that moment, and I can also still feel the remnant of doubt, the suspicion that maybe it [was] real, after all. I returned to the house almost on a run, flew up the stairs two or three steps at a jump, and rushed into that [sitting-room]--and was made glad again, for there was no casket there."

A few days later, Twain's brother left on a river boat from New Orleans. As related by Mark Twain:

> "Two or three days afterward the boat's boilers exploded at Ship Island, Memphis. I found Henry stretched upon a mattress on the floor of a great building, along with thirty or forty other scalded and wounded persons... his body was badly scalded... I think he died about dawn. The coffins provided for the dead were of unpainted white pine, but in this instance some of the ladies of Memphis had made up a fund of sixty dollars and bought a metallic case, and when I came back and entered the [dead-room] Henry lay in that open case, and he was dressed in a suit of my clothing. He had borrowed it without my knowledge during our last sojourn in St. Louis; and I recognized instantly that my dream of several weeks before was here exactly reproduced, so far as these details went--and I think I missed one [detail;] but that one was immediately supplied, for just then an elderly lady entered the place with a large bouquet consisting mainly of white roses, and in the [centre] of it was a red rose, and she laid it on his breast."

The Dream-Murder of Tanya Zachs

In a legal case investigated and reported by Joseph (2000), a beautiful young woman, Tanya Zachs, disappeared on her way home in San Jose from her job in Santa Cruz in September of 1984. Her car was found abandoned along highway 17 midway between the two cities and which courses through the Santa Cruz mountains. That night, a young woman "Sunshine" who lived in a nudist colony, Lupin Lodge, situated in the Santa Cruz Mountains, had a nightmare: A woman was being brutally murdered. The next day, Sunshine read the story of Tanya's disappearance in the local newspaper, and that night she had the dream again, but this time the victim appeared to her quite clearly. It was Tanya.

In the dream Tanya showed "Sunshine" a narrow mountain road off highway 17, one of many leading from the long and winding highway between San Jose and Santa Cruz. Tanya led the dreamer down the mountain road which was bordered by a thick canopy of redwood trees and pines, and then to an isolated spot alongside. Tanya then beckoned the dreamer to follow her down a rather steep incline leading from the mountain road into the forest and thick brush, and then along a forested trail. Finally, Tanya stopped and pointed out her naked body, lying spread eagle on a huge slab of rock surrounded by trees.

Sunshine was convinced she knew where Tanya's body lay hidden. On the morning of 9/15/84, she contacted Tanya's family, told them of her dreams, and that same day led them and the police to the mountain side road Tanya had showed her and finally to the isolated spot. The police climbed down the tree-covered steep incline, and just as Sunshine had dreamed, they found the trail leading into the forest. But, there was no body.

That night Sunshine had another dream and Tanya took her to the same spot, down the same trail, then pointed at and emphasized a little deer trail that forked off to the right between the trees, and which led directly to her body. The next day, Sunshine and the family met again, and then climbed down the incline, took the trail to the right, and there was Tanya's body laid out exactly as revealed to Sunshine when dreaming.

The murder remained unsolved, however, until four years later. Damon Wells, beset by horrible nightmares where the victim kept accusing him of her murder, sought psychiatric treatment and confessed (Joseph 2000).

Precognition Dreams Are Common

Precognition dreams are common (Fukuda 2002; Haraldsson, 1985; Lange et al. 2001; Ross & Joshi, 1992; Stowell, 1995; Thalbourne, 1994) and often involve negative, unhappy, unpleasant events such as deaths, disasters and other calamities (Ryback & Sweitzer, 1990). About 40% of precognitive dreams are linked to an event the following day (Sondow, 1988), or take place several days or weeks later. However, anecdotal evidence indicates that the dreamed events may occur just prior to waking, even triggering wakefulness.

Precognition dreams can be about mundane affairs of concern only to the dreamer. A colleague of this author admits to frequently having had precognitive dreams and relates the following:

"I dreamed that my water heater busted and that water was flooding out onto the floor. Three days later, the water heater sprung a leak." "I dreamed about getting a flat tire while driving on the freeway. It was the rear tire on the driver's side. A couple days later, the car's dashboard-computer informed me that the rear tire on the driver's side was low in air." "I dreamed that I took a girlfriend to my hot tub in the back yard, but it was empty and there was no water. When I tried to turn the water on nothing would happen. In the dream I was irritated because it would have to be replaced. A few days later the hot tub on-switch broke and after several failed attempts to fix it, I had it junked."

Several studies indicate that precognitive dreams are more common in younger than older individuals and that women report more precognitive dreams than men (Lange et al. 2001). It has also been found that those who have experienced deja vu are more likely to have precognitive dreams (Fakuda, 2002).

In large samples, anywhere from 17.8 % to 66% of individuals report that they experienced at least one precognitive dream (Fukuda 2002; Palmer, 1979; Haraldsson, 1985; Ross & Joshi, 1992; Ryback, 1988; Thalbourne, 1994) whereas over 60% of the general population believe such dreams are possible (Thalbourne, 1984; Haraldsson, 1985). However, Ryback (1988) after investigating 290 case reports of paranormal dreams, dismissed most of these precognitive dreams as coincidence and concluded that only 8.8% of the population actually have these dreams.

PreCognitive Dream Skepticism and Professional Baseball

Over 2000 years ago Aristotle wrote a book expressing his disbelieve in precognitive dreams: "On Divination in Sleep." Aristotle complained that most of those having precognitive dreams were unworthy of the honor of receiving advanced information and "are not the best and wisest, but merely commonplace persons." Aristotle argued that "the sender of such dreams should be God." According to Aristotle "most dreams are to be classed as mere coincidences..." and do not take "place according to a universal or general rule" and have no causal connection to actual events in the future.

"Coincidence" has been the major objection to claims of precognitive dream activity (Caroll 2000; Wiseman 2011). Caroll (2000) refers to the "law of large numbers" and dismisses all claims as being a function merely of coincidence. For example, the odds, are with so many dreamers having dreams about so many different themes, that a few of them will have dreams about an airplane crash or a ship that sinks. If the next day a ship sinks or a plane crashes, this is merely coincidence. According to Caroll and others, if precognitive dreams were real, they should be more commonplace, with more dreamers coming forward, and thus there should be a high "hit rate" and there is not; and as such "precognitive dreams do not exist."

However, if we applied the same reasoning to professional baseball, then professional baseball does not exist. Consider, from 2000 until 2013, the average baseball batting average ranged around 267. During regular season play during 2013, out of 750 major league players, 726 had a batting average of less than 30% (http://espn.go.com/mlb/stats/batting). Taking into considering fowl balls and hits which result in "outs", but considering that each player has at least 3 opportunities each time at bat to hit the ball, and then taking that .267 average, it could be said that the average professional baseball player actually gets a base hit less than 20% of the time. Be it a 20% or 30% hit rate, obviously this does not mean no one in professional baseball is able to hit the ball, or that when they do it is merely a coincidence. The same standard must be applied to precognitive dreams.

Precognitive dreams need not be about Earth-shaking national tragedies and it is unknown how many dreamers would ever come forward to report their dreams even if they did have national implications. In fact, most dreams are forgotten upon waking (Frank, 2012, Pagel, 2014, Stickgold & Walker, 2010). Further, many precognitive dreams may be related to mundane matters like a "flat tire," or a phone call or visit from a friend the next day; or they may be entangled with events which are about to occur just minutes or seconds into the future, i.e. backward/precognitive dreams. Since most people forget their dreams upon waking and most dreams are forgotten, how often precognitive dreaming occurs, and how many people have them, is unknown. What is known is that such dreams can be explained by quantum physics and the neurological foundations for dream activity and dream-time.

PreCognitive Backward Dreams

Precognitive experiences occurring during waking may be entangled with innocuous event which are just about to occur, such as thinking of a friend and then getting a phone call or email from that friend minutes or hours later. Just as a professional baseball player is more likely to swing and miss than hit the ball, the fact that one might think of a friend who does not call is not evidence against precognition.

During accelerated states associated with dream-time, precognitive dreams may be for events which will soon happen, or are just about to happen, perhaps seconds, or minutes away. These latter-type of dreams care best described as precognitive-backward dreams.

A case in point, "Katherine" dreams she and a friend "Sheryl" are shopping in Boston. They go from store, lugging shopping bags. "Katherine" in her dreams feels this sense of urgency to go home as if she is late for something and someone is waiting for her so she sets her bags down on the sidewalk and sits on a bench to wait for a cab or a the bus. She then realized her friend "Sheryl" is gone. Katherine looks for her, goes in and out of stores, but can't find her. "Katherine" sees a bus-like street car coming down the cobbled street and she picks up her packages and steps out onto the curb. As the street car pulls up and stops "Katherine" is surprised to see that Sheryl is driving and is ringing its bell. The sound of the bell grows louder and louder and then jolts "Katherine" from her dream. Katherine realizes that her phone is ringing. She picks it up and it is her friend "Sheryl" who is calling. Sheryl and Katherine are going shopping that day.

That "Katherine" dreamed about going shopping with "Sheryl" is not remarkable in-itself. That "Sheryl" was ringing the bell and it was Sheryl who was calling can be explained away as interesting coincidence. It is no surprise that Sheryl called. What seems paradoxical, however, is that the dream of shopping and walking down the streets of Boston laden with packages, the desire to go home, then looking for her missing friend and then seeing the bus-like street car all seemed to lead up to the ringing of the bell in a logical order of events so that its ringing made sense in the context of the dream. Hearing the ringing bell seemed to be a natural part of the dream, and it is. However, the dream did not lead up to the bell. Rather, the ringing of the bell initiated the dream. The effect (ringing bell) and the cause (the ringing phone) are identical. The effect caused itself.

There are two explanations for these quite common "backward" dreams. Dream-time and dream-consciousness does not obey the laws of physics. In dream-time, dream-consciousness may attempt to impose temporal order on a dream which has no temporal order and which may be experienced as a gestalt. In other words, in dream-time the entire dream was instantaneous and the dream was initiated by the ringing of her phone. The bell was heard and the dream was instantly produced in explanation and association. Future, present, past may be juxtaposed and experienced as a gestalt; like seeing the forest instead of the individual trees. In fact, although dreams may seem to last long time periods, they may be only seconds in length (Frank, 2012, Pagel, 2014, Stickgold & Walker, 2010).

The other explanation is that the ringing of the phone and the fact that Sheryl was calling Katherine, was perceived in dream-time, before it happened. Just as a time-machine traveling at superluminal speeds from the future into the past will pass by an observer only to be followed by its light image (which trails behind at the speed of light), information just seconds or minutes away into the future can be perceived by dream-consciousness in dream-time through entanglement. However, it is not future information traveling at superluminal speeds, but the mind and brain of the dreamer which are accelerating toward that future event in advance of those conscious minds which are still awake.

In dream time, the brain is highly active (Frank, 2012, Pagel, 2014, Stickgold & Walker, 2010), and certain regions in the limbic system are hyperactive (Joseph 1992, 1996, 2000). During dream-time, brain activity is accelerating which causes a contraction in time-space. The future comes closer to the present during dream time relative to outside observers which may include, upon waking, the conscious mind of the dreamer. However, while in dream-time, in a state of accelerated dream-consciousness, the future may be sensed and it may trigger a complex dream which then leads up to that future event when it arrives in the present thereby waking the dreamer.

Another illustrative example: French physicist Alfred Maury dreamt that he had taken part in the French Revolution and that he had been condemned to death and his head cut off at the exact moment when his bedpost broken and struck him

Quantum Physics, **Retrocausation, PreCognition, Entanglement,**

across the neck:

> "I was rather, unwell, and was lying down in my room, with my mother at my bedside. I dreamed of the Reign of Terror; I witnessed massacres, I was appearing before the Revolutionary tribunal, I saw Robespierre, Marat, Fouquier-Tinville, all the most wicked figures of that terrible era; I talked to them; finally, after many events that I only partly remember, I was judged, condemned to death, taken out in a tumbril through a huge throng to the Place de la Revolution; I mounted the scaffold; the executioner tied me to that fatal plank, he tipped it up, the blade fell; I felt my head separating from my body, I woke up, racked by the deepest anguish, and felt the bedpost on my neck. It had suddenly come off and had fallen on my cervical vertebrae just like the guillotine blade."

Certainly it would be expected that a major blow to the head and neck would cause instant waking. But in this instance, it did not. Instead, the dreamer experienced a long and convoluted dream which was initiated by what was about to happen, and which could also be considered a warning of what was about to happen; albeit in the unique dream-language characteristic of dreams. This is not a case of an instantaneous backward dream, but a precognitive dream which provided the dreamer with a glimpse of what lay in store just moments into the future.

A third example related to me by a colleague:

> "I had been working in my yard into the late Friday afternoon and was exhausted. It was hot and I stripped off my shirt and lay down in a swinging hammock in my yard to take a nap and instantly fell asleep and began to dream. In the dream I was in a nightclub and there was this exotic beautiful woman with long black hair drinking at the bar. We began drinking together and then we were dancing and kissing and then we were suddenly in my house and we were laying on the floor and I was taking off her clothes and she was getting very excited and aggressive. All at once I could see she had yellow eyes and black skin, but it wasn't skin, but resembled an insect's chitlin. Her arms and hands became claws and her teeth became razor sharp and pointed. She put her claw arms around me very tight as I struggled to escape. She had turned into some demonic insect-creature and pressed her razor sharp claw-hands into my back. I could feel her razor sharp claws knifing me and I felt I was being stabbed in the back. The pain was terrible. It seemed as if her pointed claws were going to completely pierce my back and come out my chest. The pain was so horrific I woke up. But the pain was still there. I got up from the hammock and there was a crippled black bumble bee laying there. The damn thing has stung me on my back."

Dreams that seem to paradoxically lead up to an event which wakes the dreamer are common. These dreams may be relatively brief or become lengthy complicated dreams leading up to some event which then occurs, as if on cue, waking the dreamer who discovers upon waking that someone was knocking on his door, the phone was ringing, it was the alarm clock, a kid was yelling outside the window, and so on, all of which initiated the dream which then led up to the event which caused the dream (Joseph 1992). The dream was produced, so as to explain in the unique language of the dream what was about to happen; and this is because, it already happened in the future. The only other explanation is the dream was produced as an instant gestalt and the dreamer dreamed the dream in accelerated dream-time without any temporal order, and it was upon waking that the dream was reconstructed in a temporal sequential time frame (Joseph 1992).

Be they backward dreams instantly produced as a gestalt, or examples of dream precognition, backward dreams are the most easily comprehended because the conscious mind utilizes temporal sequences to explain what is observed, and may recall the dream in reverse, so it makes temporal-sequential sense; as if the cause led to the effect, when the cause and effect were either simultaneous, or the effect was its own cause.

Joseph Dreams His Death 2000 Years Ago

In the early 1950s, when I [R. Joseph] was a boy of 3, and for many years until around age 7, I had dreams about a little boy playing by the sea shore, by the ocean. And there were crowds of people. Some lying or sitting together on the sand. Others swimming or fishing. And then in the dream the ocean began to recede... the ocean waters drew back back back... and I could see shells and fish flopping on the wet sand where moments before there had been ocean... and ships and small boats lay on their sides... and I ran to where the ocean had been, on the wet sand, picking up shells... and many other people also ran onto the wet sand picking up wiggling fish and laughing and talking in amazement that the ocean had pulled back for miles and miles leaving the sand and ocean floor completely revealed for everyone to see.... and then... and then... and then...

I walked further and further out to where the ocean had been, picking up giant shells some with wiggling living creatures still inside, and gazing in wonder at what the ocean had hidden but which was now revealed... and then I heard screams... women and men and children were screaming... and in my dream, they were all running from the wet sand where the ocean had been toward the dry shore...and people on the shore were also running... everyone was running away and screaming... and I could hear this rumbling roar from behind me... and when I looked back to see why, what they were running from and what was making that roar, I could see the ocean... it was still miles away--but it was a WALL OF OCEAN.. a WALL OF WATER looming up maybe 100 yards perhaps even miles into the sky... and in my dream the wall of ocean was rushing forward, to where the ocean had been minutes before, toward where I was standing with sea shells in my hands...

Quantum Physics, **Retrocausation, PreCognition, Entanglement,**

and I started running... like everyone else, running running running... and I could see, over my shoulder, behind me, the roaring wall of ocean water coming closer, and closer... and faster faster faster... and I kept running... everyone was running and screaming...trying to get away... and then the towering WALL OF WATER was just behind me... then looming over me... and then it crashed down upon me... and the little boy that I was, in this dream, drowned.... and then I awoke in my bed... the same boy who drowned, but a different boy...me...

I had this dream over and over for years; the same dream, the source of which was a mystery to me as I had never even imagined that the ocean could actually recede and then rush back to land as I had dreamed. It was not until 20 years later that I learned, for the first time, about Tsunamis and how characteristic it is for people to foolishly run out to where the ocean and been... and then... the ocean comes rushing back as a wall of water drowning everyone who did not immediately run away.

How could I have dreamed so vividly about something 3-year old me knew nothing about in the early 1950s when we didn't even have a television? There were clues in yet other dreams when I was a child, and they were dreams of the same little boy. But it was during ancient Roman times, and I was sitting with my mother who was dressed in royal robes typical of the Roman period. She was singing to me... And down below I could see Roman soldiers marching, and peasant women by a river, washing clothes, and the river was flowing into the ocean. The peasant women, who had with them many naked children, were dressed in clothes I associated with Biblical times, of ancient Egypt; my grandmother would often read to me from a Bible picture book. But in these river-side dreams which began so peaceful, they all ended with incredible earthquakes, like the world was turning up-side down...

How do these two dreams relate? Almost 50 years after I had these dreams, I searched the records for Tsunamis in the Mediterranean sea near Italy and Egypt. On the morning of July 21, 365 AD, an earthquake of great magnitude caused a huge tsunami more than 100 feet high and it inundated and destroyed several towns on the coasts of the Mediterranean, including Alexandria. This is how Ammianus Marcellinus, a Roman historian described it:

> Slightly after daybreak, and heralded by a thick succession of fiercely shaken thunderbolts, the solidity of the whole earth was made to shake and shudder, and the sea was driven away, its waves were rolled back, and it disappeared, so that the abyss of the depths was uncovered and many-shaped varieties of sea-creatures were seen stuck in the slime; the great wastes of those valleys and mountains, which the very creation had dismissed beneath the vast whirlpools, at that moment, as it was given to be believed, looked up at the sun's rays. Many ships, then, were stranded as if on dry land, and people wandered at will about the paltry remains of the waters to collect fish

and the like in their hands; then the roaring sea as if insulted by its repulse rises back in turn, and through the teeming shoals dashed itself violently on islands and extensive tracts of the mainland, and flattened innumerable buildings in towns or wherever they were found. Thus in the raging conflict of the elements, the face of the earth was changed to reveal wondrous sights. For the mass of waters returning when least expected killed many thousands by drowning, and with the tides whipped up to a height as they rushed back, some ships, after the anger of the watery element had grown old, were seen to have sunk, and the bodies of people killed in shipwrecks lay there, faces up or down.

Had Joseph dreamed of a previous life from nearly 2000 years ago? Or did he journey to the past, during dream-time, and visit the long ago in the time machine of consciousness?

We have been here before, we will be here again, we will always be, and this is because time and consciousness are a quantum continuum and the distinctions between past present and future are illusions.

REFERENCES

Al-Khalili (2011). Black Holes, Wormholes and Time Machines, Taylor & Francis.

Almheiri, A. et al. (2013). Black Holes: Complementarity or Firewalls? J. High Energy Phys. 2, 062

Bethe, H. A., et al., (2003) Formation and Evolution of Black Holes in the Galaxy, World Scientific Publishing.

Bilaniuk, O.-M. P.; Sudarshan, E. C. G. (1969). "Particles beyond the Light Barrier". Physics Today 22 (5): 43–51.

Bo, L., Wen-Biao, L. (2010). Negative Temperature of Inner Horizon and Planck Absolute Entropy of a Kerr Newman Black Hole. Commun. Theor. Phys. 53, 83–86.

Bohr, N., (1913). "On the Constitution of Atoms and Molecules, Part I". Philosophical Magazine 26: 1–24.

Bohr, N., (1913). "On the Constitution of Atoms and Molecules, Part I". Philosophical Magazine 26: 1–24.

Bohr, N. (1934/1987), Atomic Theory and the Description of Nature, reprinted as The Philosophical Writings of Niels Bohr, Vol. I, Woodbridge: Ox Bow Press.

Bruno, N. R., (2001). Deformed boost transformations that saturate at the Planck scale. Physics Letters B, 522, 133-138.

Blandford, R.D. (1999). "Origin and evolution of massive black holes in galactic nuclei". Galaxy Dynamics, proceedings of a conference held at Rutgers University, 8–12 Aug 1998, ASP Conference Series vol. 182.

Carroll, S (2004). Spacetime and Geometry. Addison Wesley.

Casimir, H. B. G. (1948). "On the attraction between two perfectly conducting plates". Proc. Kon. Nederland. Akad. Wetensch. B51: 793.

Chodos, A. (1985). "The Neutrino as a Tachyon". Physics Letters B 150: 431.

Einstein, A. (1915a). Fundamental Ideas of the General Theory of Relativity and the Application of this Theory in Astronomy, Preussische Akademie der Wissenschaften, Sitzungsberichte, 1915 (part 1), 315.

Einstein, A. (1915b). On the General Theory of Relativity, Preussische Akademie der Wissenschaften, Sitzungsberichte, 1915 (part 2), 778–786.

Einstein A. (1939) A. Einstein, Ann. Math. 40, 922.

Einstein, A. (1961), Relativity: The Special and the General Theory, New York: Three Rivers Press.

Einstein, A. and Rosen, N. (1935). "The Particle Problem in the General Theory of Relativity". Physical Review 48: 73.

Einstein, A., Lorentz, H.A., Minkowski, H., and Weyl, H. (1923). Arnold Sommerfeld. ed. The Principle of Relativity. Dover Publications: Mineola, NY. pp. 38–49.

Einstein A, Podolsky B, Rosen N (1935). "Can Quantum-Mechanical Description of Physical Reality Be Considered Complete?". Phys. Rev. 47 (10): 777–780.

Eisberg, R., and Resnick. R. (1985). Quantum Physics of Atoms, Molecules, Solids, Nuclei, and Particles. Wily,

Everett, A., & Roman, T. (2012). TIme Travel and Warp Drives, University Chicago Press.

Feinberg, G. (1967). "Possibility of Faster-Than-Light Particles". Physical Review 159 (5): 1089–1105.

Fuller, Robert W. and Wheeler, John A. (1962). "Causality and Multiply-Connected Space-Time". Physical Review 128: 919.

Garay, L. J. (1995). Quantum gravity and minimum length Int.J.Mod.Phys. A10 (1995) 145-166

Ghez, A. M.; Salim, S.; Hornstein, S. D.; Tanner, A.; Lu, J. R.; Morris, M.; Becklin, E. E.; Duchene, G. (2005). "Stellar Orbits around the Galactic Center Black Hole". The Astrophysical Journal 620: 744.

Geiss, B., et al., (2010) The Effect of Stellar Collisions and Tidal Disruptions on Post-Main-Sequence Stars in the Galactic Nucleus. American Astronomical Society, AAS Meeting #215, #413.15; Bulletin of the American Astronomical Society, Vol. 41, p.252.

Giddings, S. (1995). The Black Hole Information Paradox," Proc. PASCOS symposium/Johns Hopkins Workshop, Baltimore, MD, 22-25 March, 1995, arXiv:hep-th/9508151v1.

Hawking, S. W., (1988) Wormholes in spacetime. Phys. Rev. D 37, 904–910.

Hawking, S., (1990). A Brief History of Time: From the Big Bang to Black Holes. Bantam.

Hawking, S. (2005). "Information loss in black holes". Physical Review D 72: 084013.

Hawking, S. W. (2014). Information Preservation and Weather Forecasting for Black Holes.

Heisenberg, W. (1927), "Über den anschaulichen Inhalt der quantentheoretischen Kinematik und Mechanik", Zeitschrift für Physik 43 (3–4): 172–198.

Heisenberg. W. (1930), Physikalische Prinzipien der Quantentheorie (Leipzig: Hirzel). English translation The Physical Principles of Quantum Theory, University of Chicago Press.

Heisenberg, W. (1955). The Development of the Interpretation of the Quantum Theory, in W. Pauli (ed), Niels Bohr and the Development of Physics, 35, London: Pergamon pp. 12-29.

Heisenberg, W. (1958), Physics and Philosophy: The Revolution in Modern Science, London: Goerge Allen & Unwin.

Jaffe, R. (2005). "Casimir effect and the quantum vacuum". Physical Review D 72 (2): 021301.

Joseph, R (2010a) The Infinite Cosmos vs the Myth of the Big Bang: Red Shifts, Black Holes, and the Accelerating Universe. Journal of Cosmology, 6, 1548-1615.

Joseph, R. (2010b). The Infinite Universe: Black Holes, Dark Matter, Gravity, Acceleration, Life. Journal of Cosmology, 6, 854-874.

Joseph, R. (2014a) Paradoxes of Time Travel: The Uncertainty Principle, Wave Function, Probability, Entanglement, and Multiple Worlds, Cosmology, 18, 282-302.

Joseph, R. (2014a) The Time Machine of Consciousness, Cosmology, 18, In press.

Kerr, R P. (1963). "Gravitational Field of a Spinning Mass as an Example of Algebraically Special Metrics". Physical Review Letters 11 (5): 237–238.

Lambrecht, A. (2002) The Casimir effect: a force from nothing, Physics World, September 2002

Lorentz, H. A. (1892), "The Relative Motion of the Earth and the Aether", Zittingsverlag Akad. V. Wet. 1: 74–79

McClintock, J. E. (2004). Black hole. World Book Online Reference Center. World Book, Inc.

Melia, F. (2003). The Edge of Infinity. Supermassive Black Holes in the Universe. Cambridge U Press. ISBN 978-0-521-81405-8.

Melia, F. (2007). The Galactic Supermassive Black Hole. Princeton University Press. pp. 255–256.

Merloni, A., and Heinz, S., (2008) A synthesis model for AGN evolution: supermassive black holes growth and feedback modes Monthly Notices of the Royal Astronomical Society, 388, 1011 - 1030.

Minchin, R. et al. (2005). "A Dark Hydrogen Cloud in the Virgo Cluster". The

Astrophysical Journal 622: L21–L24.

Morris, M. S. and Thorne, K. S. (1988). "Wormholes in spacetime and their use for interstellar travel: A tool for teaching general relativity". American Journal of Physics 56 (5): 395–412.

O'Neill, B. (2014) The Geometry of Kerr Black Holes, Dover

Penrose, R. (1969) Rivista del Nuovo Cimento.

Preskill, J. (1994). Black holes and information: A crisis in quantum physics", Caltech Theory Seminar, 21 October. arXiv:hep-th/9209058v1.

Pollack, G. L., & Stump, D. R. (2001), Electromagnetism, Addison-Wesley.

Rindler, W. (2001). Relativity: Special, General and Cosmological. Oxford: Oxford University Press.

Rodriguez, A. W.; Capasso, F.; Johnson, Steven G. (2011). "The Casimir effect in microstructured geometries". Nature Photonics 5 (4): 211–221.

Ruffini, R., and Wheeler, J. A. (1971). Introducing the black hole. Physics Today: 30–41.

Russell, D. M., and Fender, R. P. (2010). Powerful jets from accreting black holes: Evidence from the Optical and the infrared. In Black Holes and Galaxy Formation. Nova Science Publishers. Inc.

Schrödinger, E. (1926). "An Undulatory Theory of the Mechanics of Atoms and Molecules". Physical Review 28 (6): 1049–1070. Bibcode:1926PhRv...28.1049S. doi:10.1103/PhysRev.28.1049.

Sen, A. (2002) Rolling tachyon," JHEP 0204, 048

Slater, J. C. & Frank, N. H. (2011) Electromagnetism, Dover.

Smolin, L. (2002). Three Roads to Quantum Gravity. Basic Books.

Taylor, E. F. & Wheeler, J. A. (2000) Exploring Black Holes, Addison Wesley.

Thorne, K. (1994) Black holes and time warps, W. W. Norton. NY.

Thorne, K. S. & Hawking, S. (1995). Black Holes and Time Warps: Einstein's Outrageous Legacy, W. W. Norton.

Thorne, K. et al. (1988). "Wormholes, Time Machines, and the Weak Energy Condition". Physical Review Letters 61 (13): 1446.

Wilf, M., et al. (2007) The Guidebook to Membrane Desalination Technology: Reverse Osmosis, Balaban Publishers.

2. Synchronicity, Entanglement, Quantum Information and the Psyche

Francois Martin[1], Federico Carminati[2], Giuliana Galli Carminati[3]

[1]Laboratoire de Physique Theorique et Hautes Energies, Universities Paris
[2]Physicist at CERN, Geneva, Switzerland.
[3]Mental Development Psychiatry Unit - Adult Psychiatry Service, Department of Psychiatry, University Hospitals of Geneva, Switzerland

Abstract

In this paper we describe synchronicity phenomena. As an explanation of these phenomena we propose quantum entanglement between the psychic realm known as the "unconscious" and also the classical illusion of the collapse of the wave-function. Then, taking the theory of quantum information as a model we consider the human unconscious, pre-consciousness and consciousness as sets of quantum bits (qu-bits). We analyze how there can be communication between these various qu-bit sets. In doing this we are inspired by the theory of nuclear magnetic resonance. In this manner we build quantum processes that permit consciousness to "read" the unconscious and vice-versa. The most elementary interaction, e.g. between a pre-consciousness qu-bit and a consciousness one, allows us to predict the time evolution of the pre-consciousness + consciousness system in which pre-consciousness and consciousness are quantum entangled. This time evolution exhibits Rabi oscillations that we name mental Rabi oscillations. This time evolution shows how, for example, the unconscious can influence consciousness and vice-versa. In a process like mourning the influence of the unconscious on consciousness, as the influence of consciousness on the unconscious, are in agreement with what is observed in psychiatry.

Key Words: Synchronicity, quantum entanglement, quantum information, consciousness, unconscious.

1 Synchronicity Effects

Synchronicity phenomena are characterized by a significant coincidence which appears between a (subjective) mental state and an event occurring in the (objective) external world. The notion was introduced by the Swiss psychoanalyst Carl Gustav Jung and further studied together with Wolfgang Pauli (Jung and Pauli, 1955). Jung referred to this phenomenon as "acausal parallelism" which are linked by an "acausal connecting principle." Synchronicity effects show no causal link between the two events that are correlated.

We can distinguish two types of synchronicity phenomena. The first one is characterized by a significant coincidence between the psyche of two individuals. An example of this type is when two friends at a distance simultaneously buy two identical neckties without having consulted each other beforehand. The significant coincidence appears as a correlation between the psyche of the two subjects, suggesting some type of psychic communication. There are many examples of such long range correlations between events which are causally unrelated, or subjects who engage in identical behaviors, often simultaneously: twins, relatives, members of a couple, friends, or scientists who make the same discoveries at around the same time.

For example, in March of 1951, a new comic strip appeared in over a dozen newspapers in the United States, featuring a little blond boy wearing a red and black striped shirt. The boy was called: Dennis The Menace. In the United Kingdom, a new character, a little boy wearing a red and black striped shirt was introduced in a comic book, The Beano. He was also named, Dennis The Menace. The creators of both comics claimed it was a coincidence.

The second type of synchronicity phenomena, which is closer to what was advocated by C.G. Jung, happens when the significant coincidence occurs between a mental state and a physical state. In this case the physical state is symbolically correlated to the mental state by a common meaning. They appear not necessarily simultaneously but in a short interval of time such that the coincidence appears exceptional. Jung referred to these events as "meaningful coincidences."

Another more common example goes as follows: You are sitting at home and begin thinking about an old friend who you had not seen in months, when the phone rings, and its him.

Synchronistic events between mind and matter seem difficult to explain in terms of correlations between conscious or unconscious minds. For Jung, synchronistic events are remnants of a holistic reality - the unus mundus which is based on the concept of a unified reality, a singularity of "One World" from which everything has its origin, and from which all things emerge and eventually return. The unus mundus, or "One World" is related to Plato's concept of the "World of Ideas," and has its parallels in quantum physics. Thus, the unus mundus underlies both mind and matter.

As already stressed, in a synchronicity effect, there is no causal link between correlated events localized in space and time. Synchronicity effects are global phenomena in space and time. They cannot be explained by classical physics. However, in the case of a significant coincidence appearing between the psyche of two individuals one can see an analogy with quantum entanglement (Baaquie and Martin, 2005).

Moreover one can possibly see synchronistic events between the mental and the material domains as a consequence of a quantum entanglement between mind and matter (Primas, 2003). For us mental and material domains of reality will be considered as aspects, or manifestations, of one underlying reality in which mind and matter are unseparated (Atmanspacher, 2004).

Synchronicity phenomena, especially those involving a correlation at a distance between several individuals, lead us to postulate non-localized unconscious mental states in space and time. Although different regions of the brain subserve specific functions (Joseph, 1982, 1992), mental states are not exclusively localized in the human brain. They are correlated to physical states of the brain (possibly via quantum entanglement) but they are not reducible to them.

Since we study the analogy between synchronistic events and quantum entanglement, we treat mental states (conscious and unconscious) as quantum states, i.e. as vectors of a Hilbert space (Baaquie and Martin, 2005). Moreover we treat them as vectors of a Hilbert space of information (Martin, Carminati, Galli Carminati, 2009).

2 Quantum information and the Psyche

We try to apply quantum information to some functions of the Psyche. In classical information, the memory boxes are binary system, called bits, which can take only two values: 0 or 1. A quantum bit (in a shortened form qu-bit) can take all values which are superposition of 0 and 1 (more precisely all superpositions of the states $/0>$ and $/1>$). In other words, a qu-bit can take simultaneously the values 0 and 1. Quantum information studies the monitoring of qu-bits. It studies also the transfer of quantum information from one qu-bit to another one (especially via two-qubit quantum logic gates).

As an example of a binary psychic system we have considered the phenomenon of mourning (Galli Carminati and Carminati, 2006): either mourning is achieved (qu-bit 0), either it is not (qu-bit 1). So quantum mechanics allow the existence of all superpositions of the state in which mourning is achieved with the state in which mourning is not achieved.

Quantum mechanics rests upon two fundamental properties. First it is based on the superposition principle (superposition of vector states of an Hilbert space). Second it is based on a fundamental phenomenon called quantum entanglement. This phenomenon manifests itself by the fact that a system of two, or several, quantum entangled particles is "non-separable". In technical terms this means that the wave-function of the two-particle system does not factorize into a product of a wave-function for each particle. The quantum system describing the two particle system is a global system, a non-local one. Moreover, in such a system, the particles are heavily correlated. Therefore, if we measure a certain property of one of the two particles, destroying in this way the "non-separability" of the system, we can predict with certainty the corresponding property of the other particle, even if this one is at the other extreme of the universe. However, there are caveats: the quantum specificity indicates that this property is not determined beforehand, i.e. before measurement. Quantum physics is a non-local and non-realistic theory. Quantum entanglement and the property of "nonseparability" are properties that are fundamentally quantum, that do not exist in "classical physics".

Assuming, with Belal Baaquie (Baaquie and Martin, 2005), the existence of quantum entanglement between the unconscious of two or several persons, we have proposed an explanation of correlations at a distance that appears between two (or several) individuals having affective links. This would constitute an explanation of synchronicity of the first type. It would be interesting to measure in a quantitative manner those unconscious correlations at a distance. May be those correlations could activate neural circuits that could be visible in nuclear magnetic resonance imaging (NMRI).

We propose to measure quantitatively the existence (or the non-existence) of such correlations during group therapies or group training, via "absurd" tests (Galli et al., 2008; Martin et al., 2007, 2009). Those experiments are currently in progress.

Quantum information applied to Psyche allows to explain a certain number of mental processes (Martin, Carminati, Galli Carminati, 2009). We suppose that the mental systems first proposed by Freud (1900, 1915a,b), i.e. the unconscious, pre-consciousness, consciousness, are made up of mental qu-bits. They are sets of mental qu-bits.

Specifically, Freud (1900, 1915ab), saw the mind as consisting of three mental realms; the unconscious, preconscious, and conscious mind, with the unconscious being the deepest, most inaccessible region of the psyche and which contains repressed memories and unacceptable feelings, thoughts, and ideas. The preconscious serves as a bridge, or passageway between the unconscious and conscious mind, and lies just below the surface of consciousness. Freud believed that unconscious impulses must pass through the preconscious which acts as a double doorway; one door leading from the unconscious to the preconscious, and the other from the preconscious to the conscious mind. In this way, the preconscious can censor information and prevent unacceptable impulses and ideas from becoming conscious. However, the preconscious is also the depository of information which has been pushed out of consciousness, and which may be shoved so deeply underground, so to speak, that the information becomes completely unconscious. Therefore, although separated, these mental realms interact and can influence one another.

Inspired by the theory of nuclear magnetic resonance (NMR), we have built a model of handling a mental qu-bit with the help of pulses of a mental field. Starting with an elementary interaction between two qu-bits we build two-qubit quantum logic gates that allow information to be transferred from one qu-bit to the other. For example, we build the controlled-NOT (CNOT) gate in which, under certain circumstances, the information is transferred from the control qu-bit to the target qu-bit. We also build swapping in which there is a complete exchange of information between two qu-bits. In those manners we build quantum processes that permit consciousness to "read" the unconscious and vice-versa. The most elementary interaction, e.g. between a pre-consciousness qu-bit and a consciousness one, allows us to predict the time evolution of the pre-consciousness + consciousness system in which pre-consciousness and consciousness are quantum entangled. This time evolution exhibits Rabi oscillations that we name mental Rabi oscillations. This time evolution shows how, for example, the unconscious can influence consciousness and vice-versa.

The pulses of the mental field can be emitted either by consciousness (effects of will or freewill) or by the unconscious (individual, group or collective). As we said, together with quantum entanglement, they can explain the awareness of unconscious components. In this case we can say that consciousness measures the unconscious like an experimental physics device records a microscopic process. As we said, quantum entanglement explains also the influence of the unconscious on consciousness and the reciprocal influence of consciousness on the unconscious. We have studied these two types of influences in the case of mourning and we have seen how they could allow mourning to be achieved with time.

A third mental process, already mentionned above, is the quantum entanglement between two (or several) unconscious psyches. The evolution in time of the state of the two quantum entangled unconscious shows the reciprocal influence of each unconscious on the other one. Then through the interaction of their two unconscious, a psychoanalyst named Alice can help Bob to achieve relief from his mourning.

The fundamental characteristic of the most elementary interaction between two mental qubits, e.g. between a qu-bit of pre-consciousness and a qu-bit of consciousness, is to highlight, as a function of time, oscillations between two quantum states made of two correlated qu-bits; i.e. the states $/I1>/C0>$ and $/C1>/I0>$ (I for "Insight" or pre-consciousness and C for "Consciousness").

Let us notice that at the level of the brain, there is evidence of an alternating activity of the two hemispheres (Joseph 1982, 1988). This oscillation expresses itself in the phenomenon of binocular rivalry (Blake, 1989). When two images are presented to each of the two eyes of a subject, they enter in "competition" so that one image is visible while the other is not. The same happens when the subject is presented with two superposed images, a nice metaphor to represent the superposition of two quantum states.

3. The Right and Left Hemisphere

Joseph (1982, 1988) proposed that this oscillating activity and the different functions controlled by the right and left half of the brain, could explain some mental phenomenon associated with the conscious and unconscious mind, with the corpus callosum, a major cord of nerves, linking the two brain halves and thus acting as a bridge between these two mental realms. This is similar to Freud's concept of the preconscious linking the conscious and unconscious mind. Further, Joseph (1982, 1988) linked this oscillating activity to dream recall vs dream forgetting, with the right hemisphere (the domain of visual-spatial imagery and social-emotion) producing the dream during high levels of oscillating activity, and the left (the domain of language and verbal thought) forgetting the dream due to low levels of activity.

Mental (Rabi) oscillations are still to be studied. In particular in the case of the "asleep" consciousness, the unconscious + consciousness system (or at least a part of this system) constantly oscillates between the states $/U1 > /C0 >$ and $/C1 > /U0 >$. A pendulum does not measure time. For this we need a system that keeps the memory of the number of the oscillations of the pendulum. This is a function of a clock, which measures time. In a clock the oscillations of the pendulum have a cumulative effect that allows us to keep the memory of the number of oscillations. In the case of the Rabi oscillations of the unconscious (pre-consciousness)+ consciousness system, we have to imagine a system, correlated to the first one, that is subject to cumulative effects and that allows to memorize the mental Rabi oscillations. In this case, it is only thanks to the storage of the mental Rabi oscillations that consciousness or preconsciousness or the unconscious can be modified.

4. The Limbic System

At the level of the brain this memorization can be actuated by the limbic system, and in particular by the hippocampus (Joseph 1992). To be more specific, the limbic system is the domain of our emotions, and is classically associated with the "four F's:" feeding, fighting, fleeing, and sexual behavior. The amygdala and the hippocampus are some of the main structures of the limbic system, with the amygdala playing a major role in emotional memory, and the hippocampus in storing non-emotional memories. So, in some respects, the limbic system could be compared to the unconscious (Joseph 1992). The limbic system is part of the old brain, and is buried beneath the new brain, which consists of "new cortex," i.e. neocortex. It is the neocortex, particularly that of the left hemisphere, which we associate with human thought, language, rational behavior, and the conscious mind. However, it is the limbic system and the right hemisphere which become most active during dream-sleep (Joseph 1988, 1992).

For an asleep consciousness the perturbations coming from the environment are weak. In these conditions, the mental Rabi oscillations may extend over a time that can be long, probably of the order of several minutes, or more. The situation is totally different for an awaken consciousness. Its interaction with the environment, which operates via the sensory system, perturbs the interaction between pre-consciousness and consciousness and therefore interferes with the oscillations that, as a consequence, cannot last very long. The time for the awaken consciousness to receive an external stimulus being of the order of half a second. The Rabi mental oscillations cannot last more than that.

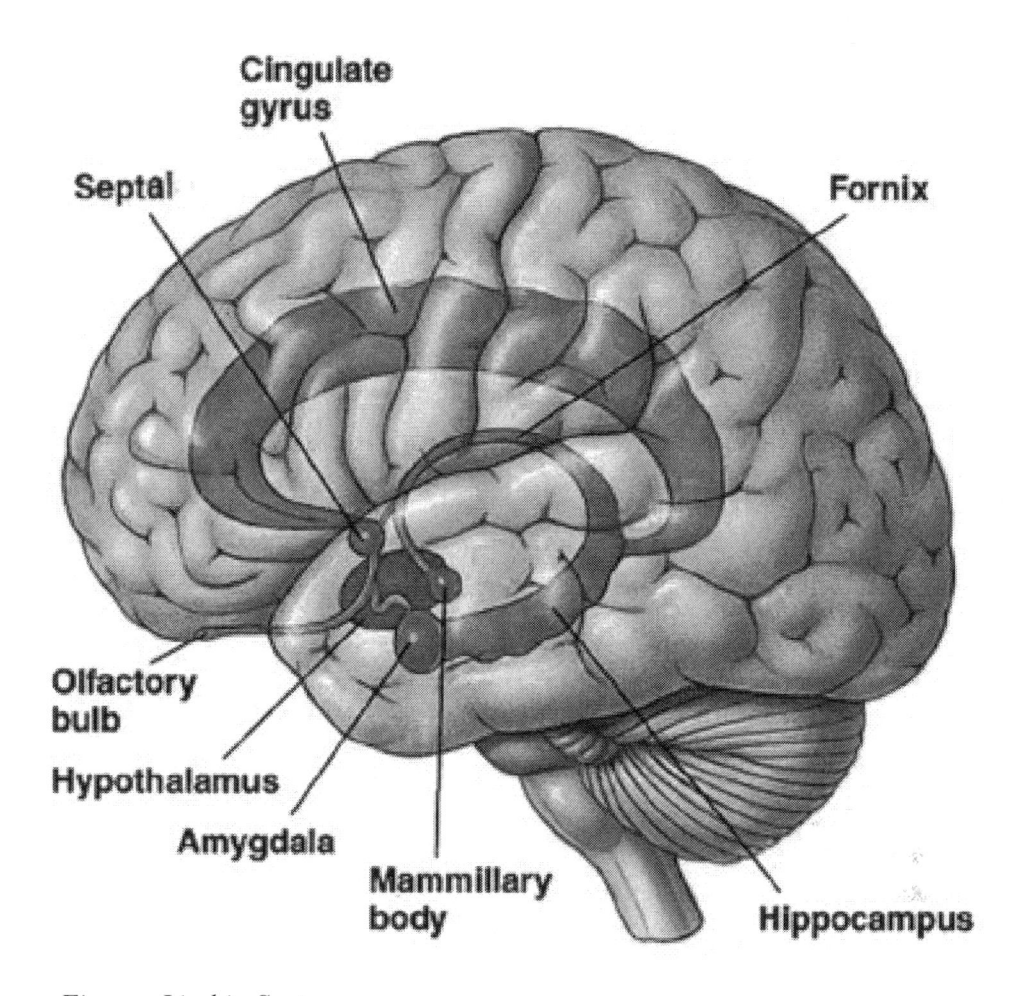

Figure: Limbic System

From a neurological perspective, this can be explained as follows: During dream sleep the left half of the brain is at a low level of activity and it cannot perceive or respond to outside sensory impressions unless they are sufficiently arousing they trigger wakefulness. By contrast, because the right half of the brain and limbic system are at a high level of unconscious-activity during dream-sleep, they can respond to external stimuli (Joseph 1982, 1988). This explains how external stimuli can become incorporated into a dream. A typical example, the dreamer is walking down a strange street when a little boy on a bike rides by ringing a bell. The bell is so loud the dreamer wakes up to discover the alarm clock is ringing. However, upon achieving full awake consciousness, the awake-consciousness may forget the dream; that is, the awake part of the psyche (the left hemisphere) forgets the dream which may remain stored in the limbic system of the right hemisphere. However, later that day, the awake dreamer sees a little boy on a bike ringing his handle-bar bell, and the awake dreamer experiences synchronicity; he remembers his dream.

This same framework can be applied to individuals who engage in the same behaviors. They may respond, unconsciously, to the same stimuli which, for a variety of reasons may not be perceived by the conscious mind. Moreover, these unconscious realms are more closely attuned to the collective unconscious and most probably to theunus mundus, the unified reality from which everything has its origin, and from which all things emerge. Although not perceived by the conscious mind, the unconscious does respond, and then influences the conscious mind of different individuals, who then engage in the same behaviors or come up with the same thoughts or ideas.

5 Conclusions

In summary, some mental phenomena are not explainable in the framework of what we call "classical" mechanics. Let us cite, among others, the phenomenon of awareness, the correlations at a distance between individuals, and more generally the synchronicity phenomena. These three types of phenomena can be explained in the framework of quantum mechanics, particularly, thanks to quantum entanglement for the correlations at a distance between individuals (synchronicity phenomena of type I) and thanks to what we can call the classical illusion of the collapse of the wave-function for the synchronicity phenomena of type II.

Let us notice that the existence of synchronicity phenomena prevents the mental states to be reducible to physical states of the brain. The mental states are correlated to such states, probably via quantum entanglement, but they are not reducible to those states. Therefore this invalidates the materialistic hypothesis.

The projection of our subjectivity in the environment in which we live (synchronicity phenomena of type II), in agreement with quantum mechanics, refutes the local hypothesis ("each individual is in his parcel of space-time") as well as the realistic hypothesis ("the object has a reality well defined independent of the subject who observes it").

As an end let us mention a quantum effect that can have important consequences in mental phenomena, for example for awareness (for the emergence of consciousness). It is the Bose-Einstein condensation, in which each particle looses its individuality in favour of a collective, global behaviour.

References

Atmanspacher, H. (2004). Quantum theory and consciousness: an overview with selected exam- ples, Discrete Dynamics in Nature and Society, 1, 51-73 (2004)

Baaquie, B.E., and Martin, F. (2005). Quantum Psyche - Quantum Field Theory of the Human Psyche, NeuroQuantology, 3, 7-42.

Blake, R. (1989). A neural theory of binocular rivalry, Psychological Review, 96, 145-167

Freud, S. (1900). The Interpretation of Dreams. Standard Edition, 5, 339-622.

Freud, S. (1915a). Repression. Standard Edition, 14, 141-158.

Freud, S. (1915b). Repression. Standard Edition, 14, 159-204.

Galli Carminati, G., and Martin, F. (2008). Quantum mechanics and the Psyche, Physics of Particles and Nuclei, Vol. 39, 560-577.

Galli Carminati, G., and Carminati, F. (2006). The mechanism of mourning: an anti-entropic mechanism, NeuroQuantology Journal, 4, 186-197.

Joseph, R. (1982). The Neuropsychology of Development. Hemispheric Laterality, Limbic Language, the Origin of Thought. Journal of Clinical Psychology, 44, 4-33.

Joseph, R. (1988). The Right Cerebral Hemisphere: Emotion, Music, Visual-Spatial Skills, Body Image, Dreams, and Awareness. Journal of Clinical Psychology, 44, 630-673.

Joseph, R. (1992). The Limbic System: Emotion, Laterality, and Unconscious Mind. The Psychoanalytic Review, 79, 405-456.

Jung, C. G., and Pauli, W. (1955). The Interpretation of Nature and the Psyche, Pantheon, New York, translated by P. Silz (1955); german original: Naturekluarung und Psyche, Rascher, Zurich (1952).

Martin, F., and Galli Carminati, G. (2007). Synchronicity, Quantum Mechanics and Psyche, talk given at the Conference on "Wolfgang Pauli's Philosophical Ideas and Contemporary Science", May 20-25, 2007, Monte Verita, Ascona, Zwitzerland; published in Recasting Reality, pp. 227-243, Springer-Verlag, (2009).

Martin, F., Carminati, F., and Galli Carminati, G. (2009). Quantum information, oscillations and the Psyche, to be published in Physics of Particles and Nuclei (2009).

Primas, H. (2003). Time-Entanglement Between Mind and Matter, Mind and Matter, Vol. 1, 81-119.

II. Uncertainty Principle, Multiple Worlds, Wave Functions, Entanglement, Violations of Causality, and Paradoxes of Time Travel

3. Classical Anthropic Everett Model: Indeterminacy in a Preordained Multiverse
Brandon Carter

LuTh, Observatoire de Paris, France.

Abstract

Although ultimately motivated by quantum theoretical considerations, Everett's many-world idea remains valid, as an approximation, in the classical limit. However to be applicable it must in any case be applied in conjunction with an appropriate anthropic principle, whose precise formulation involves an anthropic quotient that can be normalised to unity for adult humans but that would be lower for infants and other animals. The outcome is a deterministic multiverse in which the only function of chance is the specification of one's particular identity.

Key Words: Consciousness, quantum entanglement, Many Worlds, Multiverse

1. Introduction

Before the twentieth century, classical probabilistic models - such as those developed by Maxwell and Boltzmann for the treatment of many particle systems - were commonly considered as approximations of an objective de-terministic reality of which the details were unknown or at any rate too complicated to be tractable. However since the advent of quantum theory, it has come to be widely recognized that - as Berkeley had warned - such an objective material reality may not exist. A purported refutation of the bishop's scepticism had been provided by Johnson's famous stone kicking experiment (Deutsch, 97), but the learned doctor might not have remained so cockily confident if, instead of a tamely decoherent stone, he had tried kicking the closed box containing Schroedinger's superposed live-and-dead cat.

According to our modern understanding, classical probabilistic models should be considered as approximations, not of illusory material reality, but of more elaborate quantum theoretical models, whose interpretation is to a large extent subjective rather than objective. A complete understanding would therefor require a theory of the sentient mind - as distinct from, though correlated with, the physical brain.

The question of the relationship between our physical brains - the object of study by neurologists - and the thoughts and feelings in our "conscious" minds was already a subject of philosophical speculation long before the development of quantum theory. As very little substantial progress had been achieved, it was natural that some people should wonder whether a resolution of the mystery of quantum theory might provide a resolution of the mystery of the mind. A more common opinion has however been expressed by Steven Weinberg (1995), who wrote "Of course everything is ultimately quantum mechanical: the question is whether quantum mechanics will appear directly in the theory of the mind, and not just in the deeper level theories like chemistry on which the theories of the mind will be based ... Penrose may be right about that, but I doubt it."

I am inclined to share this common opinion, and will proceed here on that basis, not just because of the relatively macroscopic (multiparticle) nature of the neurons constituting the brain, but because quantum theory is not really essential for what is commonly considered to be the crux of the mind to matter relationship, namely what is known as the "collapse" of the "wave function" which is supposed to result from an observation of the kind exemplified by Schroedinger's gedanken experiment in which a cat in a box is liable to be killed by a pistol triggered by a Geiger counter.

2. The Trouble with the Traditional Doctrine

According to the "Copenhagen" interpretation, the relevant "wave func tion" collapses either to a pure state in which the cat is unambiguously alive, or else one in which it is unambiguously dead, when a human "observer" opens its box. The trouble with the Copenhagen interpretation is that it denies "observer" status to the occupant of the box, which is questionable even in the case of a humble cat, and would clearly be quite inadmissible if the cat were replaced by another human.

However as well as the underlying symmetry between the person at risk and the person who observes, the point I want to emphasize here is that the issue is not essentially quantum mechanical, because it subsists even if one goes over to the (decoherent) classical limit. In the human case, an analogous classical experiment can be - and historically has been - done with the Geiger triggering mechanism replaced by use of an old fashioned Russian roulette revolver. The classical analogue of the "collapse of the wave function" would be the Bayesian reduction of the corresponding classical probability distribution, from an a priori configuration, in which the outcome is uncertain, to an a posteriori configuration in which the subject of the experiment is either unambiguously alive or else unambiguously dead. To the question of which protagonist has the privilege of making the observation whereby the definitive "collapse" occurs, it is traditionally presumed that Bishop Berkeley's reply would have been been "God!". However physicists (since the time of Laplace) have tried to avoid such ad hoc invocation of a "deus ex machina", and (in the spirit of Ockham's razor) are therefore inclined to prefer the alternative reply that is expressible succinctly as "None!".

Such negation was originally proposed by Everett, and was advocated - but not adequately elucidated - first by Wheeler and subsequently by DeWitt (1973). By thus denying the Copenhagen doctrine of the occurrence of "collapse" as an objective physical process - rather than merely a subjective allowance for new information as in the familiar classical case of Bayesian reduction - Everett got off to a good start. However his attempt to provide a positive interpretation of the meaning of the "wave function" was not entirely successful.

Part of the trouble arose merely from misunderstanding, due to injudicious choice of wording, whereby what I would prefer to refer to as alternative "channels" were called "branches", thereby conveying the misleading idea of a continual multiplication of worlds (Leslie, 1996), whereas (since Everett's idea was that evolution remains strictly unitary) the "worlds" in question are strictly conserved, having neither beginning nor end: what changes is only the resolution of distinction between different "channels", which may become finer (or coarser!) as observational information is acquired (or lost!). A more serious - since not merely semantic - problem by which many people have been puzzled is what Graham (1973) has called the "dilemma" posed by Everett's declaration that the alternative possible outcomes of an observation are all "equally real" though not (if their quantum amplitudes are different) "equally probable". As I have argued previously (Carter, 2004), and will maintain here, the resolution of this dilemma requires the invocation of an appropriate anthropic principle.

3. The Concept of Reality

It was recognised long ago by Berkeley, and has been emphasized more recently by Page (1996), that the only kinds of entities we know for sure to be real are our mental feelings and perceptions (including dreams). The material world in which we have the impression of living is essentially just a theoretical construct to account for our perceptions. In the dualist (Cartesian) picture that used to be widely accepted, this material world was supposed to have a reality of its own, on par with the realm of feelings and perceptions. However under closer scrutinary such separate material reality has turned out to be illusive, so we find ourselves glimpsing a more mysterious but apparently unified quantum picture. Following the approach initiated by Everett (DeWitt, 1973; Graham, 1973), diverse attempts to sketch the outlines of such a unified picture have been made, albeit with only rather limited success so far, by various people (Deutsch, 1999; Wallace, 2003; Reaves, 2004), and in particular - from a point of view closer to that adopted here - by Page (1996) and the present author (Carter, 2004).

Assuming, as remarked above, that mental processes have an essentially classical rather than quantum nature, this essay has the relatively modest purpose of attempting to sketch the outlines of a simpler, more easily accessible, classical unification that may be useful pedagogically and, in appropriate circumstances, as an approximation to a more fundamental quantum unification that remains elusive. The picture proposed here is based on the use of an appropriate anthropic principle in conjunction with the Everett approach, which is relatively well defined in the classical limit, so that the notions of "equal reality" and "unequal probability" can be clarified in a coherent manner.

Deutsch (1999), Wallace (2003), and Greaves (2004) have developed an alternative approach that attempts to do this in terms of the kind of probability postulated in decision theory, on the debatable supposition that the relevant observations are performed by "rational agents". The essentially different approach advocated here is based on probability of a kind proportional to the amount of perception that is "real", in the sense not of Deutsch (1997) but of Page (1996) - as based on sentience rather than rationality. Following a line of thought originated by Dyson (1979), I have suggested (Carter, 2004, 2007) that the relevant amount of perception should in principle be measured by the corresponding Shannon type information content, but in practice that does not tell us much, as it leaves us with the unsolved question of which of the many processes going on in the brain are the ones that actually correspond to sentient perception. This fundamental question does not matter so long as we are concerned only with the standard, narrowly anthropic, case of adult humans, for whom (as in the example of the next section) it can reasonably be assumed that such processes go on at roughly the same average rate. However for more general applications it would be necessary to face the intractible problem of estimating the relevant anthropic quotient q, meaning an appropriate correction factor that might be larger than unity for conceivable extraterrestrials, but that would presumably be smaller for extinct hominids, and much smaller for other animals as well as for infants of our own species. The easiest non-trivial case to deal with would presumably be that of ordinarily senile members of our own species, as their mental processes are similar to those of adults in their prime except for a reduction in speed that can be allowed for by a factor q that should be clinically measurable (and of practical interest for therapeutic purposes).

4. Russian Roulette: A Historical Example

It is customary (Greaves, 2004) to demonstrate the application of such principles by idealized gedanken experiments in which, if there are just two protagonists, their initials are commonly taken to be A for Alice and B for Bob (while to illustrate merely logical, rather than physically conceivable possibilities, it is common (Bostrom, 2002) to consider examples that are not just idealised but frankly fantastic, in which case the protagonists are referred to as "Adam and Eve"). However to emphasise that I am concerned with what is "real" I shall take as a (simplified and approximate, but not artificially idealized) example an experiment that is not merely hypothetical, but that really occurred as a historical event during the XXth century, with a principle protagonist whose initial was actually not A but G.

Consciousness, Neuroscience, **Time Travel**

To illustrate the basic idea, I propose to consider a modified Schroedinger type experiment in which G - an unbalanced adolescent at the time - voluntarily and crazily took the role of the cat, in a solitary game of Russian roulette. The role of the external observer was taken by his big brother (the owner of the revolver) to whom I shall refer by the letter B. Having first heard about it privately from someone who had been neighbor at the time, I read about it many years later in published memoirs of G, who not only survived the experiment but recovered his mental equilibrium and lived to a ripe old age - at least in our particular branch-channel of the multiverse. To keep the arithmetic simple, I shall postulate that the revolver was just a five-shooter, of the compact kind that is most convenient as a concealed weapon. (In reality it may well have been a six-shooter of the kind familiar in cowboy movies, but it can safely be presumed that it was not what was originally used by the reputed inventors of Russian roulette, namely Czarist officers, whose standard service revolvers were actually seven-shooters.) The protocol of the potentially suicidal game is to load just one of the cartridge chambers and then to whirl it to a random position before pulling the trigger. In such a case, starting from initial conditions that are imperceptibly different, there will be five equally likely outcomes, of which four will be indistinguishable for practical purposes, whereas the other one will be fatal.

According to the traditional single-world doctrine of deterministic classical physics (as still taken for granted at the time of the incident in question) only one of the five possible outcomes would have actually occurred. However according to the Everett type many-world doctrine, a complete description will involve many separately conserved "strands" (commonly but misleadingly referred to as "branches") meaning single worlds of all the five types, in a multiverse consisting of five equally numerous sub-ensembles or "channels", one for each qualitatively distinct possibility. Such sub-ensembles will be characterized by a physical measure given by the fraction p of the total number of strands, which in this case is $p = 1/5$ for each one. Since the four possibilities in which G survives would have been effectively distinguishable (by an examination of the weapon) only for a very short time after the experiment, it will in practice be sufficient for most subsequent purposes to use a coarser representation in which they are regrouped into a single larger multistrand "channel", which will thus have measure $p = 4/5$. When Everett refers to things as "equally real" it is clear that he should be understood to have in mind the individual (single world) strands, rather than their weighted groupings into broader "channels".

The stage at which the original presentation of the Everett approach becomes unclear is when it is suggested that the physical weighting introduced as described should somehow be interpreted as a probability, despite the fact that (as the classical limit of evolution that is strictly unitary in the quantum case) the behavior of the many worlds involved is entirely deterministic, so that when their initial configurations have been specified no uncertainties remain.

To give a meaning to the concept of probability in this context, the purely materialistic framework of the classical many-world system described so far needs to be extended to include allowance for the role of mind. For the simplified classical model considered here, it will be good enough to do this in the usual way, by supposing that mental feelings and perceptions correspond to physical states of animate brains that are roughly localizable on time parametrized world lines of the animals concerned within the single world "strands".

5. Anthropic Quotient

Within the foregoing framework, the incorporation of probability into the model is achieved by an appropriate application of the anthropic principle. In the simple (weak) version that is adequate for the present purpose, the anthropic principle (Carter, 2005, 2010) prescribes that the probability of finding oneself on a particular animate world line on a single strand within a small time interval dt is proportional to q dt, where the "anthropic quotient" q, is normalized to unity in the average (adult) human case. This coefficient q is interpretable as a measure of the relative rate of conscious sentient thought (which might be very low compared with the rate of subconscious but perhaps highly intelligent information processing, such as could be performed by an insentient computer). Whereas it might be higher than unity for conceivable extraterrestrials, q would presumably be lower for other terrestrial species (such as chimpanzees) as well as for infants and senile members of our own species. On short (diurnal) time scales the anthropic quotient of an individual would fluctuate between high waking levels and low dreaming values, and it would of course go to zero at and after the instant of death, as also before conception (though perhaps not before the instant of birth).

In the CAE (classical anthropic Everett) model set up in this way, the meaning of the weighting fraction p of a channel constituted by an ensemble of very similar single-world strands is now clear. It does not directly determine the total probability of finding oneself in that channel, but it does determine the probability dP of finding oneself within a time interval dt on a world line of a particular kind (such as that of G, or alternatively that of B in the example described above) within the channel in question, according to the specification dP p q dt (with the proportionality factor adjusted so that the total probability for all possibilities adds up to unity).

Let us see how this works out in the simple example of the roulette gamester G and his brother B, as shown in the figure, on the assumption that both can be considered as average adults characterized by $q = 1$. To keep the figures round, let us take it that in the first channel, with $p1 = 4/5$, both roulette gamester G and his brother B survived 6 times longer (to an age of about 90) than G did in the second (fatal) channel, with $p2 = 1/5$, where the life of B would have been unaffected (while that of G would have been truncated at about age 15). This can be seen to imply that one is 20 per cent more likely to find oneself to be B than G. In the former case, one will have a 20 per cent chance of being in the fatal channel, and thus of witnessing the death of one's younger brother. In the latter case, that is to say conditional on being G, one will have a 20 per cent chance of finding oneself in the time interval before the game, and thus with only a 4 per cent chance of being in what will turn out to be the fatal channel.

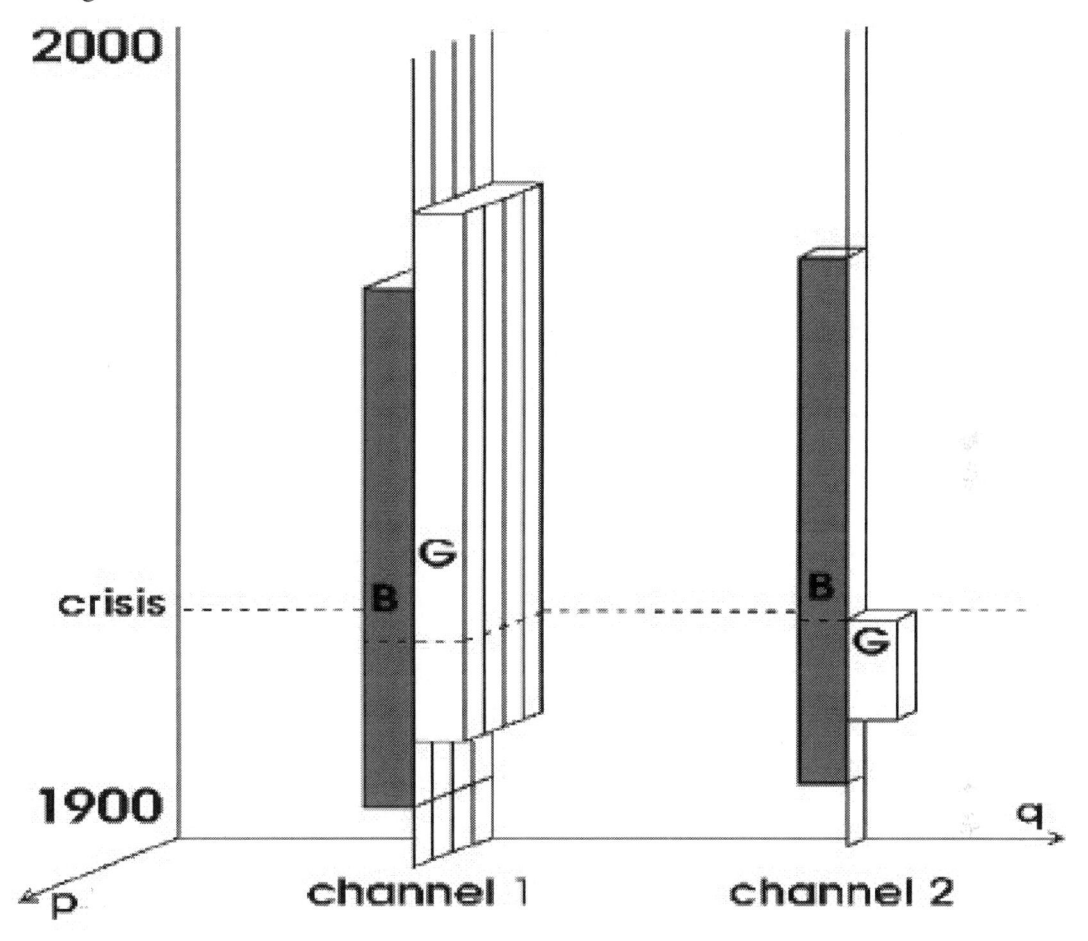

Figure 1: Crude anthropic biograph of XXth century roulette gamester G (pale shading) and his brother B (dark shading) using the vertical direction for time, while the thickness of a worldline in the sideways direction measures subjective anthropic probability weighting per unit time, as specified by the anthropic quotient q which is set to zero before birth and after death, and is taken here to have a uniform unit (average) value during life - whereas in a less crude version it would taper off at the beginning (infancy) and the end (senility). For each of the (Everett type) channels, the 3rd dimension - out of the page or screen - measures the number of "strands", representing the objective physical probability p (the square of a corresponding quantum amplitude) which is conserved. In such an anthropic diagram, the probability of finding oneself to be in a particular state

of a particular person during a particular time interval is proportional to the relevant volume (the time integral of the product pq of the anthropic and physical probability measures). In channel 1 - the one we know about historically, because we are on it ourselves - both brothers survived through a complete life span until old age, the younger naturally outliving the elder. In channel 2, for which the life of G was truncated after only one 6th of its natural span, it is supposed that the subsequent life of B would not have been substantially affected.

If B and G had been the only sentient inhabitants of the world it can be seen that the a priori odds against channel 2 would have been 48 to 7 which is almost 7 to 1. However when account is taken of all the rest of the population (who would not have been significantly affected by the outcome of the roulette game) it can be seen that the a priori odds against finding oneself in channel 2 (and thus deprived of access to G's later literary output) would actually have been barely greater than 4 to 1 (the value given by the ratio $p1/p2$ of the naive physical probabilities designated by p). A more complete picture, allowing for the many other inhabitants of the world, would of course require a much finer decomposition involving far more than two qualitatively distinct channels. Indeed a complete multibiography just of G alone would probably require many more channels to allow for the vicissitudes of his later life, which extended not just through the Second World War but even through the Cold War. In particular - to be fully realistic - an adequate multi-history of the latter would presumably require the inclusion of non-negligibly weighted channels in which an incident such as the Cuban missile crisis terminated in the catastrophic manner envisaged by Shute (1957).

6. Commentary

Although the interpretations - and perhaps the ethical implications - are different, there is no effectively observable distinction between what is predicted by the deterministic many-world CAE model presented here (in which only one's identity is unforeseeable) and what is predicted by the corresponding classical model of the ordinary single-world type (in which the material physical outcome depends on chance). It might therefore be argued that the traditional single-world model should be preferred on the grounds that it is simpler, or less ontologically "bloated". It is however to be recalled that a classical model cannot claim to represent ultimate reality, but merely provides what is at best an approximation to a more accurately realistic quantum model, a purpose for which the traditional single world-model is not so satisfactory.

Another point to be emphasized is that the ontology in question involves only mental feelings and perceptions. As foretold by Berkeley, but contrary to what used to be taught by "positivists" such as Mach, matter, as incorporated in physical fields over spacetime, should not be considered to have objective "reality", but has the status merely of mathematical machinery (that might be replaced for predictive purposes by an equivalent action at a distance formulation based on Green functions).

Having recognized that the relevant ontology does not involve matter but only mind, one is still free to entertain different opinions about how extensive or "bloated" (Leslie, 1983) it may be. The anthropic measure characterized by the coefficient q merely determines the relative probability of the perceptions in question, but but not the absolute number of times they occur. If ontological economy is a desideratum, it might seem preferable to postulate the actual occurrence only of a fraction of the perceptions admitted by the theory. On the other hand for those concerned with economy only in the sense of Ockham's razor, and particularly for those who are unhappy with the concept of probability except when it can be prescribed in terms of relative frequencies, the most attractive possibility would presumably be to suppose that all the perceptions admitted by the theory actually occur (in the indicated proportions). Although it is more ontologically extravagant, the latter alternative has the advantage of conforming to the requirement that was expressed in metaphorical language by Einstein's edict that "God does not play dice". However that is not for us mere mortals to judge: as far as scientific observation is concerned there is no way of telling the difference.

A more mundane issue (with ethical implications concerning protection from inhumane treatment) is the evaluation of the appropriate anthropic quotient q for non-human terrestrial animals (such as the cat considered by Schroedinger) and particularly for infants of our own species. It is to be presumed that q should be of the order of unity for extinct hominids such as homo erectus, whose integrated population time is at most comparable with our own (Carter, 2010). However the observation (Standish, 2008) that we do not belong to the far more numerous populations of animals of other, less closely related, kinds suggests that their anthropic quotients should be much lower, and hence, by analogy, that the same may apply to infants.

Acknowledgement The author is grateful for stimulating discussions with John Wheeler, Roger Penrose, and Bryce DeWitt on various past occasions, and also for more recent discussions with John Leslie and Don Page.

References

Bostrom, N. (2002). Anthropic bias: obsrvation selection effects in science and philosophy (Routlege, New York).

Carter, B. (2004). Anthropic interpretation of quantum theory, Int. J. Theor. Phys. 43, 721-730. [hep-th/0403008]

Carter, B. (2007). Microanthropic principle for quantum theory, in Universe or Multiverse, ed B. Carr (Cambridge U. Press. 285-319.

Carter, B. (2010). Hominid evolution: genetics versus memetics. [arXiv:1011.3393]

Deutsch, D. (1997). The Fabric of reality (Penguin, London).

Deutsch, D. (1999). Quantum theory of probability and decisions, Proc. Roy. Soc. A455, 3129-3137 [quant-ph/9906015].

DeWitt, B.S. (1973). The many-universes interpretation of quantum theory, in The many worlds interpretation of quantum theory, ed. B.S. De Witt, N. Graham (Princeton University Press, 1973) 167-218.

Dyson, F.J. (1979). Time without end: physics and biology in an open system, Rev. Mod. Phys. 51, 447-460.

Graham, N. (1973). The measurement of relative frequency, in The many worlds interpretation of quantum theory, ed. B.S. De Witt, N. Graham (Princeton University Press, 1973) 229-253.

Greaves, H. (2004). Understanding Deutsch's probability in a deterministic multiverse, Studies in the history and philosophy of modern physics 35, 423-456. [quant-ph/0312136].

Leslie, J. (1983), Cosmology, probability, and the need to understand life, in Scientific Explanation and Understanding, ed. N. Rescher (University Press of America, 1983) 53-82.

Leslie, J. 1996, A difficulty for Everett's many-worlds theory, International studies in the philosphy of science 10, 239-246.

Page, D. (1996), Sensible quantum mechanics: are probabilities only in the mind?, Int. J. Mod. Phys. D5, 583-596. [gr-qc/9507024]

Shute, N. (1957), On the beach (Ballantine, New York).

Standish, R. K. (2008). Ants are not Conscious. [arXiv:0802.4121]

Wallace, D. (2003). Everettian rationality: defending Deutsch's approach to probability in the Everett interpretation, Studies in the history and phi- losophy of modern physics 34, 415-439 [quant-ph/0303050].

Weinberg, S. (1995). Reductionism Redux, New York Review of Books XLII Oct. 5, 39-42.

4. Quantum Paradoxes of Time Travel: The Uncertainty Principle, Wave Function, Probability, Entanglement, and Multiple Worlds
R. Gabriel Joseph

Cosmology.com / BrainMind.com

Abstract:

In quantum mechanics the cosmos as a whole can be likened to a quantum continuum which is continually in flux and is thus indeterminate except at the moment of perception and registration by an observing consciousness or measuring apparatus. Because of this continual fluctuation and the limitations of conscious perceptual capacities and through a phenomenon known as "entanglement," it is only possible to make predictions about what may be observed; and these predictions can only be based on probabilities and a probability distribution. When quantum mechanics are applied to the concept of "time" then what is conceptualized as "past" "present" and "future" is also best described in terms of probabilities. Time is uncertain and not deterministic. Causes may occur simultaneously with or even after effects become effects, as demonstrated by entanglement. The experience of time or the existence of an object in space, are also manifestations of the wave function. As the wave propagates through space it effects the continuum both locally and at a distance simultaneously as demonstrated by entanglement, where even choices made in the future can effect the present. If time has a wave function, then the present can effect not just the future, but the future can effect the present and the past, as time is a continuum. Time is entangled. If considered as a unity with no separations in time and space except at the moment of conscious observation, then to effect one point in time-space is to effect all points which are entangled in spacetime. Time and events occurring in time, through entanglement, and as a manifestation of the quantum continuum, can therefore change the future and the past and events occurring in time simultaneously. The present and the future may change the past as all are interconnected, thereby giving rise to paradoxes where the past may be changed such that it becomes a different past. This paradox, however, can be resolved through Everett's conception of Many Worlds. The past which is changed, is just one past among many. Hence, in terms of the "grandfather" paradox, for example, one may travel back in time but the "grandfather" they kill would not be their "grandfather," but the "grandfather" of their doppelganger who exists in an alternate world as there are innumerable worlds each with their own probable existence and space-time.

Quantum Physics, **Retrocausation, PreCognition, Entanglement,**

Keywords: Time Travel Paradoxes, Everett's Multiple Worlds, Uncertainty Principle, Wave Function, Probability Function

TimeSpace in Relativity

In 1904, Lorentz introduced a hypothesis that moving bodies contract in their direction of motion by a factor depending on the velocity of the moving object. Time can therefore also contract such that the future and the present come closer together. He also argued that in different schemes of reference there are different apparent times which differ from and replace "real time." He also argued that the velocity of light was the same in all systems of reference. In 1905 Albert Einstein seized on these ideas and abolished what Lorentz called "real time" and instead embraced "apparent time." In his theories of special relativity, Einstein promoted the thesis that reality and its properties, such as time and motion had no objective "true values" but were "relative" to the observer's point of view (Einstein, 1905a,b,c). Einstein's conceptions of reality and time, therefore, differed significantly from that of Newton.

Time is relative to the observer (Einstein 1905a,b,c, 1906, 1961). Since there are innumerable observers, there is no universal "past, present, future" which are infinite in number and all of which are in motion. There is more than one "present" and this is because time is not the same everywhere for everyone, and differs depending on gravity, acceleration, frames of reference, relative to the observer (Einstein 1907, 1910, 1961). Time is relative and there is no universal past. No universal future. And no universal now. The "past" in another galaxy overlaps with the "present" on Earth. The "present" in another galaxy will not be experienced on Earth until the future. There is no universal now (Einstein 1955). Time is relative, and so too are the futures, presents, and pasts, which overlap and exist simultaneously in different distant regions of space-time. Time is relative, and the "present" for one observer, in one location, may be the past, or the future, for a second observer on another planet.

Time has energy. As defined by Einstein's (1905b) famous theorem $E=mc^2$, and the law of conservation of energy and mass, mass can become energy and energy can become mass. Space-time is both energy and mass which is why it can be warped and will contract in response to gravity and acceleration (Einstein, 1914, 1915a,b; Parker & Toms 2009; Ohanian & Ruffini 2013). Time and space are linked, thereby forming a fourth dimension, timespace. Time, and conceptions about the past, present of future are therefore illusions, as there is no "future" or "past" but rather there are different locations in space which relative to an observer appear far away or nearby. However, when considered from the perspective of quantum mechanics, timespace is a continuum, a unity, and time does not exist independent of this continuum, except as an act of perceptual registration by consciousness or mechanical means.

Einstein's theories did not replace Newtons. Instead Einstein came up with a new closed system of definitions and axioms represented by mathematical symbols which were are radically different from those of Newton's mechanics. For example, space and time in Newtonian physics are independent, whereas in relativity they are combined and connected by the Lorentz transformations. Moreover, although Newtonian mechanics could be applied to events where velocities are small relative to the velocity of light, Newtonian physics cannot be applied to events which take place near light speeds whereas Einstein's physics can.

By contrast, it is at light speed and beyond, and for objects and particles smaller than atoms where Einstein's theory breaks down and this was recognized in the early 1920s (Born et al. 1925; Heisenberg 1925, 1927). The phenomenon of electricity, electromagnetism and atomic science required a new physics and radically different conceptions of cause, effect, and time.

The Uncertainty Principle: Cause, Effects, Time, and Probability

In 1925 a mathematical formalism called matrix mechanics posed a direct challenge to Newton and Einstein and conceptions of reality (Born et al. 1925; Heisenberg 1925). The equations of Newton were replaced by equations between matrices representing the position and momentum of electrons which were found to be unpredictable. Broadly considered, atoms consist of empty space at the center of which is a positively charged nucleus and which is orbited by electrons. The positive charge of the atom's nucleus determines the number of surrounding electrons, making the atom electrically neutral. However, it was determined that it was impossible to make precise predictions about the position and momentum of electrons based on Newtonian or Einsteinian physics, and this led to the Copenhagen interpretation (Heisenberg 1925, 1927) which Einstein repeatedly attacked because of all the inherent paradoxes. Matrix mechanics is referred to now as quantum mechanics whereas the "statistical matrix" is known as the "probability function;" all of which are central to quantum theory.

As summed up by Heisenberg (1958) "the probability function represents our deficiency of knowledge... it does not represent a course of events, but a tendency for events to take a certain course or assume certain patters. The probability function also requires that new measurements be made to determine the properties of a system, and to calculate the probable result of the new measurement; i.e. a new probability function." Since time is also a property of a system, as events take place in time, then time also, is subject to the probability function.

Quantum physics, as exemplified by the Copenhagen school (Bohr, 1934, 1958, 1963; Heisenberg, 1925, 1927, 1930), like Einsteinian physics, makes assumptions about the nature of reality as related to an observer, the "knower" who is conceptualized as a singularity. As summed up by Heisenberg (1958), "the concepts of Newtonian or Einsteinian physics can be used to describe events in nature." However, because the physical world is relative to being known by a "knower" (the observing consciousness), then the "knower" can influence the nature of the reality which is being observed through the act of measurement and registration at a particular moment in time. Moreover, what is observed or measured at one moment can never include all the properties of the object under observation. In consequence, what is known vs what is not known becomes relatively imprecise (Bohr, 1934, 1958, 1963; Heisenberg, 1925, 1927). Time, therefore, including what is conceptualized as the "now" also becomes imprecise, as well as relative to an observer as predicted by special relativity.

As expressed by the Heisenberg uncertainty principle (Heisenberg, 1927), the more precisely one physical property is known the more unknowable become other properties. The more precisely one property is known, the less precisely the other can be known and this is true at the molecular and atomic levels of reality. Therefore it is impossible to precisely determine, simultaneously, for example, both the position and velocity of an electron at any specific moment in time (Bohr, 1934, 1958, 1963).Time, itself, becomes relativity imprecise even when measured by atomic clocks which slow or speed up depending on gravity and velocity (Ashby 2003, Chou et al. 2010; Hafele & Keating 1972a,b,)--exactly as predicted by Einstein and Lorenz.

Heisenberg's principle of indeterminacy focuses on the relationship of the experimenter to the objects of his scientific scrutiny, and the probability and potentiality, in quantum mechanics, for something to be other than it is. Time, too, therefore, would have potentiality, including what is believed to have occurred in the past (Joseph 2014). Einstein objected to quantum mechanics and Heisenberg's formulations of potentiality and indeterminacy by proclaiming "god does not play dice."

In Einstein's and Newton's physics, the state of any isolated mechanical system at a given moment of time is given precisely. Numbers specifying the position and momentum of each mass in the system are empirically determined at that moment of time of the measurement. Probability never enters into the equation. Therefore, the position and momentum of objects including subatomic particles are precisely located in space and time as designated by a single pair of numbers, all of which can be determined causally and deterministically. However, quantum physics proved that Einstein and Newton's formulation are not true at the atomic and subatomic level (Bohr, 1934, Born et al. 1925; Heisenberg 1925, 1927), whereas experiments with atomic clocks proves that even "moments in time" can vary (Ashby 2003, Chou et al. 2010; Hafele & Keating 1972a,b,).

According to Heisenberg (1925, 1927, 1930), chance and probability enters into the state and the definition of a physical system because the very act of measurement can effect the system. No system is truly in isolation. No system can be viewed from all perspectives in totality simultaneously which would require a god's eye view. Only if the entire universe is included can one apply the qualifying condition of "an isolated system." Simply including the observer, his eye, the measuring apparatus and the object, are not enough to escape uncertainty. Results are always imprecise. Time itself, is relatively imprecise depending on gravity, velocity and the observer's frame of reference.

As determined by Niels Bohr (1949), the properties of physical entities exist only as complementary or conjugate pairs. A profound aspect of complementarity is that it not only applies to measurability or knowability of some property of a physical entity, but more importantly it applies to the limitations of that physical entity's very manifestation of the property in the physical world. Physical reality is defined by manifestations of properties which are limited by the interactions and trade-offs between these complementary pairs at specific moments in time when those moments are also variable. For example, the accuracy in measuring the position of an electron at a specific moment in time requires a complementary loss of accuracy in determining its momentum; and momentum can contract time and the distance between the present and the future. Precision in measuring one pair is complimented by a corresponding loss of precision in measuring the other pair (Bohr, 1949, 1958, 1963); which in turn may be related to variations and fluctuations in time. The ultimate limitations in precision of property manifestations are quantified by Heisenberg's uncertainty principle and matrix mechanics. Complementarity and Uncertainty dictate that all properties and actions in the physical world are therefore non-deterministic to some degree--and the same applies to time and even what is considered cause and effect.

Bohr (1949) holds that objects governed by quantum mechanics, when measured, give results that depend inherently upon the type of measuring device used, and must necessarily be described in classical mechanical terms since the measuring devices functions according to classical mechanics. The measuring device effects the outcome and the interpretation of that outcome as does the observer using that device. "This crucial point...implies the impossibility of any sharp separation between the behaviour of atomic objects and the interaction with the measuring instruments which serve to define the conditions under which the phenomena appear...." (Bohr 1949). Time, however, is also determined by measuring devices, which may fluctuate depending on gravity and velocity, including the velocity of the object being measured--exactly as predicted by relativity.

Evidence obtained under a single or under different experimental conditions cannot be reduced to a single picture, "but must be regarded as complementary in the sense that only the totality of the phenomena exhausts the possible information about the objects." In consequence, the results must be viewed in terms of probabilities when applied to the nature of the object under study and its current and future behaviors in time. Bohr (1949) called this the principle of complementarity, a concept fundamental to quantum mechanics and closely associated with the Uncertainty Principle. "The knowledge of the position of a particle is complementary to the knowledge of its velocity or momentum." If we know the one with high accuracy we cannot know the other with high accuracy at the same time (Bohr, 1949, 1958, 1963; Heisenberg, 1927, 1955, 1958); and this is also because, there is no such thing as "the same time."

Central to the Copenhagen principle is the wave function and the probability distribution, i.e. the results of any experiment can only be stated in terms of the probability that the momentum or position of the particles under observation may assume certain values at a specific time. The probability distribution is a prediction for what may occur in the future, that is, within a predicted range of probabilities. When the experiments are performed many times, and although subsequent observations may differ, they are expected to fall within the predicted probability distribution. This also means that nothing is precisely determined at any particular moment in time (Bohr, 1949, 1963; Heisenberg, 1927, 1930, 1955).

Time and the measuring devices used to calculate time, are relative, and even moments in time may be stretched or contracted relative to an observer's frame of reference. There is no universal now. Thus, even what is described as "now" or the future or the past, must also be subject to a probability function. Time cannot be known precisely, even when measured by atomic clocks (Ashby 2003, Chou et al. 2010; Hafele & Keating 1972a,b,). Thus, even what is considered cause and effect" must be subject to a probability function as the moments embracing the "cause" may overlap and occur simultaneously with or even preceded the "effect" due to the stretching and contraction of local time.

These are not just thought experiments. There is considerable evidence of what Einstein (1955) referred to as "spooky action at a distance" and what is known in quantum physics as "entanglement" (Plenio 2007; Juan et al. 2013; Francis 2012). It is well established that causes and effects can occur simultaneously and ever faster than light speed (Lee et al. 2011; Matson 2012; Olaf et al. 2003); a consequence of the connectedness of all things in the quantum continuum.

For example, photons are easily manipulated and preserve their coherence for long times and can be entangled by projection measurements (Kwiat et al. 1995; Weinfurter 1994). A pump photon, for example, can split light into two lower-energy photons while preserving momentum and energy, and these photons remained maximally entangled although separated spatially (Goebel et al 2008; Pan et al. 1998). However, entanglement swapping protocols can entangle two remote photons without any interaction between them and even with a significant time-like separation (Ma et al., 2012; Megidish et al. 2013; Peres 2000). In one set of experiments entanglement was demonstrated even following a delayed choice and even before there was a decision to make a choice. Specifically, four photons were created and two were measured and which became entangled. However, if a choice was then made to measure the remaining two photons, all four became entangled before it was decided to do a second measurement (Ma et al., 2012; Peres 2000). Entanglement can occur independent of and before the act of measurement. "The time at which quantum measurements are taken and their order, has no effect on the outcome of a quantum mechanical experiment" (Megidish et al. 2013).

Moreover, "two photons that exist at separate times can be entangled" (Megidish et al. 2013). As detailed by Megidish et al (2013): "In the scenario we present here, measuring the last photon affects the physical description of the first photon in the past, before it has even been measured. Thus, the "spooky action" is steering the system's past. Another point of view...is that the measurement of the first photon is immediately steering the future physical description of the last photon. In this case, the action is on the future of a part of the system that has not yet been created."

Hence, entanglement between photons has been demonstrated even before the second photon even exists; "a manifestation of the non-locality of quantum mechanics not only in space, but also in time" (Megidish et al 2013). In other words, a photon may become entangled with another photon even before that photon is created, before it even exists. Even after the first photon ceases to exist and before the second photon is created, both become entangled even though there is no overlap in time. Photons that do not exist can effect photons which do exist and photons which no longer exist and photons which will exist (Megidish et al. 2013); and presumably the same applies to all particles, atoms, molecules (Wiegner, et al 2011).

As demonstrated in quantum physics, the act of observation, measurement, and registration of an event, can effect that event, causing a collapse of a the wave function (Dirac 1966a,b; Heisenberg 1955), thereby registering form, length, shape which emerges like a blemish on the face of the quantum continuum. Likewise, a Time Traveler or particle/object speeding toward and then faster than light and from the future into the past will affect the quantum continuum. By traveling into the future or the past, the Time Traveler will interact with and alter every local moment within the quantum continuum and thus the future or the past.

Entanglement proves that effects may precede causes, and causes and effects may also take place simultaneously. In the quantum continuum, determinism and causes and effects do not always exist and this is because, as Einstein proclaimed: "The distinction between past, present and future is only an illusion."

In quantum mechanisms, although every deterministic system is a causal system, not every causal system is deterministic (Heisenberg (1925, 1927; 1958). Rather, causality is the relationship between different states of the same object at different times whereas what is "deterministic" relates to what may occur, and is better described in terms of probabilities.

According to the Copenhagen interpretation (Bohr, 1949, 1963; Heisenberg, 1958), it is the act of measurement which collapses the wave function. It is also the measurement and observation of one event which triggers the instantaneous alteration in behavior of another event or object at faster than light speeds; i.e. entanglement (Plenio 2007; Juan et al. 2013; Francis 2012). For example, two particles which are far apart have "spin" and they may spin up or down. However, although they are far apart, an observer who measures and verifies the spin of particle A will at the same time effect the spin of particle B, as verified by a second observer. Measuring particle A, effects particle B and changes its spin. Likewise observing the spin of B determines the spin of A. There is no temporal order as the spin of one effects the spin of the other simultaneously, faster than the speed of light. Even distant objects are entangled and have a symmetrical relationship and a constant conjunction (Bokulich & Jaeger, 2010; Plenio 2007; Sonner 2013).

Because the future can effect the past or present, the relationship of cause and effect and energy or mass over time is uncertain and can be described only by probabilities (Born et al 1925, Heisenberg 1925, 1927). Time is uncertain. Temporal succession may have no probable connection with what precedes or follows (Heisenberg 1958). In quantum mechanics, one can know the connection between two events only by knowing the future state--thus one must wait for the future to arrive, or look back upon the future state of similar systems in the past. If one knows the properties of an acorn at an earlier time t1 one still cannot deduce the properties of the oak tree at time t2. This may be possible only in isolated systems (Bohr, 1949; Heisenberg 1958). Thus time must also be isolated. However, unless the entire universe is included in the measurement, then the system, which includes time, is not truly isolated.

The Probability and Wave Function

Quantum mechanics is mechanical but not deterministic and causal relationships are never teleological and not always deterministic. In quantum physics, nature and reality are represented by the quantum state. The electromagnetic field of the quantum state is the fundamental entity, the continuum that constitutes the basic oneness and unity of all things. The physical nature of this state can be "known" by assigning it mathematical properties and probabilities (Bohr, 1958, 1963; Heisenberg, 1927). Therefore, abstractions, i.e., numbers and probabilities become representational of a hypothetical physical state. Because these are abstractions, the physical state is also an abstraction and does not possess the material consistency, continuity, and hard, tangible, physical substance as is assumed by Classical (Newtonian) physics. Instead, reality, the physical world, is a process of observing, measuring, and knowing and is based on probabilities and the wave function (Heisenberg, 1955).

Consider an elementary particle, once its positional value is assigned, knowledge of momentum, trajectory, speed, and so on, is lost and becomes "uncertain." The particle's momentum is left uncertain by an amount inversely proportional to the accuracy of the position's measurement which is determined by values assigned by measurement and the observing consciousness at a specific moment in time relative to that observe and the measuring device. Therefore, the nature of reality, and the uncertainty principle is directly affected by the observer and the process of observing, measuring, and knowing, all of which are variable thereby making the results probable but not completely certain (Heisenberg, 1955, 1958):

"What one deduces from an observation is a probability function; which is a mathematical expression that combines statements about possibilities or tendencies with statements about our knowledge of facts....The probability function obeys an equation of motion as the coordinates did in Newtonian mechanics; its change in the course of time is completely determined by the quantum mechanical equation but does not allow a description in both space and time" (Heisenberg, 1958).

"The probability function does not describe a certain event but a whole ensemble of possible events" whereas "the transition from the possible to the actual takes place during the act of observation... and the interaction of the object with the measuring device, and thereby with the rest of the world... The discontinuous change in the probability function... takes place with the act of registration, because it is the discontinuous change of our knowledge in the instant of registration that changes the probability function." "Since through the observation our knowledge of the system has changed discontinuously, its mathematical representation has also undergone the discontinuous change and we speak of a quantum jump" (Heisenberg, 1958).

Einstein ridiculed these ideas: "Do you really think the moon isn't there if you aren't looking at it?"

Heisenberg (1958), cautioned, however, that the observer is not the creator of reality: "Quantum theory does not introduce the mind of the physicist as part of the atomic event. But it starts from the division of the world into the object and the rest of the world. What we observe is not nature in itself but nature exposed to our method of questioning." Nevertheless, the act of knowing, of observing, or measuring, that is, interacting with the environment in any way, creates an entangled state and a knot in the quantum continuum described as a "collapse of the wave function;" a knot of energy that is a kind of blemish in the continuum of the quantum field. This quantum knot bunches up at the point of observation, at the assigned value of measurement and can be entangled.

The same principles would also apply to time, and to time travel. The act of moving through time would effect time and all local and even more distant events. Traveling through the past or the future would effect every moment of that future; however, exactly what those changes may be, are indeterministic and can only be described by a probability function.

In the Copenhagen model, objects are viewed as quantum mechanical systems which are best described by the wave function and the probability function. "The reduction of wave packets occurs when the transition is completed from the possible to the actual" (Heisenberg, 1958).

The measuring apparatus and the observer also have a wave function and therefore interact with what is being measured. The effect of this is obvious when its a macro-structure measuring a micro-structure vs a macro-structure measuring a macro-structure.

Moreover, according to the uncertainty principle, it is not possible to restrict any analysis to position or moment without effecting the other, and this is because the very act of eliminating uncertainty about position maximizes uncertainty about momentum (Heisenberg 1927). Uncertainty implies entanglement. Likewise, eliminating uncertainty about momentum maximizes uncertainty about position. Instead, one must assign a probability distribution which assigns probabilities to all possible values of position and momentum.

Therefore, no object, or particle, or quanta, or quantum, or moment in time, has its own eigenstate (inherent characteristic). Although every object appears to have a definite momentum, a definite position, and a definite time of occurrence, the object is in flux and it can't have a position and momentum at the same time as there is no such thing as "the same time." Time is also in flux. Therefore, when applied to time, then time, including the future and the past, can only be defined by a probability function. This means, the future and the past may change and that whatever is believed to have taken place or which will take place is best described in terms of probabilities.

Time and Quantum Physics: The Future Can Lead to the Past

In contrast to Newton and Einstein, quantum mechanics concerns itself with the dynamical change of state and its probability coupled with the Schrödinger (1926) time equations which are both time dependent and time independent for particles and waves. The state-function specifies the state of any physical system as a specific time t. The Schrödinger time equations relates states at a series time t1 to a later time t2. In quantum mechanics, the Schrödinger (1926) equation is a partial differential equation that describes how the quantum state of a physical system changes with time. Like Newton's second law ($F = ma$), the Schrödinger equation describes time in a way that is not compatible with relativistic theories, but which supports quantum mechanics and which can be easily mathematically transformed into Heisenberg's (1925) matrix mechanics, and Richard Feynman's (2011) path integral formulation.

Therefore, time, in quantum physics, is not necessarily relative or even a temporal sequence, and the same is true of future and past. As summed up by Heisenberg (1958), "in classical theory we assume future and past are separated by an infinitely short time interval which we may call the present moment. In the theory of relativity we have learned that the future and past are separated by a finite time interval the length of which depends on the distance from the observer..." and where the past always leads to the future. However, "when quantum theory is combined with relativity, it predicts time reversal;" i.e. the future can lead to the past.

Time Is Entangled

Time cannot be separate from the continuum except when perceived as such by an observing consciousness or measuring device, thereby inducing a collapse of the wave function of time; experienced as the present, past, or future.

Time, be it considered a dimension known as timespace, or as a perceived aspect of the quantum continuum, is also subject to entanglement, as all aspects of time are interconnected and indistinguishable until perceived thereby inducing a collapse of the wave function. "A" future can therefore effect "a" past and change it through entanglement and by effecting the wave function.

Given faster than light entanglement, spooky action at a distance and the reality of the wave function, then the laws of physics must allow for information and effects to be conveyed faster than light speed and from the future to the past. If time is considered as a gestalt and a continuum and not a series of fragments, then the future and past are coexistensive.

The quantum continuum is without dimensions and encompasses space and time in its basic unity of oneness. Everything within the quantum continuum can be effected by local effect and distant effects simultaneously at and beyond light speeds. Therefore, the future, and the "present" being part of this continuum can effect the past by effecting the wave function of the past, present, future, and thus, the space-time continuum, as all are entangled.

Light can travel to the future and from the past relative to the observer's frame of reference. However, light and time are not the same. The speed of light, and time, be it past or future, are not synonymous, though both may be affected by gravity (Carroll 2004; Einstein 1961). Even the ticking of atomic clocks is effected by gravity as well as velocity. Time is subject to change, including what is described as "now" as there is no universal "now." Moreover, just as light has a particle-wave duality and can physically interact with various substances, time also can be perceived and therefore must have a wave function if not a particle-wave duality. Time, be it "past" "present" or "future" can be changed.

Time-space is interactional, and can contract to near nothingness and then continue to contract in a negative direction such that the time traveler can journey into the past.

Gravity, Acceleration, Relativity, and the Quantum Mechanics of Time Contraction

Time has energy. As defined by the law of conservation of energy and mass and Einstein's (1905b) theorem $E=mc^2$, mass can become energy and energy can become mass. Space-time is both energy and mass which is why it will contract in response to gravity and acceleration (Einstein, 1914, 1915a,b; Parker & Toms 2009; Ohanian & Ruffini 2013).

Time is perceived. Time is experienced. Time is "something," it exists, and therefore it must have energy and a wave function which is entangled with motion, velocity, gravity, the observer, and the quantum continuum which encompasses space-time.

Time is associated with light (Einstein 1961). Light has a particle-wave duality and travels at a maximum velocity of 186282 miles per second. However, time is not light, and light is not time. Rather, light can carry images reflected by or emitted from innumerable locations in space-time and can convey or transport information from these locations which may be perceived by an observer and experienced as moments in time. For much of modern human history time has been measured by celestial clocks such as the phases of the moon, and the tilt and rotation of Earth and Earth's orbit around the sun which marks the four seasons and the 24 hour day (Joseph, 2011b). Time is a circle and may be segmented into years, months, weeks, days, hours, minutes, seconds, nanoseconds as measured by various clocks from sundials to atomic clocks. However, time, even when measured by atomic clocks, can flow at different rates and speeds such that the "future" and the "past" can overlap and exist simultaneously with the same moment in time, and this is because there is no universal now.

Atomic clocks tick off time as measured by the vibrations of light waves emitted by atoms of the element cesium and with accuracies of billionths of a second (Essen & Parry, 1955). However, these clocks are also effected by their surroundings and run slower under conditions of increased gravity or acceleration (Ashby 2003; Hafele & Keating 1972a,b) In 1971 Joe Hafele and Richard Keating placed atomic clocks on airplanes traveling in the same direction of Earth's rotation thereby combining the velocity of Earth with the velocity of the planes (Hafele & Keating 1972a,b). All clocks slowed on average by 59 nanoseconds compared to atomic clocks on Earth. Time, like the weather, is effected by local conditions. Under accelerated conditions and increased gravity, time slows down; the same conditions which would enable a time traveler to accelerate toward the future and from the future into the past.

It has been demonstrated that atomic clocks at differing altitudes will eventually show different times; a function of gravitational effects on time. The lower the altitude the slower the clock, whereas clocks speed up as altitude increases; albeit the differences consisting of increases of a few nanoseconds (Chou et al. 2010; Hafele & Keating, 1972; Vessot et al. 1980). "For example, if two identical clocks are separated vertically by 1 km above the surface of Earth, the higher clock gains the equivalent of 3 extra seconds for each million years (Chou et al., 2010). The speeding up of atomic clocks at increasingly higher altitudes has been attributed to a reduction in gravitational potential which contributes to differential gravitational time dilation.

Consciousness, Neuroscience, Time Travel

A predicted by Einstein, clocks run more slowly (time contraction) near massive objects whereas time dilates and runs more quickly as gravity is reduced. Increases in altitude and reductions in gravity speed up the clock, whereas decreases in altitude and increases in gravity slow the clock down (Hafele & Keating, 1972; Vessot et al. 1980).

Time must have energy and energy can be converted into mass. Acceleration expands mass (as energy is converted to mass) and increases gravity which contracts time and mass. Increases in gravity can squeeze space-time into smaller spaces such that there is more time in a smaller space. According to Einstein's famous equation: $E = mc^2$, where E is energy, m is mass and c is the speed of light, mass and energy are the same physical entity and can be changed into each other (Einstein 1905a,b,c 1961). Because of this equivalence, the energy an object acquires due to its motion will increase its mass. In other words, the faster an object moves, the greater the amount of energy which increases its mass, since energy can become mass. This increase in mass only becomes noticeable when an object moves very rapidly. If it moves at 10% the speed of light, its mass will only be 0.5 percent more than normal. But if it moves at 90% the speed of light, its mass will double. And as mass increases it also shrinks and its gravity increases. This is because increased mass increases gravity which then pulls on the mass making it shrink toward the center of gravity, all of which contributes to the collapsing and contraction of space time (Carroll 2004; Einstein 1913, 1914, 1915a,b).

A similar principle applies to time travel. By accelerating toward light speed, space-time contracts (Lorentz 1982; Einstein 1961; Einstein et al. 1923), and the distance between the future and the present and distant locations in space time shrinks and are closer together.

Speed, that is velocity, per se is not effected by time travel. Velocity does not contract or dilate. Hence, since space-time contracts as one accelerates (and although time slows down), and as velocity is not effected then one can traverse and journey across this shrinking space more quickly, and cover the distance between the "now" and the "future" more rapidly because they are closer together--and this would be possible only if the "future" already exists, albeit in a different location in spacetime. Distant locations in space-time are no longer so far apart; the result of increased speed and gravity.

The relationship between time dilation and the contraction of the length of space-time can be determined by a formula devised by Hendrik Lorentz in 1895. As specified by the Lorentz factor, γ (gamma) is given by the equation $\gamma =$, such that the dilation-contraction effect increases exponentially as the time traveler's velocity (v) approaches the speed of light c. Therefore, for example, at 90% light speed 2.29 days on Earth shrinks to just one day in the time machine and 7 days in the time machine at this speed, would take the time traveler 16 days into the future. The distance between the present and the future has contracted so that the future arrives in 7 days instead of 16.

Consider for example, 30 feet of space which contracts to 10 feet. Those inside the time machine need only walk 10 feet whereas those outside the time machine must walk 30 feet. Likewise because the time traveler's clock runs more slowly, and since more time is contracted into a smaller space, it might take him 10 minutes to get 30 minutes into the future. By contrast, it takes those outside the time machine longer to get to the future because it is further away and as their clocks are running faster and it takes more time. At 99.999999% the speed of light, almost two years pass for every day in the time machine. At 99.99999999999 % of c, for every day on board, nearly twenty thousand years pass back on Earth. However, upon reaching light speed, time stops. It is only upon accelerating beyond light speed, that time runs backwards and the contraction of space-time continues in a negative direction. One must accelerate toward the future to reach the past.

The shrinkage of space-time has given rise to the famous "twin paradox" (Langevin 1911; von Laue 1913). If one twin leaves Earth and accelerates toward light speed, that twin will arrive in the future in less time than the twin left behind on Earth. Because it took less time, the time traveling twin does not age as much whereas the twin left on Earth ages at the normal rate. Because time-space has contracted, and since it takes less time to get to distant locations which are now closer together, the time traveling twin arrives in the future in less time than her twin on Earth. Hence, the time traveling twin will be younger.

Not just spacetime, but the mass of the object traveling toward light speed also contracts. The amount of length contraction can be calculated and determined by the Lorentz Transforms (Einstein 1961). For example, a 100 foot long time-space ship traveling at 60% the speed of light would contract by 20% and would become 80 feet in length. Presumably, its diameter would remain the same, though the likelihood is that all surrounding space including the diameter of the time machine would contract. If the time-space ship accelerates to 0.87 light speed, it will contract by 50%.

"Length contraction" can be expressed mathematically by the following formula: $E = mc^2/\sqrt{(1-v^2/c^2)}$, which is similar to the equation for time dilation (if one replaces the value of v for 0). As the value of v (velocity) increases, so does an object's mass which requires more energy to continue at the same velocity or to accelerate. Since energy can become mass, mass increases even as the object shrinks and contracts, thereby increasing its gravity which exerts local effects on the curvature of space-time. Not just the time machine, but space-time in front and surrounding the time machine also contracts. Eventually, the time traveler may shrink to the less than the width of a hair--at least from the perspective of outside observers. At near light speed, the time traveler's length would contract to the size of an atom. Once it shrinks in size smaller than a Planck Length, it will have so much mass and energy that it can blow a hole in spacetime and be propelled at superluminal speeds (Joseph 2014)--however once it exceeds light speed, length contraction and the contraction of time continues in a negative direction. Time reverses, and the direction of travel is into the past. One must accelerate to light speed, which takes the time traveler far into the future, and then to superluminal speeds to journey backwards in time, and this means the future leads to the past.

Although seemingly paradoxical, Einstein's theories of relativity (despite his posting of a cosmic speed limit) predicts that the only way to travel into the past is to exceed the speed of light. Upon accelerating toward light speed, space-time contracts and the space-time traveler is propelled into the future. However, it is only upon accelerating into the future and then beyond light speed that the contraction of space-time continues in a negative direction and time flows in reverse. It is only at superluminal speeds that time reverses and one can voyage backward in time. Einstein's general theory of relativity predicts that the future leads to the past. Likewise, as shown by Gödel 1949a,b), Einstein's field equations predict that time is a circle; and this violates the laws of causality (Buser et al. 2013).

Because the present leads to the future which leads to the past, past, present and future are linked in spacetime. The future can therefore effect the past and effects may take place before the cause.

Time, and time-space are embedded in the quantum continuum and can effect as well as be effected by other particle-waves even at great distances; a concept referred to as "entanglement." Time and space-time are entangled.

Probabilities and The Wave Function of the Time Traveler

According to quantum mechanics the subatomic particles which make up reality, or the quantum state, do not really exist, except as probabilities (Born et al. 1925; Dirac 1966a,b; Heisenberg 1925, 1927). These "subatomic" particles have probable existences and display tendencies to assume certain patterns of activity that we perceive as shape and form. Yet, they may also begin to display a different pattern of activity such that being can become nonbeing and thus something else altogether.

The conception of a deterministic reality is rejected and subjugated to mathematical probabilities and potentiality which is relative to the mind of a knower which registers that reality as it unfolds, evolves, and is observed (Bohr 1958, 1963; Heisenberg 1927, 1958). That is, by measuring, observing, and the mental act of perceiving a non-localized unit of structural information, injects that mental event into the quantum state of the universe, causing "the collapse of the wave function" and creating a bunching up, a tangle and discontinuous knot in the continuity of the quantum state.

Therefore, quantum mechanics, as devised by Niels Bohr, Werner Heisenberg, Dirac, Born and others in the years 1924–1930, does not attempt to provide a description of an overall, objective reality, but instead is concerned with quanta, probabilities and the effects of an observer on what is being observed. The act of measurement causes what is being measured to assume one for many possible values at specific moments of time, and yields the probability of an object or particle to be moving at one speed or direction or to be in one position or location, vs many others at a specific moment in time. Thus, it could be said that the act of observation causes a wave function collapse, a discontinuity in the continuum which is interpreted as reality and cause and effect. However, time too, is subject to measurement and can therefor yield different values by being measured. Observing and measuring time causes time to have certain values.

Central to quantum mechanics is the wave function (Bohr, 1963; Heisenberg, 1958). All of existence has a wave function, including light and time. However, quantum physics is also based on the fact that matter appears to be a duality, and can be both a wave and a particle; that is, to have features of both, i.e. particle-like properties and wave-like properties (Niel Bohr's complementary principle). Therefore, every particle has a wave function which describes it and which can be used to calculate the probability that a particle will be in a certain location or in a specific state of motion, but not both at certain moment of time. Again, however, time also has a wave function. Every aspect of existence can be described as sharing particle-like properties and wave-like properties and this would necessarily have to include the experience of time. Time can be perceived, therefore time must have energy, and energy has a particle wave duality.

The wave function is the particle spread out over space and describes all the various possible states of the particle. Likewise, the wave function would describe all the various possible states of time, including past, present, and future. According to quantum theory the probability of findings a particle in time or space is determined by the probability wave which obeys the Schrodinger equation. Everything is reduced to probabilities, including time. Moreover, these particle/waves and these probabilities are entangled.

Reality and the experience of time, are manifestation of wave functions and alterations in patterns of activity within the quantum continuum which are entangled and perceived as discontinuous, and that includes the perception of past, present, future. The perception of a structural unit of information is not just perceived, but is inserted into the quantum state which causes the reduction of the wave-packet and the collapse of the wave function. It is this collapse which describes shape, form, length, width, and future and past events and locations within space-time (Bohr, 1963; Heisenberg, 1958).

In quantum physics, the wave function describes all possible states of the particle and larger objects, including time, thereby giving rise to probabilities, and this leads to the "Many Worlds" interpretation of quantum mechanics (Dewitt, 1971; Everett 1956, 1957). That is, since there are numerous if not infinite probable outcomes, each outcome and probable outcome represents a different "world" with some worlds being more probable than others and each of which may be characterized by their own unique moments in time. "Many Worlds" must include "Many Times."

For example, an electron may collide with and bounce to the left of a proton on one trial, then to the right on the next, and then at a different angle on the third trial, and another angle on the fourth and so on, even though conditions are identical with one exception: they occur at different moments in time. This gives rise to the Uncertainty Principle and this is why the rules of quantum mechanics are indeterministic and based on probabilities. The state of a system one moment cannot determine what will happen the next moment, because moments in time, and thus time itself has a wave function and a probability function. The wave function describes all the various possible states of the particle (Bohr, 1963; Heisenberg, 1958) and that includes the experience of time, including the eternal now.

Wave Functions: The Past, Present and Future Exist Simultaneously

Only when the object can be assigned a specific value as to location, or time, or moment, does it have possess an eigenstate, i.e. an eigenstate for position, or an eigenstate for momentum, or an eigenstate for time; each of which is a function of the "reduction of the wave function;" also referred to as wave function collapse (Bohr, 1934, 1958, 1963; Heisenberg, 1930, 1955, 1958). Wave function collapse, which is indeterministic and non-local is a fundamental a priori principle of the Copenhagen school of quantum physics and so to is the postulate that the observer and the observed, and the past, present, and future, become entangled and effect one another.

Wave function collapse has also been described as "decoherence" which in turn leads to the "many-worlds" interpretation and the thought experiment known as "Schrödinger's Cat'" i.e. is a cat in a sealed box dead or alive? According to the Copenhagen interpretation, there is a 50% chance it will be dead and 50% chance it will be alive when it is observed, but one cannot know if it dead or alive until observed (measured). However, if there are two observers, one in the box with the cat the other outside the box, then the observer in the box knows if the cat is dead or alive, whereas the observer outside the box sees only a 50-50 probability (Heisenberg 1958).

The wave function describes all the various possible states of the particle. Rocks, trees, cats, dogs, humans, planets, stars, galaxies, the universe, the cosmos, past, present, future, as a collective, all have wave functions.

Waves can also be particles, thereby giving rise to a particle-wave duality and the Uncertainty Principle. Particle-waves interact with other particle-waves. The wave function of a person sitting on their rocking chair would, within the immediate vicinity of the person and the chair, resemble a seething quantum cloud of frenzied quantum activity in the general shape of the body and rocking chair. This quantum cloud of activity gives shape and form to the man in his chair, and is part of the quantum continuum, a blemish in the continuum which is still part of the continuum and interacts with other knots of activity thus giving rise to cause and effect as well as violations of causality: "spooky action at a distance."

Since mass can become energy and energy mass, the "field" is therefore a physical entity that contains energy and has momentum which can be transmitted across space. Likewise, since time can be perceived it must have energy and energy mass as well as momentum which can be transmitted across space. Therefore, "action at a distance" may be both distant and local, a consequence of the interactions of these charges within the force field they create in conjunction with the force field know as "time."

Because time has a wave function which interacts with the continuum which includes time, then effects can be simultaneous, even at great distances, and occur faster than the speed of light (Plenio 2007; Juan et al. 2013; Francis 2012; Schrödinger & Dirac 1936), effecting electrons, photons, atoms, molecules and even diamonds (Lee et al. 2011; Matson 2012; Olaf et al. 2003; Schrödinger & Born 1935). Since time has a wave function and is entangled, then effects may precede the cause since time is a continuity, and this explains why effects may take place faster than light.

If considered as a unity with no separations in time and space, then to effect one point in time-space is to effect all points which are entangled; and those entangled connections includes time and consciousness (Joseph 2010a). And this gives rise to the uncertainty principle because all are interactional (Heisenberg, 1927) and there is no universal "now." Everything effects everything else and thus time in the "future" can effect "time" in the "past" via the wave function which propagates instantaneously throughout the continuum.

Likewise, the intrepid time traveler, journeying into the past, is also a wave function; consisting of particles and waves which interact locally with other local waves and creating additional blemishes in the quantum continuum. By traveling into the past or the future, the time traveler would come into contact with and change and alter the wave function of other blemishes in space-time. Hence, speeding into the past would therefore change the past, or rather, local events in that past, even if the Time Traveler sat still and did nothing at all except go with the flow. The wave function of the observer effects the wave function of what is observed and the wave function of immediate surroundings. The backward traveling time traveler effects each moment of "local" space-time as she travels through it. The wave function of the time traveler moving through time would spread out over space, becoming vanishingly small until disappearing.

By traveling into the past, the time traveler changes the past locally and perhaps even at a distance, depending on his actions. Likewise, since the future, present, and past are entangled, events taking place in the future, can effect and alter the past, thereby violating causality such that the past the time traveler visits may no longer be the past he was familiar with.

The probability function and entanglement when applied to the space-time continuum indicates that the, or rather "a" past may be continually changed and altered to varying degrees. This may also explain why memories of the past do not always correspond with the past record (Haber & Haber, 2000; Megreya & Burton 2008). Although blamed on faulty memory, perhaps the past has been and is continually and subtly being altered through entanglement.

As demonstrated in quantum physics, the act of observation, measurement, and registration of an event, can effect that event, causing a collapse of a the wave function (Dirac 1966a,b; Heisenberg 1955),. Likewise, a Time Traveler or particle/object speeding toward and then faster than light and from the future into the past will affect the quantum continuum. By traveling into the future or the past, the Time Traveler will interact with and alter every local moment within the quantum continuum and thus the future or the past. However, the past which is changed, always existed, albeit, as a probability; one past world among infinite worlds each with their own past, presents, and futures.

Everett's Many Worlds

Since the universe, as a collective, must also have a wave function, then this universal wave function would describe all the possible states of the universe and thus all possible universes, which means there must be multiple universes which exist simultaneously as probabilities (Dewitt, 1971; Everett 1956, 1957). And the same would be true of time. Why shouldn't time have a wave function?

The wave function of time means there are infinite futures, presents, pasts, with some more probable than others.

As theorized by Hugh Everett the universal wave function is "the fundamental entity, obeying at all times a deterministic wave equation" (Everett 1956). Thus, the wave function is real and is independent of observation or other mental postulates (Everett 1957), though it is still subject to quantum entanglement.

In Everett's formulation, a measuring apparatus MA and an object system OS form a composite system, each of which prior to measurement exists in well-defined (but time-dependent) states. Measurement is regarded as causing MA and OS to interact. After OS interacts with MA, it is no longer possible to describe either system as an independent state. According to Everett (1956, 1957), the only meaningful descriptions of each system are relative states: for example the relative state of OS given the state of MA or the relative state of MA given the state of OS. As theorized by Hugh Everett what the observer sees, and the state of the object, become correlated by the act of measurement or observation; they are entangled.

However, Everett reasoned that since the wave function appears to have collapsed when observed then there is no need to actually assume that it had collapsed. Wave function collapse is, according to Everett, redundant. Thus there is no need to incorporate wave function collapse in quantum mechanics and he removed it from his theory while maintaining the wave function, which includes the probability wave.

According to Everett (1956) a "collapsed" object state and an associated observer who has observed the same collapsed outcome have become correlated by the act of measurement or observation; that is, what the observer perceives and the state of the object become entangled. The subsequent evolution of each pair of relative subject–object states proceeds with complete indifference as to the presence or absence of the other elements, as if wave function collapse has occurred. However, instead of a wave function collapse, a choice is made among many possible choices, such that among all possible probable outcomes, the outcome that occurs becomes reality.

Everett argued that the experimental apparatus should be treated quantum mechanically, and coupled with the wave function and the probable nature of reality, this led to the "many worlds" interpretation (Dewitt, 1971). What is being measured and the measuring apparatus/observer are in two different states, i.e. different "worlds." Thus, when a measurement (observation) is made, the world branches out into a separate world for each possible outcome according to their probabilities of occurring. All probable outcomes exist regardless of how probable or improbable, and each outcome represent a "world." In each world, the measuring apparatus indicates which of the outcomes occurred, which probable world becomes reality for that observer; and this has the consequence that later observations are always consistent with the earlier observations (Dewitt, 1971; Everett 1956, 1957).

Predictions, therefore, are based on calculations of the probability that the observer will find themselves in one world or another. Once the observer enters the other world he is not aware of the other worlds which exist in parallel. Moreover, if he changes worlds, he will no longer be aware that the other world existed (Everett 1956, 1957): all observations become consistent, and that includes even memory of the past which existed in the other world.

The "many worlds" interpretation (as formulated by Bryce DeWitt and Hugh Everett), rejects the collapse of the wave function and instead embraces a universal wave function which represents an overall objective reality which consists of all possible futures and histories all of which are real and which exist as alternate realities or in multiple universes. What separates these many worlds is quantum decoherence and not a wave form collapse. Reality, the future, and the past, are viewed as having multiple branches, an infinite number of highways leading to infinite outcomes. Thus the world is both deterministic and non-deterministic (as represented by chaos or random radioactive decay) and there are innumerable futures and pasts.

As described by DeWitt and Graham (1973; Dewitt, 1971), "This reality, which is described jointly by the dynamical variables and the state vector, is not the reality we customarily think of, but is a reality composed of many worlds. By virtue of the temporal development of the dynamical variables the state vector decomposes naturally into orthogonal vectors, reflecting a continual splitting of the universe into a multitude of mutually unobservable but equally real worlds, in each of which every good measurement has yielded a definite result and in most of which the familiar statistical quantum laws hold."

DeWitt's many-worlds interpretation of Everett's work, posits that there may be a split in the combined observer–object system, the observation causing the splitting, and each split corresponding to the different or multiple possible outcomes of an observation. Each split is a separate branch or highway. A "world" refers to a single branch and includes the complete measurement history of an observer regarding that single branch, which is a world unto itself. However, every observation and interaction can cause a splitting or branching such that the combined observer–object's wave function changes into two or more non-interacting branches which may split into many "worlds" depending on which is more probable. The splitting of worlds can continue infinitely.

Since there are innumerable observation-like events which are constantly happening, there are an enormous number of simultaneously existing states, or worlds, all of which exist in parallel but which may become entangled; and this means, they can not be independent of each other and are relative to each other. This notion is fundamental to the concept of quantum computing.

Likewise, in Everett's formulation, these branches are not completely separate but are subject to quantum interference and entanglement such that they may merge instead of splitting apart thereby creating one reality.

Changing the Past: Paradoxes and the Principle of Consistency

Entanglement and "spooky action at a distance" prove that effects can occur faster than the speed of light (Lee et al. 2011; Matson 2012; Olaf et al. 2003), such that effects may take place simultaneously with or before the cause, such that the effect causes itself and may be responsible for the "cause;" a consequence of entanglement in the quantum continuum Likewise, a Time Travel can also effect the present and change the future of the past, or rather, "a" past or "a" future.

Since the time traveler and his time machine are comprised of energy and matter their presence and movement through time-space will also warp and depress the geometry of space-time thereby creating local and distant effects. Time travel would effect each local moment of time-space leading from one moment and location in time (e.g. the present) to another location, i.e. from the present to the future and from the future into the past, and these effects can occur simultaneously and at superluminal speeds.

Many physical systems are very sensitive to small changes which can lead to major change. Unless the past and the future are "hard wired" and already determined, then the very act of voyaging to distant locations in time will alter every local moment of that time continuum. In terms of "Many Worlds" the time traveler is continually creating or entering new worlds which exist in parallel. Each "world" becomes most probable the moment he interacts with the quantum continuum, including simply by passing through time.

As detailed by quantum mechanics (Dirac 1966a,b; Heisenberg, 1955), shape and form appear as blemishes and bundles of energy in the quantum continuum, the underlying quantum oneness of the cosmos, emerging out of the continuum but remaining part of it. According to the Copenhagen interpretation (Bohr 1934, 1963; Heisenberg, 1930, 1955), all quanta are entangled and therefore any jostling of one quanta can create an instantaneous ripple which can effect local as well as distant objects and events through intersecting wave functions.

The space-time continuum is part of that basic oneness and is the sum of its parts including what can and can't be observed. And this includes distant locations in space-time corresponding to all possible futures, presents, and pasts.

As pertaining to time travel, as the traveler journeys through the quantum continuum of space-time he will jostle and affect all the particles (or waves) he contacts as he passes through time, and these will effect particles and waves elsewhere in space-time, thus altering the very fabric of every local and more distant moments of space-time. In the "Many worlds" interpretation, the time traveler is not really changing the future or the past but is engaging in actions which cause branching and splitting, which leads him to a future and a past which exists in parallel with innumerable other futures and pasts. He is not changing the past, but entering a different past which always existed as a probability.

As predicted by the Many Worlds interpretation, if the Time Traveler did make a significant impact in the past, then the alteration of the past would effect the entire world-line of history related to that event, including the memories of everyone living since that event and all those who retain any knowledge of that event; such that no one would realize anything has changed.

Minds, consciousness, the brain, memory, are also part of the quantum continuum and can be altered by changes in it (Joseph 2010a). The act of observation can change an event and an event can alter the observing mind. If the past were changed, we would not know it had changed because everything related to that event would have changed, from the writing of books to documentary films about the event. The alteration of the quantum continuum is not limited to just that event but can alter the entire continuum, including the quantum composition of the brain and memories of everyone who has lived since that event (Everett 1956, 1957).

The Principle of Self-Consistency

Many theorists have argued that it is impossible to change the past. Igor Novikov and Kip Thorne (Friedman et al. 1990) called this the "self-consistency conjecture" and "the principle of self-consistency" and various paradoxes have since been proposed to support this contention such as: "what if you killed your grandmother before she gave birth to your mother? If you did, then you could not be born and could not go back in time to kill your grandmother! Presumably these paradoxes are supposed to prove it is impossible to travel back in time.

In some respects these "paradoxes" are the equivalent of asking: "What if you went into the past and grew wings?" And the answer is: "You can't." The time traveler can not go back in the past and grow wings, or an extra pair of hands, or develop super powers, and so on. Nor could the time traveler kill anyone in the past who, according to the past record, did not die on the date he was killed.

Just as in "real life" there are boundaries which prevent the average person from engaging in or making world-altering decisions, these same limitations would apply in the past. Therefore, according to the principle of self-consistency, it is impossible to change the past, and if any changes were made, they may be "local" rather than global, and thus completely non-significant and not the least memorable--just like daily life for 99.999999999% of the 7 billion souls who currently dwell on Earth and who live and die and are quickly forgotten except by a few other insignificant souls who are also quickly forgotten as if they never even exists. Any changes made in the past may be so insignificant as to be meaningless.

Just as it is impossible to determine position and momentum of a particle, the past may also be subject to imprecision such that by establishing certain facts, makes other facts less certain The past may also be subject to the Uncertainty Principle, which may explain why historians, eye-witnesses, and husbands and wives may not always agree about what exactly happened in the past or just moments before.

The Principle of Self-Consistency, however, holds that the past is hard wired and cannot be altered, and reverse causality is an impossibility (Friedman et al. 1990). By contrast, reverse causality (also referred to as backward causation and retro-causation) is based on the premise that an effect may occur before its cause, such that the future may effect the present and the present may effect the past. A "cause" by definition must precede the effect, otherwise the effect may negate the cause and the effect! For example, the if a man went back in time and killed his grandfather he would negate his own existence making it impossible to go back in time and kill his grandfather. On the other hand, if he did kill his "grandfather" it might turn out that his paternal lineage leads elsewhere, i.e. "grandmother" had an affair and another man fathered his own father. Thus killing his "grandfather" has no effect on his existence and does not interfere with his ability to go back in time to kill his grandfather. In this instance, the effect does not nullify the cause; which is in accordance with the principle of self-consistency. The past can't be altered and if it is, the result is not significant.

If the past is "fixed" and hard-wired and can't be altered, then although the time traveler may go back in time with the intention of killing his grandfather, or Hitler, or Lee Harvey Oswald, the result would be that he would be unable to do so; his gun would misfire, the bullet would miss, or he never got close enough to the intended victim to do the deed. The past is hard wired and can't be changed.

If the past can't be altered, then this also implies that the future may also be fixed and hard wired and is not subject to alteration. However, if the future is subject to change (as demonstrated by classical physics and the laws of cause and effect), then the future must exist in order to be altered; as predicted by quantum mechanics, entanglement, and Einstein's theories of relativity. If the future may be changed, then why not the past? According to the "Many worlds" interpretation, the past is not changed, instead one changes which past world becomes his reality.

The "Many Worlds" interpretation of quantum mechanics would allow one to kill their mother or commit a murder which had not taken place, in this "world;" but in so doing would be effecting the quantum continuum and contributing to the probability that an alternate world would become the time traveler's world once he commits these crimes.

Paradoxes and Many Worlds

Most time travel "paradoxes" are based on the premise that the time traveler some how gains powers or the will to do things he would never do, or to accomplish what others tried to do and failed. Even if the time traveler wanted to kill his mother before he was born, or assassinate Hitler before he came to power, would he be able to do it? Would he be able to get close enough to shove in that knife or fire that bullet? And if he did, maybe the victims would live. Maybe the knife or the bullet would miss the necessary organ. Maybe he would change his mind at the last moment. Maybe in the struggle someone else would shoot the Time Traveler in the head and he would die instead. Many people tried to kill Hitler and failed.

"Paradoxes" can be reduced to simple probabilities. What is the probability a time traveler would want to go back in time and kill his mother? What is the probability he would succeed? What is the probability others would intervene before he could do the deed? What is the probability that he might be killed in the attempt? ... and so on.

And if he did kill his mother, it would not be "his" mother.

An observer, object, particle, interacts with its environment, with the quantum continuum, changing and altering it. As postulated by "Many Worlds" theory, there is one ultimate reality, but many parallel realities and histories, like the branches of a tree, a hallway with infinite doors, or infinite highways all of which lead out of the city. One highway leads to a past where Hitler won the war. Another highway leads to a past where the Kennedy brothers were never killed. Yet another takes the time traveler to a world where he was never born.

According to quantum theory and the "many worlds" interpretation, a new highway, a new door, a new branch of the tree appears every time a particle whizzes by or an observer interacts with his environment, makes a decision, or records an observation. Thus, the time traveler may go back in time and kill the mother who dwells in a parallel world or universe, but he would be unable to kill his mother.

The "Many Worlds" Resolution of The Grandmother Paradox Time Traveler "A" goes back into the past and kills his grandmother when she was still a little girl. An observer, object, particle, interacts with its environment, with the quantum continuum, changing and altering it. A time traveler going into the past would change every moment leading to that past simply by traveling through it, so that the past and the grandmother he encounters would be a different past and a different grandmother. As also predicted by the "Many Worlds" interpretation of quantum physics, a time traveler can appear in different parallel worlds. Therefore by killing this grandmother in this past time, Traveler "A" would be preventing the birth of that woman's time-traveling grandson "B", thereby preventing "B" from going into the past and killing the grandmother of the Time Traveler "A."

Multiple Paradoxes. Effects Negating Causes A very wealthy scientist invents a time machine and travels 30 years back into the past to prevent the car accident which killed his very beautiful wife. He arrives in the parking lot of the business where she works and lets all the air out of her tires and disables the engine.

He visits the younger version of himself and gives him the blue print for building a time machine, and a list of 100 stocks and when to buy and sell them. The Time Traveler returns to the future.

When was the time machine invented?

His wife takes a cab to her Lover's apartment and that night they drive to her home and that of her husband (the younger version of the time traveler). The Lover discovers the blue print for the time machine and the list of 100 stocks. The Lover and the wife sneak into the bedroom where her husband (the younger version of the time traveler) is napping and shove a knife through his heart.

Who invented the time machine?

The "Lover" upon killing the younger version of the Time Traveler (with the help of Time Traveler's wife), suddenly finds himself alone with the body, still holding the bloody knife in his hand. However, the Time Traveler's wife (and the blue print for a time machine and list of stocks) have disappeared. Upon his arrest he learns the Time Traveler's wife was killed hours before in a car accident.

Information Exists Before it is Discovered Two research scientists, both bitter rivals, are competing to make a major scientific discovery. Scientist A, who is better funded, makes the discovery first, publishes the results, receives world wide acclaim and receives a Nobel Prize.

Scientist B loses all funding, does not get tenure, and is reduced to living in obscurity and working in his basement lab, where, 20 years later, he invents a time machine. Scientist B makes a copy of the article which won his rival,

Scientist A, the Nobel Prize, and goes back in time and gives it to the younger version of himself. To ensure that the true inventor, Scientist A, does not get credit, the time traveler, Scientist B, kills Scientist A.

The younger version of Scientist B publishes the discovery and receives all the credit and the Nobel prize. When the time traveler, Scientist B returns to his own time he is famous and has a Nobel prize on his shelf. When he looks at the scientific journal where the original article appeared, he sees the same article but with himself listed as the author. He no longer understands why he went back in time to kill his rival.

Who made the discovery?

Another scientist after laboring his entire life makes a major discovery which brings him wealth and world wide acclaim. However, he is old and sick and unhealthy and is unable to savor the honors, women, and riches which are now his for the asking but he is too old to enjoy. So, he invents a time machine, takes a copy of his notebook describing the discovery, goes back 50 years in time and gives it to his younger self, and explains: "here are the answers you are searching for. You are going to be rich and famous."

So where did the discovery come from?

Science is replete with examples of scientists who independently make the same discoveries although they were working independently of each other and often not knowing of the other's work (Merton, 1961; 1963; Hall, 1980). Examples include the 17th-century independent formulation of calculus by Isaac Newton, Gottfried Wilhelm Leibniz and others; the 18th-century discovery of oxygen by Carl Wilhelm Scheele, Joseph Priestley, Antoine Lavoisier and others; In 1989, Thomas R. Cech and Sidney Altman won the Nobel Prize in chemistry for their independent discovery of ribozymes; In 1993, groups led by Donald S. Bethune at IBM and Sumio Iijima at NEC independently discovered single-wall carbon nanotubes and methods to produce them using transition-metal catalysts. And the list goes on.

What this could imply, if the past and future are a continuum, is that the discovery exists before it is discovered, albeit in a distant location of space-time. Or, in terms of multiple worlds theory, one branch leads to a world where the discovery is made by scientist A, a different branch leads scientist B to the discovery. Yet another branch leads to a world where the discovery is not made until 20 years into the future, whereas a different branch leads to a world where it is discovered in just a few days.

Mozart heard his music in his head, already composed--and some have proposed there is a cosmic consciousness which contains all information, and that one need only a brain that can tap into this source to extract this information. If true, this may explain why discoveries are made simultaneously or why Mozart heard his music "already composed" in his head and then simply wrote it down.

Time Travel Through Many Worlds

As based on a Many Worlds interpretation of quantum physics, traveling backwards into the past would itself be a quantum event causing branching. Therefore the timeline accessed by the time traveller simply would be one timeline among many different branching pasts. Hence, the time traveler from one world/universe may kill his grandfather in another world/universe. Likewise, in the past of some worlds, Hitler won the war, the Kennedy brothers were never killed, the dinosaurs did not become extinct, mammals and humans never evolved, and so on. All quantum worlds, many worlds, all exist as there is an infinity of possible universes and worlds, each of which differs in some manner from the other, from the minute to the major.

However, by changing (or choosing) his past, the Time Traveler would not just be making this past "World" more probable, but may cause all pasts to become unified. That is, the other pasts disappear as they are subsumed by and merge to become this one unified past.

Therefore, according to the Many Worlds interpretation, by changing the past, and by creating a single unified past, then once the merging occurs, all "memories" of earlier branching events will be lost. No one will ever remember that there was any other past and no observer will even suspect that there are several branches of reality. As such, the past (and the future) becomes deterministic and irreversible, and this effects the wave function of time, such that the past shapes the future, and conversely, the future can shape the past.

Therefore, if a time traveler journeys to the past, his passage will either change the past so that those in the future can only remember the past that has been altered since this past is the past which leads up to them. Or, the past was never really altered and always included the Time Traveler's journey into the past. That is, this altered past has always existed even before he journeyed to it and this is because he traveled to and arrived in the past before he left. Thus everything he does, from the moment he left for the past, has already happened. The past, like the future is irreversible and has been hard wired into the fabric of space-time.

According to the Copenhagen model, one may predict probabilities for the occurrence of various events which are taking place or which will take place. In the many-worlds interpretation, all these events occur simultaneously. Therefore, the time traveler is not changing the past, but choosing one past among many: "new worlds" which always existed as probabilities.

REFERENCES

Bohr, N., (1913). "On the Constitution of Atoms and Molecules, Part I". Philosophical Magazine 26: 1–24.

Bohr, N., (1913). "On the Constitution of Atoms and Molecules, Part I". Philosophical Magazine 26: 1–24.

Bohr, N. (1934/1987), Atomic Theory and the Description of Nature, reprinted as The Philosophical Writings of Niels Bohr, Vol. I, Woodbridge: Ox Bow Press.

Bohr. N. (1949). "Discussions with Einstein on Epistemological Problems in Atomic Physics". In P. Schilpp. Albert Einstein: Philosopher-Scientist. Open Court.

Bohr, N. (1958/1987), Essays 1932-1957 on Atomic Physics and Human Knowledge, reprinted as The Philosophical Writings of Niels Bohr, Vol. II, Woodbridge: Ox Bow Press.

Bohr, N. (1963/1987), Essays 1958-1962 on Atomic Physics and Human Knowledge, reprinted as The Philosophical Writings of Niels Bohr, Vol. III, Woodbridge: Ox Bow Press.

Born, M. Heisenberg, W. & Jordan, P. (1925) Zur Quantenmechanik II, Zeitschrift für Physik, 35, 557-615.

DeWitt, B. S., (1971). The Many-Universes Interpretation of Quantum Mechanics, in B. D.'Espagnat (ed.), Foundations of Quantum Mechanics, New York: Academic Press. pp. 167–218.

DeWitt, B. S. and Graham, N., editors (1973). The Many-Worlds Interpretation of Quantum Mechanics. Princeton University Press, Princeton, New-Jersey.

Dirac, P. (1966a) Lectures on Quantum Mechanics.

Dirac, P. (1966b). Lectures on Quantum Field Theory .

Einstein, A. (1905a). Does the Inertia of a Body Depend upon its Energy Content? Annalen der Physik 18, 639-641.

Einstein, A. (1905b). Concerning an Heuristic Point of View Toward the Emission and Transformation of Light. Annalen der Physik 17, 132-148.

Everett , H (1956), Theory of the Universal Wavefunction",Thesis, Princeton University.

Everett, H. (1957) Relative State Formulation of Quantum Mechanics, Reviews of Modern Physics vol 29, 454–462.

Friedman, J. et al. (1990). Cauchy problem in spacetimes with closed timelike curves". Physical Review D 42 (6): 1915.

Haber, R. N., Haber, L. (2000). Experiencing, remembering and reporting events. Psychology, Public Policy, and Law, 6(4): 1057-1097.

Heisenberg, W. (1925) Über quantentheoretische Umdeutung kinematischer und mechanischer Beziehungen, ("Quantum-Theoretical Re-interpretation of Kinematic and Mechanical Relations") Zeitschrift für Physik, 33, 879-893, 1925

Heisenberg, W. (1927),"Über den anschaulichen Inhalt der quantentheoretischen Kinematik und Mechanik", Zeitschrift für Physik 43 (3–4): 172–198,

Heisenberg. W. (1930), Physikalische Prinzipien der Quantentheorie (Leipzig: Hirzel). English translation The Physical Principles of Quantum Theory, University of Chicago Press.

Heisenberg, W. (1955). The Development of the Interpretation of the Quantum Theory, in W. Pauli (ed), Niels Bohr and the Development of Physics, 35, London:

Pergamon pp. 12-29.

Heisenberg, W. (1958), Physics and Philosophy: The Revolution in Modern Science, London: Goerge Allen & Unwin.

Joseph, R. (2010) Quantum Physics and the Multiplicity of Mind: Split-Brains, Fragmented Minds, Dissociation, Quantum Consciousness. "The Universe and Consciousness", Edited by Sir Roger Penrose, FRS, Ph.D., & Stuart Hameroff, Ph.D. Science Publishers, Cambridge, MA.

Juan Y., et al. (2013). "Bounding the speed of `spooky action at a distance". Phys. Rev. Lett. 110, 260407.

Lee, K.C., et al. (2011)."Entangling macroscopic diamonds at room temperature". Science 334 (6060): 1253–1256. Matson, J. (2012) Quantum teleportation achieved over record distances, Nature, 13 August.

Megidish, E., Halevy, T. Shacham, A., Dvir, T., Dovrat, L., Eisenberg, H. S. (2013) Entanglement Swapping Between Photons that have Never Coexisted. ArXiv.1209.4191v1, 19, Sep, 2012. Physical Review Letters, 110, 210403.

Megreya, A. M., & Burton, A. M. (2008). Matching faces to photographs: Poor performance in eyewitness memory (without the memory). Journal of Experimental Psychology: Applied, 14(4): 364–372.

Olaf, N.. et al. (2003) "Quantum interference experiments with large molecules", American Journal of Physics, 71 (April 2003) 319-325.

Plenio, V. (2007). "An introduction to entanglement measures". Quant. Inf. Comp. 1: 1–51.

Schrödinger E; Born, M. (1935). "Discussion of probability relations between separated systems". Mathematical Proceedings of the Cambridge Philosophical Society 31 (4): 555–563.

Schrödinger E; Dirac, P. A. M. (1936). "Probability relations between separated systems". Mathematical Proceedings of the Cambridge Philosophical Society 32 (3): 446–452.

****
****
****

III. Time Travel Through Black Holes and Worm Holes

****
****
****

5. Time Travel Through Black Holes and Worm Holes in the Fabric of Space-Time
Lan Tao and R. G. Joseph

Cosmology.com

Abstract:

Relativity and the quantum physics of time travel through black holes, worm holes, and quantum holes in the fabric of space time are detailed and discussed. As first predicted by Einstein and Rosen, these tunnels through time may lead to a mirror universe. Black holes can also serve as gravitational sling shots to propel a space craft or time machine toward the speed of light. Because of length contraction, as a time traveler accelerates, once they reach light speed they will have shrunk in size to smaller than a Planck Length (10-33 cm). Concentrated mass in spaces this small generates tremendous gravity and energy which can blow infinitely small black holes in space-time and propel a time traveler to superluminal speeds and from the future into the past. It is only possible to voyage to the past by first traveling to the future and to exceed light speed. Because of length contraction, at superluminal speeds time space continues to contract in a negative direction, thereby resulting in a time reversal. However, be it quantum holes, worm holes, or super massive black holes, because of negative energy density within these holes, the time traveler may become composed of negative energy and negative mass and emerge from the hole into a mirror universe where time runs backwards. Because the mirror universe also consists of negative energy and negative mass, there is no violation of the laws of thermodynamics.

Keywords: Time Travel, Black Holes, Worm Holes, Quantum Holes, Length Contraction, Negative Energy, Negative Mass, Quantum Physics, Relativity, Einstein Rosen Bridge.

Curvatures And Holes in Space Time

If you drop two balls at the same time from the roof of a building, one weighing 10 pounds and the other weighting 100, they will fall at the same rate and hit the ground at the same time due to gravity. However, the heavier ball will form a larger crater in the soil and this is because of their differential mass and gravity. The same can be said of planets, stars, and galaxies in space-time which can, theoretically, create cavities and even holes in space time and form what has been referred to as Einstein-Rosen bridges which lead to a mirror universe on the other side (Einstein & Rosen 1935). It is these holes which may make time travel possible.

According to Einstein and Rosen, at the top of the hole is a "mouth" and at its center is a "throat." The hole does not lead to a bottom but opens at the other end, forming another "mouth" and which leads to a mirror universe. However, these theories were eventually abandoned and the Einstein-Rosen bridge dismissed as a mathematical anomaly.

Rosen and Einstein's ideas and theorems were resurrected in the 1960s by Robert Fuller and John Wheeler (1962) who saw them as a mathematical requirement for proving the existence of supermassive black holes. They are also a key component to the Reissner-Nordstrom solution which describes an electrically charged black hole. John Wheeler in fact coined the term "black hole" to symbolize its two central characteristics: emptiness and blackness. Many now believe these "holes" can serve as tunnels through time.

The cosmos is believed to be littered with holes of varying size and magnitude, from those smaller than an atom to super-massive black holes with the gravity and mass of entire galaxies (Al-Khalili 2011; Hawking 1988; Joseph 2010a; Thorne 1994).

Gravitational fields are not uniform and differ in strength in various regions of this galaxy (Carroll 2004); just as the weather is different in distant localities on Earth. Galaxy distribution is asymmetric with great walls of galaxies clustering together. The planets and stars orbiting these galaxies, the galaxies themselves, and the clustering of these galaxies all differentially effect and torque the geometry of space-time such that it is littered with pockets of varying size which can form holes in the fabric of space-time. Theoretically, it is these holes, which range from super-massive black holes, to those smaller than a Planck Length, which could make time travel possible, including journeying at superluminal speeds thereby making possible a voyage into the past (Joseph 2014).

The effects of gravity differ throughout the cosmos and even on the same planet. On Earth, gravity is stronger at sea level than at higher elevations, and even stronger toward the center of the planet. Hence, objects are pulled down toward the center of gravity.

All falling objects are drawn toward the center of Earth's gravitational attraction--the center of Earth. If two identical objects are placed hundreds of miles apart miles above Earth and allowed to fall, they will come closer and closer together as they fall in a pattern similar to an inverted triangle, until they end up side by side and pointing directly toward Earth's center. Because of Earth's curvature, the balls are pulled together, toward the center of gravity which is at the center of Earth. Once they strike the ground, they would create adjacent holes and craters. By tunneling through the walls of these holes, the distance between the two objects would be considerably shorter than going up and over and then down into these adjacent holes--and the same principles apply to time travel through holes in space time.

Earth is comprised of layers. The outermost layer is referred to as the crust, beneath which is the mantle which is made up of heated rock under high pressure, and below that is believed to be exceedingly hot liquid metal (with an estimated temperature of around 7,000 kelvin) and a compressed metal core. Earth has a mass of 5.9736 x 1024 kg (5973600000000000000000000 kg), an equatorial diameter of 12,756.1 km and radius of 6,378 km. If the inner layers of Earth were to collapse and the planet imploded to half it size, the outer surface would be pulled toward the Earth's center, the planet's gravity would double, and time would slow. All the clocks and survivors on Earth's surface would also move very slowly from the perspective of those aboard the International Space Station (ISS) in orbit above the shrunken planet. However, if astronomers on Earth were looking back at the crew of the ISS, then everything taking place within the ISS would seem to have speeded up; a predicted by Einstein's (1961) theories of relativity.

The gravity and mass of Earth deforms and creates a deep pocket in the geometry of surrounding space-time (Carroll 2004). If Earth were to implode, the increased gravity would create a deeper pocket, or pit in space-time and drag surrounding space time toward it. If Earth continued to implode, thereby concentrating its mass, Earth's center of gravity would increase it's gravitational power, pulling more of the outer surface toward it, along with space-time. Those on Earth's surface would begin to shrink and sink into the ground and their movements (and time) would be so slow they would appear to be frozen in place. They would also shrink in size due to the pull of gravity.

If its radius and diameter continued to shrink the collapse and implosion of Earth would accelerate due to the increasingly powerful pull of gravity toward its center. Soon what had been a planet with a radius of 6400 km and a mass of 5.9736 x 1024 kg, would become the size of a golf ball with a radius of just a few centimeters while retaining the mass and gravity of an entire planet. Moreover, because of the energy involved in its acceleration toward miniaturization, it may increase in mass and gravity even as it grows smaller becoming molecular in size.

Einstein's and Newton's theories of gravity both predict that if mass is shrunk to a subatomic space, its gravity will become increasingly powerful (Einstein 1915a,b, 1961). In consequence, anything on the surface would be crushed to atomic size, no reflected light would be able to escape, and time would stop due to the forces of gravity which would prevent the movement of time.

The Earth would likely shrink to the size of an atom and would no longer be visible. Instead there would be a tiny black hole in the fabric of space time; and this atom sized planet would be at the center, halfway between the top and the bottom of the hole. Passing light may also be pulled down into this hole and to the surface of this miniature planet. Because no light can escape, the hole would be black. However, if the shrinkage continued, and the Earth became smaller than an atom and less than 10-13cm in size, then, as summed up by Heisenberg (1958), the combined mass and energy would blow a tiny hole into the tissues of space time and the result would "time reversal...The phenomenon of time reversal...belongs to these smallest regions." Time travels in reverse in these tiny spaces because of the tremendous energy released which blows a hole through the tissues of space-time which tunnels from the present to the past at superluminal speeds. Gravity is so powerful in spaces smaller than 10-13cm, that, theoretically, as predicted by quantum mechanics, it can suck time backwards at such incredible speeds that time itself would exceed the speed of light thereby propelling everything in its wake into the past (Joseph 2014).

If a time traveler were subject to the same increasing forces of gravity, a consequence of acceleration, they would also shrink in size as they are propelled toward the future. At superluminal speeds, they may also blow a hole in space time and experience time reversal. Therefore, if a time traveler wished to journey into the future or past he would have to first accelerate toward light speed and in so doing would shrink in size to smaller than 10-13cm (impossible with current technology) or dive his time machine into a super massive black hole--because according to Einstein and his colleague Nathan Rosen, these hole have no bottom and lead to mirror planets, galaxies, and universes (Einstein & Rosen 1935) and where tine may run in reverse (Joseph 2014).

Tunneling Through Time

Just as a planet is curved into the geometry of a circle, space-time is curved (as predicted by Einstein's theories of relativity) and, theoretically, this can allow for "short cuts" between planets, solar systems, and entire galaxies. For example, China and Argentina are antipodal, on opposite sides of the planet. The distance in a curving "straight line" between Beijing China and Buenos Aires Argentina, is 12,326 miles. The distance can be reduced by 35% however, if one were to journey through the center of the Earth, by tunneling from Argentina to China to the other side. By traveling through the Earth, the distance would only be about 7,900 miles, a reduction by 3,426 miles (35%). Theoretically, just as one might drill a hole tunneling downward from Argentina, and end up on the other side of the planet in China, a hole in the curvature of space-time may also lead to a galaxy or a universe on the other side. The holes formed by two galaxies or universes essentially create two pockets, or holes, which become linked thereby allowing a time travel to travel through the hole, instead of taking the curved route which would involve much more time and distance.

If a sufficiently massive object such as a super-sized star or galaxy were to collapse and implode, they may not simply sink deeper into the original pocket or pit which their gravity had already carved into the tissues of space-time, but they may punch a hole that has no bottom but opens at the other end, creating a passageway (Einstein & Rosen 1935). Holes forming an Einstein-Rosen Bridge are believed to have such immense gravity that not just space-time, but light is sucked down into the hole (Bethe et al., 2003; Melia 2007; McClintock 2004).

Although Einstein's theory of special relativity erected a cosmic speed limit which proclaimed nothing can exceed the speed of light (1905b, 1906a), his theory of general relativity, of gravity, abolished the speed limit (1915a,b). Under the influence of tremendous forces of gravity, the speed of light can be exceeded; and this is what may happen to light sucked into the mouth of a super massive black hole (Melia 2007; McClintock 2004; Thorne & Hawking 1995). And when light, or any object exceeds light speed, they are flung into a mirror universe where time runs backwards into the past.

Accelerating toward light speed collapses and shrinks space-time, compacting more time into a smaller space; shrinking the distance between the present and the future (Einstein et al. 1961). At light speed, time stops. However, once the cosmic speed limit is surpassed, the compression of space-time implodes and turns inside out continuing in a negative direction (Joseph 2014). Time and space-time are reversed, like looking into a mirror, except that the time traveler has entered the mirror and is looking back. However, the only way to reach the mirror universe is to speed faster than light and into the future, and in so doing shrink to a size smaller than 10-13cm--a function of Lorenz length contraction. Theoretically those who accelerate beyond the speed of light journey into the future and then through the looking glass into a mirror universe which leads from the future backwards in time.

Black Holes: Mirror Universe

The mirror universe is not fancy, but a mathematical fact based on the Schwarzschild solution of Einstein's equations; commonly used to calculate the gravitational field of a massive star (Einstein 1915a). The collapse of super massive stars creates super massive black holes in space-time.

Stars are born and they die, and the larger stars have a spectacular death, literally going out with a big bang, a supernova explosion, at which point they begin to contract. Those which are three times the size of our sun, are believed to collapse into black holes. Smaller stars collapse into what are referred to as neutron stars (Becker 2009). Neutron stars are the remnants of collapsed stars similar to or a few times larger than the mass of the sun (Becker 2009). Consider for example the "Crab Nebula" deep in the constellation of Taurus which exploded in a vast supernova in 1054 and then collapsed. Gravity is so powerful that the atoms which made up this star have been crushed into neutrons and light is unable to escape its surface, meaning that time has stopped.

Stars collapse after they burn up their internal hydrogen and then their helium fuel causing them to expand and becoming red giant as they eject mass into space; and then they begin to implode with the increasing concentration of mass and gravity exacerbating and accelerating the implosion until shrinking and becoming compressed to a singularity, perhaps as small as a single atom, and with a density of about 5×10^{93} grams per cubic centimeter (Bethe et al. 2003; McClintock, 2004).

Because of its incredible concentrated mass, gravity, and density, the singularity forms a huge depression, or cavity in space-time. Just as a man weighing 500 pounds would sink deeper into the mud than a woman weighing 100 pounds, collapsed stars also sink into the fabric of space time; and those with the most mass sinking so deep they form a huge hole, at the center of which sits that star's remains, which could be the size of a marble but with the gravity and concentrated mass of a billion suns.

As based on Einstein's general theory of relativity, once a star has collapsed it will create an intense super massive gravitational field. In consequence, according to conventional wisdom, anything which falls into this gravity-laden cavity, including nearby stars and even light, can not escape such that surrounding this hole is blackness; i.e. a black hole.

It is believed that there are tens of millions of "stellar mass" black holes lurking in the inner and outer galactic arms and on the outskirts of the Milky Way galaxy, the gravitational remnants of dead stars, each with a mass anywhere from 10 to 25 more massive than the sun (McClintock, 2004; Schödel, et al., 2006). Then there are the super massive black holes which have the mass of a million billion suns, one of which appears to sit at the axial center of this galaxy (Melia 2007). Then there are yet others which may have the concentrated mass and gravity of entire galaxies and perhaps others with the concentrated mass and gravity of an millions of galaxies (Joseph 2010a).

For example, and as detailed by Joseph (2010a) VIRGOHI21 (Minchin, et al., 2005) has swallowed all the stars of its galaxy and has the gravity of a small galaxy; an estimated total mass of about 1/10th the Milky Way; ten times more dark matter than ordinary matter; and is surrounded by vast clouds of hydrogen. Because of its galaxy-in-mass gravity, VIRGOHI21 has pulled up to 2000 galaxies toward it, creating the Virgo Cluster (Fouqué, et al., 2001). Thus, thousands of galaxies have been caught up in the vortex of this galaxy-in-mass gravity hole and now cluster about it.

The billion-light-years across "Eridanus black hole" may be typical of black holes which have the gravity-mass of millions of entire galaxies (Joseph 2010a). The Eridanus black hole sits like a giant black spider in an ocean of nothingness, having swallowed up all surrounding galaxies, gas, and light, including radiation from the Cosmic Microwave Background. Based on an analysis of the NRAO VLA Sky Survey data, Rudnick et al. (2007) in fact discovered that there was a significant and rather remarkable absence of galaxies even in the distant space surrounding this hole, in the constellation of Eridanus. Thus, the billion-light-years across "Eridanus black hole" must have consumed the gravity-mass of millions of entire galaxies all of which have been collapsed and concentrated into the singularity of this super-galactic hole.

Super massive black holes not only suck up light, but may serve as mirrors into the past or act as windows into mirror galaxies. Smaller holes lurking in the arms of the galaxy may serve as smaller mirrors, mirroring nearby stars and their planets. Space-time may be littered with trillions of billions of mirrors into the past.

Theoretically, to journey into the past, the time traveler could dive her time machine into the hole, at which point she will be accelerated to faster than light speed and then enter the mirror at the bottom of the hole which leads to the long ago. Such a journey, of course, is not without its perils. The time traveler may be crushed to death, disintegrate into particles, or be irradiated into ash (Almheiri et al., 2013), with any remnants spewed out as disembodied particles (Hawking 2014).

Black Hole's and Length Contraction

The existence of black holes and their formation was detailed in 1939, by Robert Oppenheimer who argued that stars above approximately three solar masses would collapse into black holes and create singularities in the fabric of space-time. Adjacent to and surrounding the singularity would be empty space and enclosing that empty space would be a spherical event horizon; though some scientists, such as Hawking (2014) have proposed that "gravitational collapse produces apparent horizons but no event horizons."

Event horizons very depending on the mass of the singularity which punctured the hole into space-time. For example, the apparent black hole at the center of galaxy M87, has an estimated event horizon with a circumference of 56 billion kilometers (52 light hours).

Stars, objects, particles and astronauts just outside the event horizon are drawn toward it and their ability to escape from it become exceedingly difficult as they approach due to the incredible amount of gravity and the vortex that sucks in all surrounding space-time including light (Melia 2007, McClintock, 2004). As they approach the event horizon they will accelerate toward light speed giving off a red-shift light pattern indicating their incredible velocity. As their velocity increase, time-space contracts, and they are propelled into the future.

Specifically, and if we accept that time and the speed of light are related, then if the time traveler journeys at 80% speed of light, then one day (1.197 days) from the perspective of the time traveler would be the equivalent of 2 days back on Earth. If he achieves 99% light speed, then 104 days in the time machine (time contraction) would be the equivalent of about 2 years on Earth. At 99.9% the speed of light, then 1 day (26 hours) in the time machine would be the equivalent of 6 years on Earth. If the time traveler wished to experience a future 2190 years distant she would have to spend one year in the time machine traveling at 99.999% light speed.

The amount of space-time contraction significantly increases as velocity nears light speed such that at lower speeds the effects are negligible even for velocities at 50% the speed of light. By contrast the amount of contraction becomes dramatic as velocities approach light speed. For example, at 99.999999% the speed of light, almost two years pass for every day in the time machine. At 99.99999999999 % of c, for every day on board, nearly twenty thousand years pass back on Earth. However, upon reaching light speed, time stops. It is only upon accelerating beyond light speed, that time runs backwards and the contraction of space-time continues in a negative direction. One must accelerate toward the future to reach the past.

According to conventional wisdom, whatever is captured at the event horizon will spin and circle round and round the circumference of the horizon at light speed; and as such, time will stop (Dieter 2012). Particles and objects that pass beyond the event horizon into the hole at first continue to circle at the speed of light, but as they are drawn down into the hole the forces of gravity accelerate their movement until they exceed light speed, and as such they are sucked into the past and experience a time-reversal; a function of the contraction of time-space in a negative direction as predicted by the Lorenz transforms.

As summarized by Joseph (2014) "when an object is in motion and it accelerates, its mass increases as it absorbs energy, and it shrinks in the direction of its motion. Contraction in the direction of motion has been referred to as Lorentz contraction (Lorentz 1892, 1995; Einstein et al. 1923) and is predicted by Einstein's theory of relativity (Einstein 1961). Length contraction can be expressed mathematically by the following formula: $E = mc^2/\sqrt{(1-v^2/c^2)}$, which is similar to the equation for time dilation (if one replaces the value of v for 0). As the value of v (velocity) increases, so does an object's mass which requires more energy to continue at the same velocity or to accelerate. Since energy can become mass, mass increases even as the object shrinks and contracts, thereby increasing its gravity which exerts local effects on the curvature of space-time. Not just the time machine, but space-time in front and surrounding the time machine also contracts. Everything shrinks and contracts in the direction of motion, as velocity increases. For example, a 100 foot long time-space ship traveling at 60% the speed of light would contract by 20% and would become 80 feet in length. Presumably, its diameter would remain the same, though the likelihood is that all surrounding space including the diameter of the time machine would contract. If the time-space ship accelerates to 0.87 light speed, it will contract by 50%. However, at light speed, it would contract to a size less than a Planck Length and become all mass but with no length or width. Upon exceeding light speed, the contraction continues, albeit in a negative direction; and this is what gives rise to time reversal. Time travels into the past at superluminal speeds and leads to a mirror universe where everything runs in reverse."

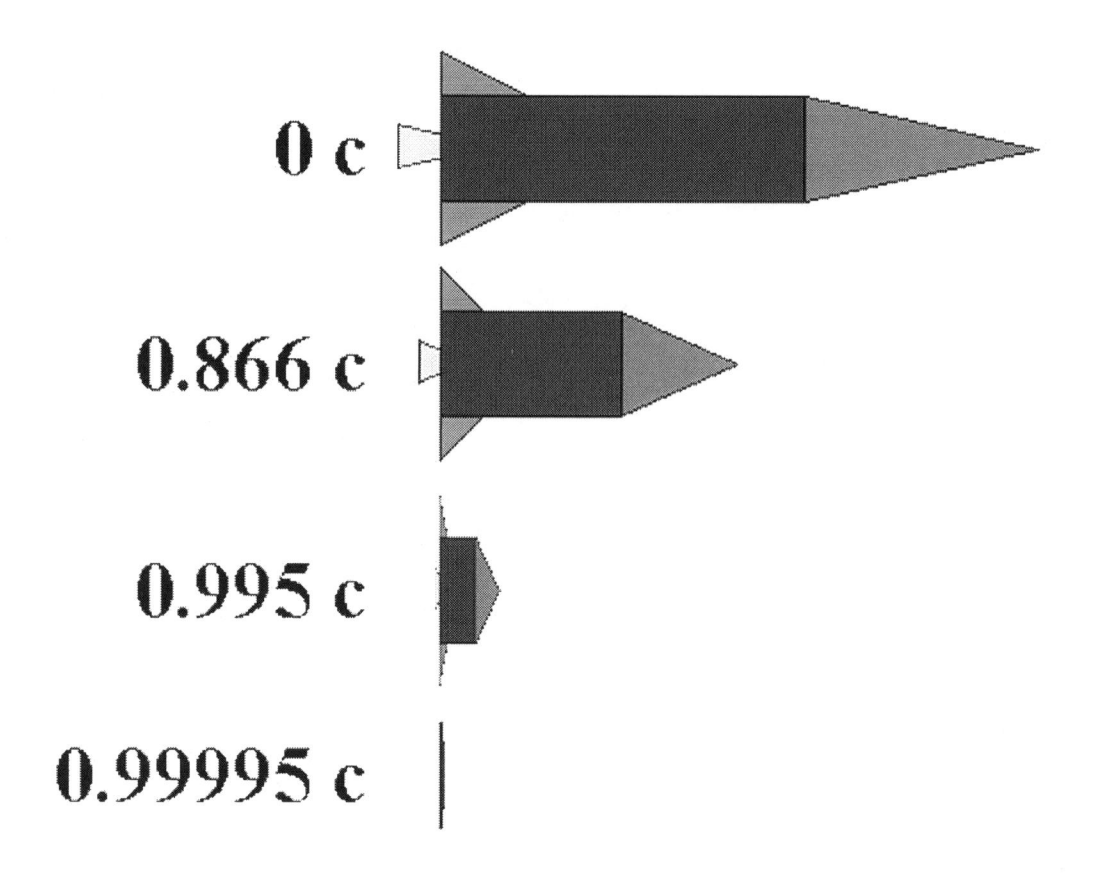

Event Horizons and the Physics of a Black Hole

Broadly considered, there are several major types of super massive black holes, including those with an electric charge, spin, and angular momentum, and those without (referred to as Schwarzschild black holes, named after karl Schwarzschild). General relativity predicts that any rotating mass will "drag" and pull space-time which also begins to circle around it (Einstein 1915a,b, 1961). Therefore, black holes with spin and momentum exert an organizational and gravitational effect on matter, just like a vortex in the ocean will drag surrounding water into the vortex. Almost every spiral galaxy is believed to have a super massive black hole at its center (Bethe et al. 2003; Melia 2007, McClintock, 2004).

Signature of piece of matter falling into black hole Cygnus XR-1

Disk of spiraling hot gas

Black hole event horizon

1000-mile gap

Blob of gas breaks off disk to spiral toward event horizon

Super massive black holes are believed to attract and organize stars which then circle and orbit around them, much in the same manner that water is drawn toward and then circles 'round a drain before disappearing inside. Therefore, just as the sun creates a depression and pocket in space time, thereby curving and dragging space-time into the hole it occupies, which in turn causes the planets to move along the curvatures, super massive black holes at the center of galaxies have the same effect on all the stars of those galaxies. Stars littered throughout the spiraling arms of the galaxy, are like roulette balls, and orbit along the grooves in the fabric of space-time created by the curvature of space. However, the super massive gravity of these super-massive black holes is so powerful, that not just space-time but entire stars are dragged into the hole.

Stars closest to the black hole have a greater orbital velocity compared to those further away (Ghez et al., 2005; Petrovskaya, 1994; Teerikorpi, 1989). Their velocity will increase as they come even closer to the hole. The light of the star will also begin to dim as light is suck toward the hole, and become red-shifted.

The point of no return is called the "event horizon." The "event horizon" is like the lips of a mouth, and completely encircles the outer rim of the black hole. Tidal forces are believed to be smaller at the event horizon and to decrease as the size of the hole increases, such that the bigger the hole, the more likely someone can survive.

If a space-time machine were to approach the event horizon its velocity would begin approaching the speed of light and it would be flung into the future. Upon reaching the event horizon, the space-time machine would have a velocity at light speed and time would stop. Those inside the time machine would experience an infinite "now." At some point the time machine, if still intact, would be sucked inside the hole and its velocity would increase well beyond the speed of light. And the same fate befalls all stars which are caught in the gravitational grip of a black hole.

From the perspective of an outside observer, the light associated with a star (or time machine) approaching and then falling into a black hole, would become dimmer and red shifted as it accelerates toward the event horizon until reaching an infinite red shift at the horizon. If the observer did not know there was a black hole, it would appear as if that distant star (based on the dimness of light) was accelerating (based on its red shift) and given its dimness (due to the black hole's capture of light) there would be an illusion that it is rapidly increasing its distance, becoming further and further away as it speeds up. However, although it is accelerating, it is not speeding further away but speeding into the future, due to the contraction of space-time; and the light associated with the star simultaneously becomes redder and dimmer. This is known as "gravitational red shift." Finally, the light would disappear, sucked down into the hole along with the star (and the space-time machine).

The Event Horizon of Eternal Now
As a thought experiment Einstein imagined that if he flew away from a big clock in the town square precisely at 12 noon, and traveled at the speed of light, the clock would appear to stop and would remain 12 noon forever--and this is because Einstein would be traveling at the same rate of speed as the light coming from the clock, in tandem and in parallel with it. Time would also essentially stop for Einstein, for if he were looking at the light beams on either side of him, they would look like stationary waves of electromagnetic activity consisting of crests and valleys--and this is because he would be moving in tandem and relative to these light beams; like two trains traveling at exactly the same speed, side by side and the only view is of the other train. At light speed Einstein would be captured in an "eternal now" with the future on one side and the past on the other.

All observers in uniform motion (like two trains traveling side by side) view themselves as at rest (so long at they can only see the two trains). If traveling at the speed of light, a light from a flashlight held in that time traveler's hand will never escape from the flashlight. The light from the flashlight will be frozen in place, in an eternal now.

If a star, astronaut, or space-time machine were to approach a super massive black hole, they would accelerate toward the "event horizon" at light speed (Dieter 2012; McClintock, 2004). The Time Traveler's clock would tick increasingly slower and light trailing behind would become redder (red shifted). However, for the time traveler, time continues as before.

Once caught by the gravitational grip of the vortex spinning round the event horizon, the star, astronaut, or space-time machine would have a velocity of light speed (Dieter 2012). Time stops. They would be captured and held in the grip of what could best be described as an "eternal now." Light could not escape, and the outside of the hole would appear black, whereas the event horizon would be blazing brightly illuminated with light.

However, once the time traveler falls inside the black hole, their space craft may accelerate toward and then beyond light speed. If they survived, they would be hurtled back form the future into the long ago.

Gravitational Sling Shots Into the Future And The Past

Space is permeated by holes: those smaller than a Planck Length; Worm holes of varying size which presumably open up between folded layers of space-time; Holes created by the compressed mass of four or more sun-like stars, millions of which litter the arms of the Milky Way Galaxy; and super-massive black holes with the concentrated gravity-mass of millions of billions of stars and which sit at the center of spiral galaxies (Blanford 1999; Melia, 2003a,b; Jones et al., 2004; Ruffini & Wheeler 1971). Super massive black holes have such incredibly powerful gravitational tidal forces that they suck entire stars and their planets into its depths, never to be seen again (Giess, et al., 2010; Melia, 2003a,b; Merloni & Heinz, 2008).

The gravity of massive black holes sprinkled throughout the outer arms of the galaxy, and those at the center of galaxies, offer the time traveler a means of accelerating to near light speed and into the future, and beyond light speed into the past. A black hole can be employed as a gravitational slingshot.

A seeming paradox of time travel into the past is that one must first journey to the future to reach the past; and this also calls for superluminal Lorentz transformations which many believe is not possible. In fact, the major objections to time travel into the past are based on Lorentz transformation equations and Einstein's special theory of relativity which decreed, by law, a cosmic speed limit and which erected a cosmic stop sign which proclaims: Nothing can travel faster than the speed of light. Einstein also proclaimed "God does not play dice." However, Einstein's cosmic speed limit is man-made and not dictated by a "god" whereas Einstein's theory of gravity allows for superluminal speeds.

As predicted by Einstein's theories of relativity, to exceed the speed of light one must first accelerate toward light speed and thus, to the future. For example, at 90% light speed the time traveler can reach the future in half the time its takes those back on Earth; one day in the time machine leading to 2.29 days in the future. At 99% light speed, one can leap 6 days into the future in just 26 hours. At 99.999999999999 the speed of light, one could leave Earth in the year 2050, and arrive in the year 2,2050 in 24 hours; two thousand years in a day. Time continues to shrink and the distance between the present and the future continues to contract with velocities approaching light speed, until at 100% light speed, time stops and one sits upon a event horizon with no future, no past, and an infinite "now." On one side of the horizon lies the future, on the other, the past.

At a speed of 99.99999999% c, one is hurtled far into the future. At light speed one arrives in the eternal present, the horizon of eternal "now." At 100.1% light speed the time traveler crosses over the horizon and continues into the past.

It is only upon accelerating toward light speed, which compresses time allowing the future to arrive more quickly, and then accelerating beyond light speed which triggers a time reversal; a consequence of the contraction of length, time-space, and time continuing in a negative direction. The time traveler must first journey to the future before continuing into the past.

Currently, there is no know means of achieving speeds sufficient to make even a few seconds of time travel possible. Due to the limitations of current technology one can only accelerate to only a fraction of light speed. For example, one of the fastest vehicles on Earth is the Bugatti Veyron Super Sport's car, which has attained a velocity of 429.69 km/h (266.99 mph). Andy Green rode the ThrustSSC vehicle to a speed of 1228 km/h (763 mph). The New Horizons space craft shot into space in January 2006 and reached a velocity of 36,373 mph when its engines shut down. The Helios space probe satellites launched in the 1970s attained a velocity of 240,000 km/h (157,000 mph). However, the fastest manned vehicle was the Saturn V rocket which reached a velocity of 24,000 mph.

Since the speed of light is 186,282 miles per second (670,616,629 mph) the ability to fly even a few hours into the future will require new technologies, based perhaps, on matter / anti-matter engines or propulsion systems designed to harness the repulsion powers of negative energy (Joseph 2014). However, if a space craft were to locate a black hole of sufficient size, it could use the energy and gravity of the hole, while avoiding the pitfalls of falling within, to propel itself to light speeds and beyond.

Innumerable massive black holes which are believed to populate the spiral arms of the Milky Way Galaxy by the hundreds of millions (McClintock, 2004; Schödel, et al., 2006). Essentially, the gravity of a black hole can be borrowed to increase the space-time machine's velocity. The space-time machine would fly toward a spinning black hole which would cause the craft to accelerate to near light speed (Penrose 1969). While remaining a safe distance from the hole's event horizon, the time traveler could then ride piggy back on the movement and energy of the spin to be flung around the exterior of the hole at a speed equivalent to the craft's velocity plus the near light speed velocity of the outer rim of the vortex.

General relativity predicts that a rotating black hole will drag space-time around it and eventually into the hole. Surrounding the "event horizon" of a black hole is a vortex of space-time referred to as the ergosphere (Bethe et al. 2003; Taylor & Wheeler 2000). Objects, particles, space-time machines, theoretically, would be able to escape from the ergosphere. Through the Penrose process (1969), objects can emerge from the ergosphere with more energy than they entered. The energy and accelerated velocity gained by the space-time machine would be equal in magnitude to that lost by black hole; which, given the relative differences in size would be negligible insofar as these stellar objects are concerned. However, the combined velocities could propel the space-time machine to well beyond the cosmic speed limit. At superluminal speeds, the time traveler would the return from the future and voyage into the past.

For example, if a time traveler were standing on a railroad track with a tennis ball in hand, and tossed the ball at 25 mph at a train approaching at 100 mph, passengers in the train would see the ball approaching at 125 mph, and then bouncing off the train at nearly the same speed. Likewise, if a professional baseball pitcher was riding in the front end of the lead locomotive racing along the tracks at 100 mph, and he threw a base ball at 100 mph to someone standing in front of the train, the initial speed of the ball would be 200 mph relative to the ground. The additional energy and acceleration in both instances are borrowed from the train. The same principles can be applied to black holes so as to gain speed.

General relativity predicts that any rotating mass will "drag" and pull space-time which begins to circle around it. Objects closest to the black hole have a significantly higher velocity that those further away. Therefore, using the Penrose process the time-space machine can ride on the spin of a spinning black hole, and by employing a parabola trajectory, circle round at an accelerated rate of speed and leave the vicinity in the opposite direction at beyond light speed. The ergosphere of the black holes becomes a gravitational sling shot.

By using a parabola trajectory the time traveler could avoid the Schwarzschild radius and event horizon where the gravity and curvature of space is so severe that the space-time machine would be unable to escape. The point of no-return is the event horizon (Bo & Wen-Biao, 2010; Melia, 2003b; Hawking 1990; Thorn 1994; Thakur, 1998). Just as a ship can't be seen after it crosses over the horizon on Earth, everything that passes over the "event" horizon of a black hole can't be seen--though it may be remembered.

The size of the event horizon is given by the "Schwarzschild radius" (Brill et al. 2003) and anything which falls through the radius, and over the horizon, is cut off from the outside universe (Bethe et al. 2003; McClintock 2004). What happens next is anyone's guess. Possibly everything decays and is torn apart due to the incredible tidal forces, with even atoms becoming ripped apart and crushed to near infinite density; unless what falls in is already atom-in-size or smaller than a Planck length--a consequence of length contraction as velocities approach light speed.

Tidal forces are believed to be smaller at the event horizon, even less so at the ergosphere, and to decrease as the size of the hole increases (Bethe et al. 2003; Brill et al. 2003; McClintock 2004), Therefore, the bigger the hole, the more likely someone can survive, especially if the time machine has shrunk to the size of a Planck length, at which point it may also blow a hole in space time and be propelled to the other side.

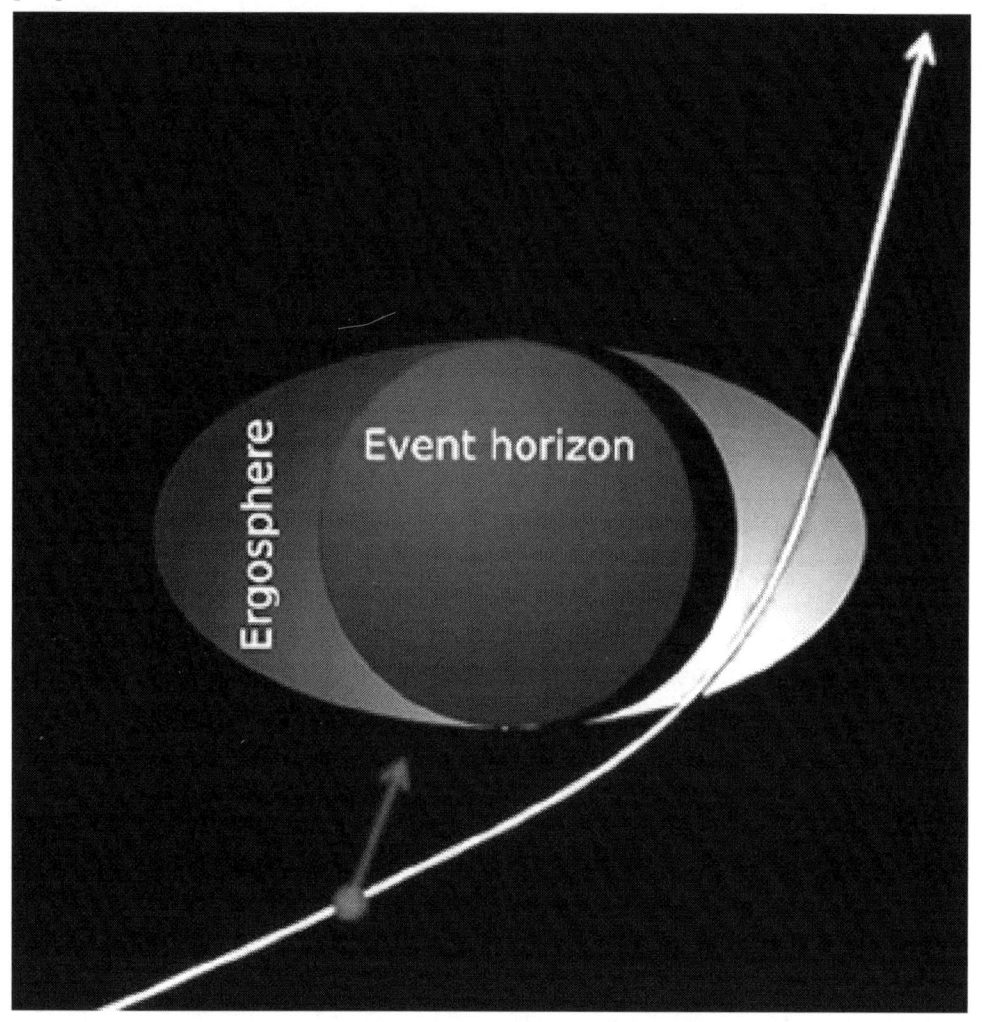

The best way to avoid the many perils which may await the intrepid time-traveler within the darkening depths of a black hole, is to use its gravity to increase the speed of the time machine and then go around it.

A parabola orbital path would enable the time traveler to avoid the event horizon and borrow the speed, gravity and the spin velocity of the black hole which would accelerate the time-space machine as it approaches the hole. Combined with the near light speed of the outer vortex of the hole and the craft's near light speed velocity, the time machine would be flung into the past at superluminal speeds.

Gravitational slingshots, also known as gravity assist maneuvers, have been employed to increase velocity by numerous interplanetary missions beginning with the Mariner 10 as well as the Voyager probes which used the combined gravity of Jupiter, Saturn, Uranus and Neptune which were closely aligned at the time. An alien space-craft approaching from another solar system could use our sun or this entire solar system for the same purposes.

Thus, a spinning black hole could also serve as a slingshot; though if superluminal speeds can be achieved is open to debate and would be dependent on numerous variables. For example, if the time traveler misjudges the gravitational grip of the black hole and is caught up in the vortex of surrounding space, she will not only accelerate to near light speed but she may be pulled to the hole's event horizon where she will spin round and round the event horizon at the speed of light. It is only upon falling, or diving into the hole that she may exceed light speed and be hurtled toward the past.

Hence, aiming directly for the heart of a black hole is yet another means of achieving superluminal speeds. This course of action, however, also poses many perils including the possibility the Time Traveler will emerge no larger than an atom, and with negative mass and negative energy.

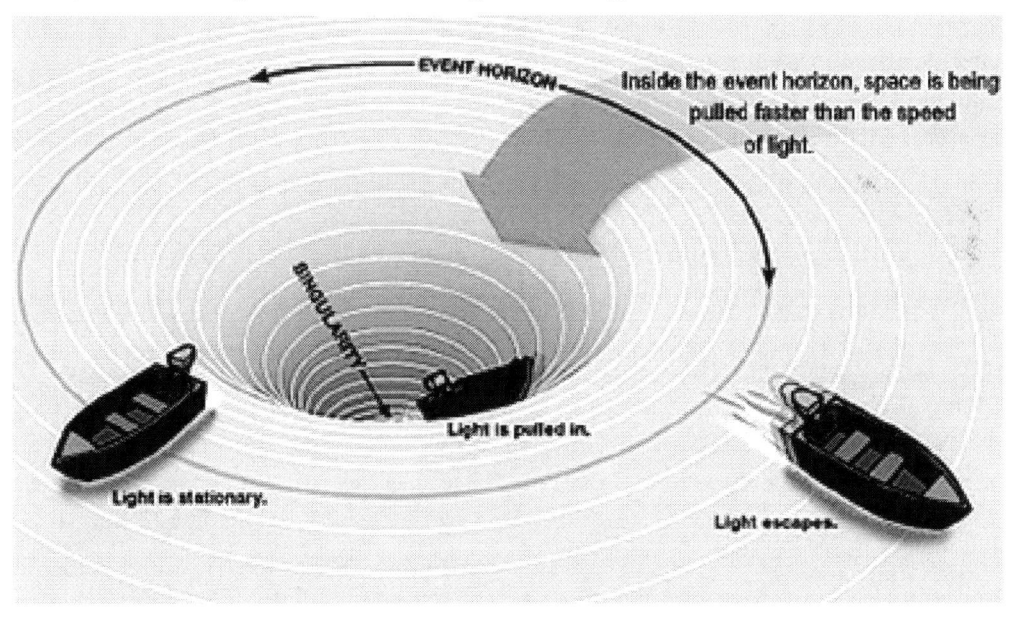

The Interior Of A Black Hole: Vacuums and Negative Energy

There is considerable disagreement about the nature of black holes and what would become of a star or a time machine that happened to fall inside (Almheiri et al., 2013; Hawking 2014); with some scientists arguing that black holes do not even exist. For example, in addition to being ripped to pieces by tidal forces the Time Traveler may experience Unruh thermal radiation; energetic photons which derive their positive energy from the vacuum within the hole and this gives the hole an energy density lower than zero; i.e. negative energy. These conditions create what is referred to as a Rindler vacuum which has negative energy density and negative pressure (Rindler 2001). A time traveler or space craft which falls or dives into the hole will lose all its positive energy and will have a negative energy and negative mass--and this is why time travel into the past will not violate the laws of thermodynamics (Joseph 2014).

An example of this is negative density quantum vacuum is illustrated by the Casimir force, a vacuum which becomes populated by virtual photons and negative energy (Casimir 1948; Jaffe 2005; Lambrecht 2002). Likewise, be it a super massive hole or a worm hole, the vacuum of the hole may create its own quantum vacuum which generates its own negative energy (Everett & Roman 2012). And negative energy has negative mass. This negative energy would be repellent whereas the positive energy would act as propulsion, thereby propelling the time traveler to beyond light speed; albeit, with negative energy and negative mass; a consequence of positive energy and mass being drained off to counterbalance the negative energy of the hole, thereby creating equilibrium as well as propulsion.

Specifically, if the time traveler were to enter the hole (and depending on the negative density of the hole or surrounding vacuum) she may lose mass and energy, become miniaturized due to length contraction, and emerge at the other end of the hole with negative mass and negative energy and then either grow larger in size, remain the same, or appear ghostlike, disintegrate or even self-annihilate if there is a collision with her positively charged double which is heading for the future while she is heading from the future for the past.

Even if there is no danger of inner hole negative energy density and thus no danger of positive to negative energy conversion, the time traveler and her time machine may still be stripped of electromagnetic energy which will radiate out into space (Hawking, 2005, 2014). However, this electromagnetic force (i.e. the basic unit of which is the quantum, the photon) being stripped of its particle, would have no mass, consisting only of a wave, thereby violating the particle-wave duality believed to characterize all matter; and leaving only a wave function as predicted by the Copenhagen interoperation of quantum physics (Bohr 1963; Heisenberg 1927). That is, upon exiting the hole and voyaging into the past, the time traveler may consist of negative mass (no mass) and a wave function.

Black holes spew particles and radiation (Giddings, 1995; Hawking, 2005; Preskill 1994; Russell & Fender, 2010). This raises the possibility that what is radiating out of one end of the hole, are the remnants of whatever entered at the opposite end of the hole. If, as Einstein and Rosen (1935) predicted, a mirror universe is at the bottom of the hole, stars, objects, time machines, etc. may enter from "the bottom" and emerge at the top as radiation. Black holes may be two way streets leading to the future and the past.

Worm Holes in Space-Time

The Einstein-Rosen bridge and the holes they visualized are also sometimes referred to as "worm holes." "Worm holes" differ from super massive "black holes" but also share many of the same features, such as negative energy densities (Everett & Roman 2012).

Wheeler and colleagues came to the conclusion that these "holes" would be unstable and would collapse so quickly that even a light beam would never make it through but would be pinched off at the throat (Fuller & Wheeler, 1962; Taylor & Wheeler, 2001). Wheeler and others have suggested that the curvatures of the walls inside these worm holes may smack up against one another, leaving no space between them.

"Worm holes" are not necessarily worm-size but can be planet-size or larger. It has also been proposed that worm holes (like "black holes") have an event horizon and any object approaching a worm hole would get caught at the horizon and could never make it inside. The "Einstein-Rosen Bridge" was thus a bridge to nowhere and no one could get across.

Others have proposed various worm hole and black hole geometries which could make it possible to tunnel from one galaxy or universe to another (Al-Khalili 2011; Hawking 1988; Thorne 1994). In 1963, Roy Kerr proposed a solution to Einstein's equations which predicted that a collapsing star would be rotating as it collapsed. Because of the rotation it would not collapse to a singularity, but become compressed into a ring which allows passage from one end of the hole to the other (Kerr 1963; O'Neill 2014). Therefore, anything passing through the throat would not be subject to infinite curvature or infinite gravity, but finite gravity. A time traveler falling into the hole would be shot right though to the other side. The Kerr black hole is a hole that acts as a gateway to the mirror universe. Unfortunately, Kerr holes are so unstable they might collapse before the time traveler could exit.

Mike Morris and Kip Thorne (1988) proposed that "traversable wormholes" would have no event horizon, no infinite curvatures, no tidal forces, and could be easily traversed if propped open and lined with "exotic matter" which would have a repulsive gravitational effect--meaning, other mass would fly away from it. The inner walls of the worm hole would repulse one another and remain open. Their proposals, however, violate the so called "weak energy" hypothesis: neither matter or energy can ever be negative and can never have less than the mass or energy density of empty space. The weak energy "hypothesis" however, is just that and it has been proved wrong (Jaffe, 2005; Lambrecht, 2002).

Morris and Thorne (1988) proposed that their "exotic matter" should be in the throat of the worm-hole, forming, presumably, a symmetric ring of exotic matter which would repulse the walls of the hole, keeping the worm hole open and the walls far apart. Unfortunately, any space-time traveler would have to pass through this ring, and suffer the unknown effects of this exotic matter. As a "solution" Thorne and Morris suggested inserting a vacuum tube through the throat which could act as a shield. Thorn has also suggested towing these holes to different locations in space-time thereby making it easier to journey from one location to another in faster than light speed (Thorn & Hawking 1995). Unfortunately, Thorn has never explained what this "exotic matter" is, or how it would be possible to cart these holes from place to place.

The "worm hole" theories and proposals that have received the most attention from the scientific community and the media are in many respects a generalized modification of the standard "black hole" theory (Fuller & Wheeler 1962; Taylor & Wheeler 2000); i.e. the gravity of a mass compacted to a singularity punches a hole in space-time; or, a spinning vortex of a collapsing star drills the hole and never forms any singularity; or exotic matter creates and maintains the hole.

Nevertheless, "worm holes" should be distinguished from "black holes" and although rather fantastical scenarios have been published in various books and scientific journals about their nature and creation, only a few scientists have provided explanations as to how and why a "worm hole" would form between planets, solar systems or galaxies and how and why the hole would link them.

For example, John Wheeler proposed what he calls "spacetime foam" which can form a labyrinth of tunnels and holes the size of a Plank length filled with positive and negative (virtual) particles (Wheeler 2010). Virtual particles are continually popping into existence, permeating space with quantum activity such that there is no emptiness, but only fluctuations involving frenzied activity. An example of this is a quantum vacuum as illustrated by the Casimir force which becomes populated by virtual photons and negative energy (Casimir 1948; Jaffe 2005; Lambrecht 2002).

These quantum holes and the virtual (negative mass) particles within them live on borrowed time. Because they disappear so quickly, lasting no longer than Planck time, they have been referred to as "virtual" (Wheeler 2010). There would not be much time to travel through a tunneling hole in space-time foam.

How Holes Form Between Layers in Extreme Curvatures of Space-Time

Space-time is not a solid but an expression of the quantum continuum which consists of a frenzy of particles and waves that are perceived and experienced as substance, form, and the various dimensions of the universe (Bohr, 1934, 1958, 1963; Dirac 1966a,b). Because it is not a solid but consists of space, time, and vacuum, space-time may contract and can be stretched, twisted, ripped apart, and may be permeated with perforations and holes which tunnel to distant locations (Fuller & Wheeler, 1962; Wheeler 2010).

When a single layer of space-time folds up and over or curves around itself, thereby forming bilayers with an empty space (vacuum) in between, differences in concentration gradients in gravity and energy may create pockets of pressure and forces similar to those which drive osmosis and diffusion (Wilf et al. 2007). As to how long these holes remains open depends on the pressure, concentration, temperature, gravity, electrical charges and other factors involving not just the folded layers (which form a bilayer) but the vacuum in between the bilayers and differential energy gradients and negative versus positive energy fluxes within and between them.

Differential gravity and energy densities, and the consequence of dragging space-time into respective gravity-bubble cavities can create so much tension that the tissues of space time becomes weakened making specific areas subject to ripping, tearing, and hemorrhaging; like when too much air in a rotating tire heats up and expands and a hole bursts open allowing the air to escape. When negative vs positive energy levels within the vacuum and the pressures above or beneath one layer or both become extreme, any weak points that have been formed in space-time hemorrhage and holes pop open thereby allowing passage through these layers and the vacuum between them; similar in some ways, to osmosis (Wilf et al. 2007).

Consider, for example, Andromeda and the Milky Way Galaxies, each of which causes extreme curvatures as they sink deep down into space-time--just as two fat men standing in a pool of mud will sink into the wet ground. However, unlike the two men where some of the mud beneath their feet plops out onto the surface of the pool of mud, these two galaxies drag space-time down into the cavities they create. In consequence, the distance between them also shrinks as surrounding space is pulled down into the depressions they've made. These spherical balloon-like cavities may be nearly side by side, separated by lengths of the layer of space-time dragged down into each of these gravity pockets, with the layer forming the wall of one cavity separated from the wall of the other cavity by the empty space between them; like two adjacent pockets woven from the same cloth. One layer of space-time stretched down into and lining the two cavities which are relatively side by side, creates a bilayer and between the bilayer: a vacuum of empty space.

However, when a vacuum forms between layers, a negative pressure density develops and the vacuum comes to be permeated by negative energy (Casimir, 1948; Bressi et al., 2002; Jaffe 2005; Rodriguez et al., 2011). Exterior to the layers surrounding the vacuum, the opposite sides of space-time will be subject to a positive energy density, which may be greater on one side than the other. These differential pressure gradients can exert so much stress that holes may open up and tunnel between those layers thereby releasing pressures building up within the vacuum and between them.

Moreover, the layer of time-space lining one gravity-cavity may also be differentially effected by gravity and electromagnetic activity triggering changes in polarity and negative vs positive energy fluxes which propagate between layers, thereby effecting the vacuum between them. For example, one layer may have a greater positive energy density than the other, or the vacuum (empty space) between the bilayers may be permeated by pockets of positive and negative energy (Everett & Roman 2012). In consequence, holes, or channels, may pop open to release this pressure and to equalize energy gradients.

Differential gravity and electromagnetic activity also creates an imbalance in electric fields which can lead to electromagnetic diffusion and the conductance of electrical charges (Pollack & Stump 2001; Slater & Frank 2011) and the opening up of holes in the bilayers of space-time so as to equalize charges and polarity (Joseph 2014). Different positive energy densities on either side of a folded layer of space may trigger electromagnetic permeability and the conductance of charged particles from one side to the other to establish electromagnetic equilibrium (Pollack & Stump, 2001; Slater & Frank 2011). Holes may be forced open from the outside, as well as from within the vacuum due to the interactions of different electric fields and energy densities on either side of the folded space-time layer.

The Casimir Vacuum: Negative vs Positive Energy

The vacuum between a folded up layer of space-time will most likely have a negative energy density, or, perhaps pockets of negative and positive energy (Casimir, 1948; Bressi et al., 2002; Jaffe 2005). In a series of experiments first performed by Hendrik Casimir in 1948, it was found that the "vacuum" between two uncharged silver plates aligned together developed a negative energy density, i.e. less than zero. Thus energy must be added to reach zero. By contrast, in a vacuum that was not surrounded on two sides by plating there was no negative energy and the energy density was already zero.

Casimir (1948) found that these two uncharged mental plates, when aligned close together with the vacuum between them, demonstrated an attractive force due to vacuum fluctuations; the plates were drawn together. Further, the negative energy density increased as the two plates were pulled closer together.

The same "Casimir Effect" may also characterize the vacuum between folded or curled up layers of space-time; i.e. there is a buildup of negative energy and the layers of space-time are drawn closer together.

In the Casimir "vacuum" there is positive pressure in two directions parallel to the plates, but a large negative pressure inside which sucks the two together--and the closer the plates (or layers) the greater the negative energy density between them (Bressi et al., 2002; Jaffe 2005; Rodriguez et al., 2011). Normally, a layer or vacuum with positive energy density would exert pressure outward (like air in a tire). A negative energy vacuum generates inward pressure or suction. Thus, in response to extreme curvatures of space-time, adjacent layers (be they like the crests or valleys of two waves, or bilayers formed when two depressions are drawn together) would also be pulled closer together by the suction of the negative energy vacuum. When adjacent layers are pulled closer together this puts more pressure on the vacuum between them which in turn increases the negative energy density within (Casimir, 1948; Bressi et al., 2002; Jaffe 2005; Lambrect 2002).

Space-time is not a solid. Suction from a vacuum between layers would suck positive energy outside the layers into the vacuum, thereby creating holes large enough to allow the passage of positively charged particles and mass; acting similar to a semi-permeable membrane. As negative energy density increases it would cause holes to pop open in the outer layers thereby allowing negative energy to escape and decreasing the vacuum state within; just as a garden hose with a kink sprouts several holes as the pressure builds up thereby allowing water to burst from the holes. Holes open up so as to establish equilibrium; again, similar to osmosis.

For example, if a container of water is divided by a semi-permeable membrane, and 6 ounces of salt were poured on one side and 1 ounce on the other, it would create an imbalance with osmotic pressure which is higher on one side and lower on the other. High osmotic pressure provides a high energy density and thus the energy to force salt across the membrane, from the side with the higher concentration to the side with lower, until the different concentrations would equalize thereby driving the osmotic pressure to zero (Wilf et al 2007). The same amount of salt eventually end up on both sides of the membrane.

Punching Holes in Space-Time: Leakage of Positive vs Negative Energy

These kinks in the curvature of space time would also give rise to vacuum fluctuations which differ in the squeezed areas vs those further apart (Everett & Roman 2012). Those areas squeezed and drawn closer together would contain pockets of increasing negative energy density whereas those further way may contain pockets of positive energy density. Positive energy is more powerful than negative energy, with positive being attracted and negative repulsive, thereby increasing pressure against the fabric of space-time. In consequence, one or both folded or curled up layers would hemorrhage thereby allowing negative energy to escape and positive energy to rush in through the holes between them.

Likewise, in regions where space-time is folded or curled together, differences in concentration gradients in gravity and positive vs negative energy density or quantum inequalities in electromagnetic activity may puncture these layers and force holes to open. Differential gravity and electromagnetic activity also creates an imbalance which can lead to electromagnetic diffusion and the conductance of electrical charges (Pollack & Stump, 2001; Slater & Frank) and the opening up of holes in the bilayers of space time. Holes may be forced open from the outside, not from within the vacuum. Once the holes are formed, energy will stream through the hole and in and out of the vacuum between the bilayers, thereby allowing for the diffusion of positively charged molecules and larger objects which can pass from one side to the other and into and out of the vacuum.

The same principles can be applied to a time traveler. The time traveler and the time machine would consist of positive energy and positive mass. Therefore a time traveler could also be sucked through these holes and deposited on the other side--though there is a danger he may remain trapped inside the vacuum if the hole closes up or he may be stripped of positive energy and mass.

For example, as positive charged particles pass through the vacuum between the bilayers, they will lose positive energy, thereby equalizing the energy density outside and inside the hole. The hole then closes up. However, positively charged particles, in losing their positive charge as they pass through the hole may have a negative charge when they emerge on the other side (Everett & Roman 2012). If positive charges enter both ends of the hole, and exit the opposite ends as negative (after having their positive charge stripped away), this will have an equalizing effect, such that the charges inside the vacuum of the hole and those outside the hole are in equilibrium; at which point, the hole may grow smaller in size and then close up.

Therefore the tube-like vacuum between folded up layers of space-time may contain pockets of positive and negative energy (Everett & Roman 2012). As the "attraction" is repulsive, and as positive energy can have mass whereas negative energy would have only negative mass, the pockets of positive energy can push against the negative. As the negative energy is repulsive and has no where to go except out, these oppositional positive vs negative forces may cause the fabric of space-time to hemorrhage and leak negative energy which is pushed out by the pockets of positive energy inside the vacuum; or as a consequence of the two layers being drawn so close together than negative energy has no where to go except out.

Thus any the particles and radiation spewed out of a worm hole may be a consequence of the negative energy density within the vacuum between bilayers. These represent the remnants of positively charged particles and mass which originate from outside the layers of space time from the opposite end of the hole, or from negative energy from within the vacuum which runs the length between the bilayers of space-time. Once equilibrium is reached, these "worm holes" close up because they have equalized the negative and positive energy densities within the vacuum which forced these hole to open in the first place (Everett & Roman 2012). However, as the nature of these vacuums is to generate negative energy, the cycle of holes opening and then closing may infinitely repeat itself.

Length Contraction and Quantum Worm Holes in Space-Time

Upon accelerating toward light speed the time-space traveler and her time machine will gain incredible mass and energy and shrink and contract in size. For example, if a 1000 foot in length time-space ship attained a velocity of 60% light speed, it would contract by 20%, to 800 feet. The amount of length contraction can be calculated, determined by the Lorentz Transforms (Einstein et al. 1923).

If her time machine has spin, rotation, and acceleration the phenomenon of length contraction will be equalized and she and her vehicle will shrink to molecular and then atom sized; otherwise she will be spaghettified and be all height yet hair-like thin as length contraction is always in the direction of motion. Moreover, although growing smaller in size she will be gaining mass as predicted by general relativity (Einstein 1915a,b, 1961).

According to this equation: $E = mc^2$, mass and energy are the same physical entity and can be changed into each other. Because of this equivalence, the energy an object has due to its motion will increase its mass. In other words, the faster an object moves, the greater its mass which at the same time is contracting. This only becomes noticeable when an object moves at great speeds. If it moves at 10% the speed of light, for example, its mass will only be 0.5% more than normal. But if it moves at 90% the speed of light, its mass will double and it will also contract by 50%. As an object approaches the speed of light, its mass rises precipitously as it shrinks to nearly nothing in size.

Upon reaching a velocity of 186,000 miles per second, time stops and the time traveler will experience an eternal present which is frozen in place. The time traveler will also have shrunk to a size perhaps smaller than a Planck Length (Joseph 2014). It has been theorized, based on Einstein's theories of relativity, that at this speed she and her time machine will have infinite mass and will have disappeared into nothingness. However, this prediction may represent the limitations of Newtonian and Einsteinian physics, the applicability of which diminishes and loses all meaning when applied to space smaller than a Planck length, the smallest unit of measurement. Space-time, within the Planck scale, is subject to extreme uncontrollable quantum fluctuations, as it is continually being bent, folded, crumpled, and torn apart by powerful gravitational forces; and the laws of physics break down (Bruno, et al., 2001).

It is accepted by most scientists that relativity breaks down and quantum theory takes over when making measurements and observations of the microscopic world. Quantum mechanics describes effects at the scale of single particles, atoms and molecules whereas classical physics and Einstein's theory of gravity are more applicable when applied to large masses. Quantum physics takes over from relativity and Newtonian physics at the atomic and subatomic level.

In quantum physics, the smallest unit of space has a Planck length which is defined as 10-33 cm (Eisberg & Resnick 1985), about 10-20 times smaller than the radius of a proton. Space smaller than a Planck length cannot be conceptualized by quantum mechanics or classical physics: Geometry ceases to exist, Cartesian coordinates, x, y and z, cannot be applied, and time ceases to have meaning (Garay 1995). Instead, a defining feature of these tiny spaces is gravity so powerful that it punches a hole in space-time (Joseph 2010b).

As predicted by Einstein's (1915a,b, 1961) theory of general relativity, any mass m has a length called the Schwarzschild radius, Sr. Compressing an object of mass m, to a size smaller than this radius Sr, generates tremendous gravity and immediately results in the formation of an almost infinitely small black hole in the fabric of space-time. Subatomic holes may be continually forming and then disappearing within spaces smaller than a Planck length (Joseph 2010b).

Two lengths of the Schwarzschild radius, Sr, become equal at the Planck length. Contained within a Planck volume is a Planck mass. Planck mass is also associated with quanta, which also refers to particles of light. Thus, in many respects Planck mass represents a division line between quantum and classical mechanics. In terms of time travel, Planck mass only becomes meaningful when applied to objects smaller than a Planck Length; objects which may then experience time reversal (Heisenberg 1958),

Time Travel in Space Smaller than a Plank Length

A Plank mass has a mass of about 22 micrograms ($mP = 2.18 \times 108$ kg). Planck mass has sometimes been described as having gravitational potential energy which is generated between two masses which are separated by the angular wavelength of a photon. The gravitational potential energy can be derived mathematically when the Compton wavelength and Schwarzschild radius are equal. The Schwarzschild radius is the radius in which a mass, if confined, would become a black hole.

If we try to combine relativity with quantum physics, neither of which can accurately describe events taking place in space smaller than a Planck Length, and accepting that an object cannot be compressed to nothingness and retain height, width and mass, then it may be safe to say that a time machine upon reaching light speed, may contract to a size smaller than a Planck Length (Joseph 2010b) and then experience time reversal (Heisenberg 1958). Instead, of infinite mass, the now smaller than a Planck length time machine would have incredible mass, energy and gravity as suggested by relativity, Newtonian physics, and quantum mechanics. It would also have incredible energy; enough energy to blow a hole through time-space, thereby using up all its mass/energy, resulting in negative mass and negative energy.

Einstein's and Newton's theories of gravity both predict that if mass is shrunk to a subatomic space, its gravity will become increasingly powerful. Quantum physics tells us that in spaces the size of the Planck length, coupled with the corresponding Planck energy (10^{19} GeV), that the gravitational forces becomes so incredibly powerful (Eisberg & Resnick 1985; Smolin, 2002) that holes are created in space time. Hence, a space-time machine or any object with a Planck mass and whose radius is less than the Planck length, would have so much gravity that it could collapse surrounding space-time and create a black hole about the size of a Planck length.

Space-time, within the Planck scale, is subject to extreme uncontrollable quantum fluctuations, as it is continually being bent, folded, crumpled, and torn apart by these powerful gravitational forces (Bruno, et al., 2001). Presumably holes in space-time and the quantum continuum are continually forming and disappearing (Joseph 2010b) and it is these holes will lead to a mirror universe where time runs in reverse.

As predicted by General Relativity, at least one hole may exist for every Planck length throughout space time. Therefore, all of space-time may be permeated by Planck length black holes which continually pop in and out of existence (Joseph 2010b).

Does this mean that particles, energy, and time travelers are continually leaking through these holes? The principles of quantum computing are based on this belief. Einstein (1939), however, argued that because of the principles governing the speed of light, particles could never enter these holes. And yet, Einstein's theories also predict that objects with sufficient mass and which are sufficiently small, will create these holes, whereas his theory of general relativity allows for superluminal speeds.

Therefore, based on general relativity and quantum mechanics, it can be predicted that at light speed, the time machine and the time traveler do not acquire infinite mass and shrink to nothing. Instead, upon approaching near light speed, the time machine and the time traveler may shrink to a size smaller than a Planck length, and they will contain so much concentrated mass, gravity and energy that they blow a hole in space-time. In other words, the time traveler and the time machine upon reaching light speed may create its own hole in space-time, and then emerge on the other side. However, the intrepid time traveler will not be journeying into the future, but into the past.

According to Heisenberg (1956), "when quantum theory is combined with relativity, it predicts time reversal" in spaces smaller than 10^{-13}cm; smaller than the radii of an atomic nucleus--a Planck length. "The phenomenon of time reversal...belongs to these smallest regions."

Although relativity says mass becomes infinite at or beyond the speed of light, quantum mechanics does not support this prediction. Infinity is possible only in an infinite universe (Joseph 2010a,b). In a finite (vs infinite) universe, the concepts of infinite mass and infinite energy may well be mathematical anomalies which represent not the impossibility of accelerating above light speed and then back below the speed of light, but the limitations and breakdown of those theories which are no longer applicable when applied to events taking place at these velocities and at these small sizes. Most physicists would agree that Newton's laws of motion are suitable for macro-objects whereas quantum mechanics must be applied for micro-structures if there is any hope of obtaining measurements and making predictions which might agree with experimental observation (Heisenberg 1927, 1958). However, at an atomic scale the energy and speed of a particle is uncertain and the smaller the size the greater the uncertainty and the greater the fluctuations. In fact, at the Planck length, the Uncertainty Principle and quantum indeterminacy becomes virtually absolute (Eisberg & Resnick 1985; Heisenberg 1958; Smolin, 2002).

If particles, objects, or time-machines contract to a size smaller than a Planck Length as they near and then reach light speed, they have so much energy they may instead create their own hole in space-time as predicted by quantum mechanics, and then be expelled at superluminal velocities with negative energy and negative mass; with positive energy acting as propulsion and negative energy as repulsive, the close proximity of which increases escape velocity. The same consequence may befall all those who dive into a super-massive black hole with a negative energy density.

As to the possibility of "infinite mass" photons travel at light speed and photons have zero mass, not infinite mass. Likewise, it can be predicted that a time machine upon reaching light speed, may also have zero mass, a consequence, in part of its release of energy when it shrinks to a size smaller than a Planck length and blows a hole through space-time (Joseph 2010b). In a finite (Big Bang) universe, and as predicted by quantum mechanics, it is more likely that at light speed the time traveler will have shrunk to a size smaller than a Planck length. Because so much energy will be released, instead of infinite mass, she would have no mass or negative mass and negative energy when she emerges from the hole (Joseph 2014) be it a worm hole or super massive black hole.

Emerging From the Worm Hole: The Mirror Universe

Worm holes presumably tunnel through a negatively charged vacuum between the layers of space time which would strip away positive energy to equalize energy gradients, such that the Time traveler would have a negative charge when they emerge. The same stripping away of positively energy may also take place when passing through a super massive black holes. It is precisely because these holes may be filled with a negative energy vacuum that the Time Traveler may be stripped of positive energy and mass while being propelled to well beyond the speed of light into the future, the past, or a mirror universe.

Moreover, due to length contraction, upon emerging from the hole, be it a super massive black hole, worm holes, or hole the size of Planck Length, she may be no larger than the smallest particle or emerge as electromagnetic energy which radiates out from the hole into space (Giddings, 1995; Hawking, 2005, 2014; Preskill 1994; Russell and Fender, 2010), possibly speeding off at superluminal speeds into the past.

Coupled with negative mass and negative energy, the time traveler upon emerging from the future (a consequence of accelerating toward light speed) and then speeding from the future into the past, may share characteristics with superluminal (hypothetical) particles such as "tachyons" (Bilaniuk & Sudarshan 1969; Chodos 2002; Feinberg, 1967; Sen 2002). The hypothetical tachyons are believed to always travel at superluminal velocities. Tachyons, which may have negative energy and negative mass, are the true time travelers, forever journeying from the future to the past.

Therefore, upon entering this mirror universe, the time traveler not only accelerates into the past but she may consist only of negative mass and energy; a phenomenon which also allows for the creation of duality without violating the laws of mass and energy conservation. There is no violation because she will consist of negative mass and energy and thus no extra mass or energy is introduced and as the negative equalizes the positive energy / mass of the time traveler who already exists in the past.

Duality is a natural consequence of traveling first to the future to get to the past. The time traveler will continually pass herself coming and going.

REFERENCES

Al-Khalili (2011). Black Holes, Wormholes and Time Machines, Taylor & Francis.

Almheiri, A. et al. (2013). Black Holes: Complementarity or Firewalls? J. High Energy Phys. 2, 062

Bethe, H. A., et al., (2003) Formation and Evolution of Black Holes in the Galaxy, World Scientific Publishing.

Bilaniuk, O.-M. P.; Sudarshan, E. C. G. (1969). "Particles beyond the Light Barrier". Physics Today 22 (5): 43–51.

Bo, L., Wen-Biao, L. (2010). Negative Temperature of Inner Horizon and Planck Absolute Entropy of a Kerr Newman Black Hole. Commun. Theor. Phys. 53, 83–86.

Bohr, N., (1913). "On the Constitution of Atoms and Molecules, Part I". Philosophical Magazine 26: 1–24.

Bohr, N., (1913). "On the Constitution of Atoms and Molecules, Part I". Philosophical Magazine 26: 1–24.

Bohr, N. (1934/1987), Atomic Theory and the Description of Nature, reprinted as The Philosophical Writings of Niels Bohr, Vol. I, Woodbridge: Ox Bow Press.

Bruno, N. R., (2001). Deformed boost transformations that saturate at the Planck scale. Physics Letters B, 522, 133-138.

Blandford, R.D. (1999). "Origin and evolution of massive black holes in galactic nuclei". Galaxy Dynamics, proceedings of a conference held at Rutgers University, 8–12 Aug 1998, ASP Conference Series vol. 182.

Carroll, S (2004). Spacetime and Geometry. Addison Wesley.

Casimir, H. B. G. (1948). "On the attraction between two perfectly conducting plates". Proc. Kon. Nederland. Akad. Wetensch. B51: 793.

Chodos, A. (1985). "The Neutrino as a Tachyon". Physics Letters B 150 (6): 431.

Einstein, A. (1915a). Fundamental Ideas of the General Theory of Relativity and the Application of this Theory in Astronomy, Preussische Akademie der Wissenschaften, Sitzungsberichte, 1915 (part 1), 315.

Einstein, A. (1915b). On the General Theory of Relativity, Preussische Akademie der Wissenschaften, Sitzungsberichte, 1915 (part 2), 778–786, 799–801.

Einstein A. (1939) A. Einstein, Ann. Math. 40, 922.

Einstein, A. (1961), Relativity: The Special and the General Theory, New York: Three Rivers Press.

Einstein, A. and Rosen, N. (1935). "The Particle Problem in the General Theory of Relativity". Physical Review 48: 73.

Einstein, A., Lorentz, H.A., Minkowski, H., and Weyl, H. (1923). Arnold Sommerfeld. ed. The Principle of Relativity. Dover Publications: Mineola, NY. pp. 38–49.

Einstein A, Podolsky B, Rosen N (1935). "Can Quantum-Mechanical Description of Physical Reality Be Considered Complete?". Phys. Rev. 47 (10): 777–780.

Eisberg, R., and Resnick. R. (1985). Quantum Physics of Atoms, Molecules, Solids, Nuclei, and Particles. Wily,

Everett, A., & Roman, T. (2012). TIme Travel and Warp Drives, University Chicago Press.

Feinberg, G. (1967). "Possibility of Faster-Than-Light Particles". Physical Review 159 (5): 1089–1105.

Fuller, Robert W. and Wheeler, John A. (1962). "Causality and Multiply-Connected Space-Time". Physical Review 128: 919.

Garay, L. J. (1995). Quantum gravity and minimum length Int.J.Mod.Phys. A10 (1995) 145-166

Ghez, A. M.; Salim, S.; Hornstein, S. D.; Tanner, A.; Lu, J. R.; Morris, M.; Becklin, E. E.; Duchene, G. (2005). "Stellar Orbits around the Galactic Center Black Hole". The Astrophysical Journal 620: 744.

Geiss, B., et al., (2010) The Effect of Stellar Collisions and Tidal Disruptions on Post-Main-Sequence Stars in the Galactic Nucleus. American Astronomical Society, AAS Meeting #215, #413.15; Bulletin of the American Astronomical Society, Vol. 41, p.252.

Giddings, S. (1995). The Black Hole Information Paradox," Proc. PASCOS symposium/Johns Hopkins Workshop, Baltimore, MD, 22-25 March, 1995, arXiv:hep-th/9508151v1.

Hawking, S. W., (1988) Wormholes in spacetime. Phys. Rev. D 37, 904–910.

Hawking, S., (1990). A Brief History of Time: From the Big Bang to Black Holes. Bantam.

Hawking, S. (2005). "Information loss in black holes". Physical Review D 72: 084013.

Hawking, S. W. (2014). Information Preservation and Weather Forecasting for

Black Holes.

Heisenberg, W. (1927), "Über den anschaulichen Inhalt der quantentheoretischen Kinematik und Mechanik", Zeitschrift für Physik 43 (3–4): 172–198.

Heisenberg. W. (1930), Physikalische Prinzipien der Quantentheorie (Leipzig: Hirzel). English translation The Physical Principles of Quantum Theory, University of Chicago Press.

Heisenberg, W. (1955). The Development of the Interpretation of the Quantum Theory, in W. Pauli (ed), Niels Bohr and the Development of Physics, 35, London: Pergamon pp. 12-29.

Heisenberg, W. (1958), Physics and Philosophy: The Revolution in Modern Science, London: Goerge Allen & Unwin.

Jaffe, R. (2005). "Casimir effect and the quantum vacuum". Physical Review D 72 (2): 021301.

Joseph, R (2010a) The Infinite Cosmos vs the Myth of the Big Bang: Red Shifts, Black Holes, and the Accelerating Universe. Journal of Cosmology, 6, 1548-1615.

Joseph, R. (2010b). The Infinite Universe: Black Holes, Dark Matter, Gravity, Acceleration, Life. Journal of Cosmology, 6, 854-874.

Joseph, R. (2014a) Paradoxes of Time Travel: The Uncertainty Principle, Wave Function, Probability, Entanglement, and Multiple Worlds, Cosmology, 18, 282-302.

Joseph, R. (2014a) The Time Machine of Consciousness, Cosmology, 18, In press.

Kerr, R P. (1963). "Gravitational Field of a Spinning Mass as an Example of Algebraically Special Metrics". Physical Review Letters 11 (5): 237–238.

Lambrecht, A. (2002) The Casimir effect: a force from nothing, Physics World, September 2002

Lorentz, H. A. (1892), "The Relative Motion of the Earth and the Aether", Zittingsverlag Akad. V. Wet. 1: 74–79

McClintock, J. E. (2004). Black hole. World Book Online Reference Center. World Book, Inc.

Melia, F. (2003). The Edge of Infinity. Supermassive Black Holes in the Universe. Cambridge U Press. ISBN 978-0-521-81405-8.

Melia, F. (2007). The Galactic Supermassive Black Hole. Princeton University Press. pp. 255–256.

Merloni, A., and Heinz, S., (2008) A synthesis model for AGN evolution: supermassive black holes growth and feedback modes Monthly Notices of the Royal Astronomical Society, 388, 1011 - 1030.

Minchin, R. et al. (2005). "A Dark Hydrogen Cloud in the Virgo Cluster". The Astrophysical Journal 622: L21–L24.

Morris, M. S. and Thorne, K. S. (1988). "Wormholes in spacetime and their use for interstellar travel: A tool for teaching general relativity". American Journal of

Physics 56 (5): 395–412.

O'Neill, B. (2014) The Geometry of Kerr Black Holes, Dover

Penrose, R. (1969) Rivista del Nuovo Cimento.

Preskill, J. (1994). Black holes and information: A crisis in quantum physics", Caltech Theory Seminar, 21 October. arXiv:hep-th/9209058v1.

Pollack, G. L., & Stump, D. R. (2001), Electromagnetism, Addison-Wesley.

Rindler, W. (2001). Relativity: Special, General and Cosmological. Oxford: Oxford University Press.

Rodriguez, A. W.; Capasso, F.; Johnson, Steven G. (2011). "The Casimir effect in microstructured geometries". Nature Photonics 5 (4): 211–221.

Ruffini, R., and Wheeler, J. A. (1971). Introducing the black hole. Physics Today: 30–41.

Russell, D. M., and Fender, R. P. (2010). Powerful jets from accreting black holes: Evidence from the Optical and the infrared. In Black Holes and Galaxy Formation. Nova Science Publishers. Inc.

Schrödinger, E. (1926). "An Undulatory Theory of the Mechanics of Atoms and Molecules". Physical Review 28 (6): 1049–1070. Bibcode:1926PhRv...28.1049S. doi:10.1103/PhysRev.28.1049.

Sen, A. (2002) Rolling tachyon," JHEP 0204, 048

Slater, J. C. & Frank, N. H. (2011) Electromagnetism, Dover.

Smolin, L. (2002). Three Roads to Quantum Gravity. Basic Books.

Taylor, E. F. & Wheeler, J. A. (2000) Exploring Black Holes, Addison Wesley.

Thorne, K. (1994) Black holes and time warps, W. W. Norton. NY.

Thorne, K. S. & Hawking, S. (1995). Black Holes and Time Warps: Einstein's Outrageous Legacy, W. W. Norton.

Thorne, K. et al. (1988). "Wormholes, Time Machines, and the Weak Energy Condition". Physical Review Letters 61 (13): 1446.

Wilf, M., et al. (2007) The Guidebook to Membrane Desalination Technology: Reverse Osmosis, Balaban Publishers.

III: Time, Reality, Temporal Non-Locality Vendanta, Upanishads and Quantum Mechanics

6. The Nature of Reality, the Self, Time, Space and Experience
Menas C. Kafatos[1] and Deepak Chopra[2]

[1]Chapman University, Orange, CA 92866
[2]Chopra Foundation, 2013 Costa Del Mar Road, Carlsbad, CA 92009

Abstract

In this article we address the nature of Reality, universal Consciousness and the subjective Self found in monistic systems of India, particularly in many sūtras of Vedanta and Kashmir Śaivism. These monistic systems do not consider time (or space) as primary but treat it as a derivative of consciousness. A brief summary of monism's process of manifestation of all objective existence, including time, greatly adds to the present special volume. Monism complements western science. Indian schools of monism complement Western science through intriguing parallels with modern quantum theory and the role of the observer. Moreover, qualia, the innumerable qualities of subjective experiences, play a fundamental role in both monistic and scientific descriptions of consciousness as it unfolds from its source. The conscious universe is an emerging view in science, in accord with ancient views about the primacy of consciousness at every level of reality.

Keywords: Consciousness, Self, time, Reality, experience, quantum mechanics, qualia

Introduction

Several of the topics explored in this text, are at the very center of what constitutes subjective experience of time. However, the problems of experience, and, ultimately, of conscious self-awareness are not addressed by modern science as science deals with the nature of the objective world, with interactions between objects. Although attempts are made in a variety of fields, including psychology, neuroscience and even philosophy of science to address the issue of experience, the basic challenge remains that by its nature, science can only talk about objective reality, while self-awareness ultimately concerns itself with subjective experience and existence. Yet, Indian perennial philosophies have covered the issue of consciousness extensively and provide the means to merge intellectual understanding with personal experiences that begin with everyday perceptions but extend far into the possibility of mind as a force for altering reality "out there." Monistic systems attribute time (and space) as being derived from experience itself, Consciousness being fundamental. This view is profoundly different from the ontology of science, wherein consciousness is a derivate of physical processes.

This introductory article offers a summary of the nature of Reality, consisting of universal Consciousness, the Self of every being and at every level of existence. The monistic explanation of manifestation and for the emergence of time and space has sufficient rigor to complement Western science and provide insights into the particularly challenging issues of experience itself.

Perennial Monistic Systems

Perennial philosophies concern themselves with the nature of Consciousness, the relationship of the individual to the universe and the relationship of the individual to Consciousness itself. Below we will illustrate with a few selected statements some of these eternal truths and examine how they relate to experience, including how space and time themselves arise.

The non-dualistic systems originating in India, specifically Advaita Vedanta and Kashmir Śaivism, give us a higher view of the individual, the universe and the nature of consciousness. The underlying premise, is that the human being is a reflection of fundamental Consciousness and that in fact there is no difference between the individual and universal Consciousness. Śaivism and Vedanta are complete systems of teachings on the nature of ultimate reality, or the Absolute. This underlying reality is called Brahman in Vedanta and Paramaśiva or Supreme Śiva in Kashmir Śaivism. Both accept Absolute undifferentiated Consciousness as the ultimate Reality, as the underlying reality of all countless objects, subjects and the processes tying them together, such as observation, sentience, understanding, dynamics, cause and effect, etc. Although agreeing on the ultimate Reality, they do differ in emphasis: Vedanta emphasizes that Brahman is the only reality and the perception of the universe as something separate from that as an illusion; while Śaivism accepts the universe as real, being itself, as we will see below, part of universal Consciousness. The universe is part of the whole, albeit being the physical, mental, subtle and in fact all objective experience, of the great underlying sea of Consciousness.

Advaita Vedanta

Vedanta accepts the authority of three sets of works, spanning several centuries: The Upanishads, the Brahma Sūtras, and, the Bhagavad Gītā (Kuiken, 2006). Its basic principles are summarized in the Viveka Chudamani (Crest-Jewel of Discrimination) by Adi Śankarā:

a) "Brahman is Reality" b) "The world is an illusion" and, 3) "The individual Self is nothing but Brahman".

Śankarā's "illusion" is taken to mean the misinterpretation of experience (and not "non-existent."). To see the world as independent and separate from the Self, is an illusion. To know reality is to experience the diversity of the universe (Brahman) as identical to one's Self (Ātman). Sutra 4 of the Isha Upanishad asks: "How can the multiplicity of life delude the one who sees its unity?" And the Nrsimhottaratapaniya Upanishad states: "All this is Ātman, all this is Brahman, all this is consciousness".

The Viveka Chudamani (Crest-Jewel of Discrimination) of Adi Śankarā states: "The Ātman is one, absolute, indivisible. It is pure consciousness. To imagine many forms within it is like imagining palaces in the air. Therefore, know that you are the Atman, ever-blissful, one without a second." The Aparokshānubhuti (Self-Realization) of Adi Śankarā states in sūtra 45: "There exists no other material cause of this phenomenal universe except Brahman. Hence this whole universe is but Brahman and nothing else". While sūtra 49 states "Inasmuch as all beings are born of Brahman, the supreme Ātman, they must be understood to be verily Brahman". In short, Vedanta's ultimate teaching is that Ātman (individual Self of any being) is identical to Brahman.

Kashmir Śaivism

The ancient system, Śaivism and its more recent specific form as developed in Kashmir, (Dyczkowski, 1992, 1994; Singh, 1980, 2006), constitutes a body of philosophical teachings, with practical implications for everyday life. Kashmir Śaivism flourished in Kashmir between the 8th – 12th centuries CE, and was developed and built in the tradition of Vedanta. Kashmir Śaivism was developed in that brief but very active period by sages such as Vamana (779 – 813 CE), Vasugupta (875 – 925 CE), Utpaladeva (disciple of Vasugupta, 900 – 950 CE), Kallata (also disciple of Vasugupta), Bhaskara, Somananda, the great Śaivite master Abhinavagupta (950 – 1020 CE), and Kṣemarāja (a great disciple of Abhinavagupta). They have left us a rich, dynamic vision of the universe and our place in it.

Śaivism is a Trika (triadic) system, consisting of Paramaśiva or supreme Śiva, the Absolute, undifferentiated Being; Śakti (universal Energy), also known as Citi (universal Consciousness, as the creative power of the Absolute); and the individual soul. The triadic teaching holds that there is no difference between Śiva and Śakti/Citi, and in fact no difference between Consciousness which is the One Paramaśiva/Citi and the individual. The monism could be also viewed as a three-fold Reality, consisting of Consciousness, the universe, and the individual; or, alternatively, the object, the subject and the processes tying them together: Paramaśiva, the supreme Being is identical to supreme Consciousness, the Self of everything in the universe. The view of the underlying Reality in Śaivism is in harmony with Vedanta (Pandit, 1977). Paramaśiva is the Absolute, undifferentiated universal Being, and as Brahman in Vedanta, is the underlying substratum of all existence. As Citi unfolds the universe, She (the Creatrix of everything) is the ultimate source of all manifestations, all objective existence, all experiences of the subjective selves and as such, is also the source of the mind. In other words, the dynamical aspect of Consciousness gives rise to countless beings and countless worlds. Citi represents the immanent part of existence, while Paramaśiva the transcendent aspect of the same identical existence. Śaivism is perhaps unique of all monistic systems in assigning reality to everything that exists. It does not deny the existence of the universe, but instead it considers the universe as real as the infinite Self, because in fact the universe arises from the Self. Countless or infinite numbers of universes and countless beings and objects are all emanations of the creative power of Citi. The creative process itself manifests in an infinite variety of vibrations (Spanda) of Ultimate Reality. Spanda derives from a term which means "subtle motion", and ultimate Reality is called Spanda because it pulsates. As such, Śaivism is based on the doctrine of vibration (Dyczkowski, 1994). Here quantum field theory (QFT) and Śaivism agree on the importance of vibration in the creative process. Whereas QFT assigns objective existence to vibrations of the quantum field, Śaivism assigns reality of objects to vibrations of the infinite field of Citi.

As in Advaita Vedanta, Śaivism accepts a triadic Self, consisting of the static (Sat) aspect of the universal Being, the dynamic (Citi or Śakti) aspect of Consciousness, and Bliss (Ānanda). These Three-in-One aspects are not different from each other, they are integral aspects of the One, the undivided sea of Consciousness, which is a dynamic, creative and intelligent Reality.

The first sutra of the Śiva Sūtras states: "Consciousness is the Self". The actual Sanskrit term for Consciousness refers to luminous awareness, the Light of Consciousness. Moreover, Consciousness means more than just (conscious) awareness, it has the absolute freedom of will, knowledge and action. In other words, Reality, which is the universal Self, is identical to Consciousness.

How does the universe manifest? What is the source of the universe? The first sutra of the Pratyabhijñā-hṛdayam, "The Secret of Self Recognition", authored by Kṣemarāja (Singh, 1980) states: "Citi, supremely independent universal Consciousness, is the cause of the manifestation, maintenance, and reabsorption of the Universe". Alternatively, "The universe is the means to attain the realization of free universal Consciousness" (Swāmī Shāntānanda, 2003). The first aphorism gives the underlying cause of the changing universe, as universal Consciousness. Here the three cosmic actions which create, maintain and re-absorb all existence are attributed to Citi. In contrast to the way scientists view the universe as being caused and driven by the laws of Nature, the Pratyabhijñā-hṛdayam states that the cause is Consciousness itself.

How does universal Consciousness unfold the universe? Sūtra No. 2 of the Pratyabhijñā-hṛdayam states: "By the power of her own will, she (Citi) unfolds the universe upon her own screen (i.e. in herself, as the basis of the universe)". Here Kṣemarāja describes the universe as being nothing other than the projection by

Consciousness onto Consciousness. What appear as differences in the objectified world, are projected differences in the universal screen of Consciousness.

But then the question would arise, what is the origin of the vast diversity of objectified existence? Sūtra No. 3 of the Pratyabhijñā-hṛdayam, explains: "That becomes diverse because of the division of reciprocally adapted objects and subjects". In other words, division in what appear as objects and subjects gives rise to all diversity. Moreover, in the Paramārthasāra Abhinavagupta states: "Just as in a mirror's reflection a town or a village appears as an image that is not separate from them, yet it appears as separate, and each separate from the other as well as from the mirror, similarly the universe appears differentiated as one thing is from another, and as well is differentiated from the awakened consciousness of the Self, most pure, though that difference too is not real".

Śaivism emphatically holds that the individual is none other than the entire existence. Sūtra No. 4 of the Pratyabhijñā-hṛdayam states: "Even the individual, whose nature is Consciousness in a contracted state, embodies the universe in a contracted form". While sūtra No. 5 holds that "Consciousness herself, having descended from her expanded state, becomes the mind, contracted by the objects of perception".

In the Īśvara Pratyabhijñā Kārikā (Pandit, 2004) Utpaladeva says: "The Great God is the real self of each and every being. He alone endures, through his undiversified Self—his awareness, I am all these". This complementary relationship between undifferentiated Consciousness and the individual operating through the mind, which itself is nothing but Consciousness, is part and parcel of Śaivism.

Manifestation that gives rise to all objects in countless worlds is referred to in many texts of Śaivism, in the Śiva Sūtras, in the Īśvara Pratyabhijñā Kārikā, in the Pratyabhijñā-hṛdayam, in the Paramārthasāra, in the great culminating work Tantrāloka of Abhinavagupta, etc. (Singh, 1980; Chatterji, 1986; Kafatos and Kafatou, 1991; Dyczkowski, 1992; Swāmī Muktānanda, 1997; Pandit, 1997; 2004; SenSharma, 2007; Singh, 2006). The universe is projected out in thirty six levels of creation, or levels of manifestation, or planes of existence, called tattvas, from Paramaśiva, to the Earth plane.

At the highest or "pure" five levels, the separation between subject and object has not occurred. It is in a potential or subtle form, the play between Aham (I am) and Idam (This). These levels are:

1) I: Paramaśiva, absolute undifferentiated existence
2) That: Paraśakti, absolute undifferentiated Consciousness or Citi.

The first two tattvas can be counted as one and are implied at every level of existence, as they are not really different from each other. They constitute the perfect I-Consciousness.

The next three levels begin the potential process of manifestation:
3) I (Am) That:

Here **I** is written in bold for emphasis. In this relationship, as the subjective part of the relationship **I** (Am) That is emphasized, it signifies the Will aspect of Consciousness. Before any knowledge (which is to follow) or subsequent to knowledge any action is undertaken, the subject has to be identifying itself with its own will.

4) That (Am) I:

The emphasis is in That, i.e. the statement is written as **That** (Am) I. Here, as the objective part of the relationship, That is emphasized. As before any action is undertaken, the object has to be identified. Here, it signifies the Knowledge aspect of Consciousness. However, in both these cases, Will and Knowledge, there is no separation, only a (latent) potential of what eventually become a separation between Subject and Object.

5) I (Am) That:

This statement shows balance between the I am and That, and it is recursive, i.e. repeated forever. The balance between Subject and Object signifies the (potential) for Action. But the Subject and Object even in this balanced state, are still One. In action, both the Subject and the Object are balanced, equally weighted.

As we move next to the level of breakdown of the above universal relationships, we have the operation of Māyā. This is the universal (and most often un-understood) freely-undertaken Power, which limits or hides the true nature of Paramaśiva/Paraśakti. Without it, no objectified experience could arise. It gives the appearance of separation, as Śaivism emphasizes. From this point on, the subject and the object seem to be separated:

Māyā and the next five tattvas which accompany it give rise to, respectively: The experience of time; experience of space; (limited ability) of will (to know and act); (limited ability to) know (and then to act); (limited) ability to act.

The same universal statements operate but now in limited form. At that point, the Subject and Object appear separated and they become (many, essentially infinite in number) subjects and objects. The subjects interacting with other subjects and objects now appear as differentiated levels of existence, willing (in a limited way) to know (in a limited way) and act (in a limited way). This in brief, is how Śaivism accounts for time, space and limited abilities of individual beings in terms of the free Power of Māyā.

All 36 tattvas are levels of reality emphasized in different levels of experiences. They are the qualia of all experiences. As such, Śaivism accounts for all Absolute and relative levels of Reality, including time and space, which manifest in the universe. Beyond this commentary, we can also show how the three universal principles which apply to the quantum, life and all fields, namely complementarity, recursion and sentience, also apply and operate at all thirty six tattvas.

From this monism, the possibility of qualia science can emerge. In qualia science, as in Indian monism, experience comes first. After all, that's how reality actually comes to us, experientially, not in quanta or in differentiated data. At the moment qualia science remains in potentia, but the tradition of Indian monism has fleshed out the absolute state of consciousness, its emergence into manifestation, and every minute gradation of existence that the human mind can conceive. What remains is to build a bridge to the worldview of contemporary science, which is what the essays in this volume of Cosmology attempt to do.

References

Chatterji, J.C. (1986). Kashmir Śaivism. SUNY Press, N.Y.

Dyczkowski M.S.G. (1994). Spandakārikā ("The Stanzas of Vibration"). Dilip Kumar Publishers, Varanasi.

Dyczkowski M.S.G. (1992). The Aphorisms of Śiva, SUNY Press, Albany.

Kafatos M., Kafatou T. (1991). Looking in, Seeing out: Consciousness and Cosmos. Quest Books, Wheaton, IL.

Kuiken, G.D.C. (2006). The Original Gita. OTAM Books, The Netherlands.

Pandit B.N. (1997). Aspects of Kashmir Śaivism. Utpal Publications, Santarasa Books, Boulder.

Pandit B.N. (2004). Īśvara Pratyabhijñā Kārikā ("Verses on the Recognition of the Lord") of Utpaladeva (translation with commentary). Muktabodha Indological Research Institute, Delhi.

SenSharma, D.B. (2007). Paramārthasāra ("The Essence of Supreme Truth") of Abhinavagupta (with the commentary of Yogaraja, translation). Muktabodha Indological Research Institute, New Delhi.

Swāmī Muktānanda (1997). Nothing Exists that is not Śiva. SYDA Foundation, South Fallsburg, NY.

Swāmī Prabhavānanda, Isherwood, C. (1975). Viveka Chudamani ("Crest-Jewel of Discrimination") of Adi Śankarā (translation). Vedanta Press, Hollywood, CA.

Swāmī Shāntānanda (2003). The Splendor of Recognition. Siddha Yoga, South Fallsburg.

Swāmī Vimuktānanda (2005). Aparokshānubhuti ("Self-Realization") of Adi Śankarā (translation). Advaita Ashrama, Delhi.

Singh J. (1980). Pratyabhijñā-hṛdayam ("The Secret of Self-recognition"). Motilal Banarsidass, Delhi.

Singh J. (2006). Śiva Sūtras ("The Yoga of Supreme Identity"). Motilal Banarsidass, Delhi.

7. Perceived Reality, Quantum Mechanics, and Consciousness

Subhash Kak[1], Deepak Chopra[2], and Menas Kafatos[3]

[1]Oklahoma State University, Stillwater, OK 74078
[2]Chopra Foundation, 2013 Costa Del Mar Road, Carlsbad, CA 92009
[3]Chapman University, Orange, CA 92866

Abstract:

Our sense of reality is different from its mathematical basis as given by physical theories. Although nature at its deepest level is quantum mechanical and nonlocal, it appears to our minds in everyday experience as local and classical. Since the same laws should govern all phenomena, we propose this difference in the nature of perceived reality is due to the principle of veiled nonlocality that is associated with consciousness. Veiled nonlocality allows consciousness to operate and present what we experience as objective reality. In other words, this principle allows us to consider consciousness indirectly, in terms of how consciousness operates. We consider different theoretical models commonly used in physics and neuroscience to describe veiled nonlocality. Furthermore, if consciousness as an entity leaves a physical trace, then laboratory searches for such a trace should be sought for in nonlocality, where probabilities do not conform to local expectations.

Keywords: quantum physics, neuroscience, nonlocality, mental time travel, time

Introduction

Our perceived reality is classical, that is it consists of material objects and their fields. On the other hand, reality at the quantum level is different in as much as it is nonlocal, which implies that objects are superpositions of other entities and, therefore, their underlying structure is wave-like, that is it is smeared out. This discrepancy shows up in the framework of quantum theory itself because the wavefunction unfolds in a deterministic way excepting when it is observed which act causes it to become localized.

The fact that the wavefunction of the system collapses upon observation suggests that we should ask whether it fundamentally represents interaction of consciousness with matter. In reality this question is meaningful only if we are able to define consciousness objectively. Since consciousness is not a thing or an external object, its postulated interaction with matter becomes paradoxical. If we rephrased our question we could ask where our personal self is located and if it is just some neural structure that exists in the classical world, why is it that its activity (observation) collapses the wavefunction?

The human observer interacts with the quantum system through apparatus devised by him and, therefore, the actual interaction is between physical systems. If the human were directly observing the system, then also the interaction is within the human brain that consists of various neural structures. The observation is associated with a time element since the apparatus must first be prepared and then examined after the interaction with the quantum system has occurred. The observer's self-consciousness is also the product of life experience, and therefore the variable of time plays a central role in it.

Neuroscience, which considers brain states in terms of electrical and chemical activity in the interconnection of the neurons, that somehow gives rise to awareness, cannot explain where the aware self is located. Since each neuron can only carry a small amount of information, no specific neuron can be the location of the self. On the other hand, if the self is distributed over a large area of the brain, what is it that binds this area together? Furthermore, since we have different inner senses of self in different states of mind, are these based on recruitment of different set of neurons, and why is it that behind each such sense of selfhood, ostensibly associated with different parts of the brain, is the certitude that it corresponds to the same individual? What integrates all these different senses into one coherent and persisting sense of self? Behind these questions lie a whole set of complementarities. Thus in some contexts it is appropriate to view the self as located at some point in the brain, whereas in other contexts it is distributed all over the body.

The question whether mental states are governed by quantum laws is answered in the affirmative by those who accept a quantum basis of mind (von Neumann, 1932/1955; Wigner, 1983; Penrose, 1994; Nadeau and Kafatos, 1999; Roy and Kafatos, 2004; Hameroff and Penrose, 2003; Stapp, 2003; Freeman and Vitiello, 2006). Recent findings in support of quantum models for biology (e.g., Lambert et al., 2013) lend additional support to this position. We add a point of caution here: mind as a random quantum machine as suggested by some theories may be an advance but not the complete picture since it will not be able to account for the individual's freedom and agency; and it cannot imply an objective reality independent of observation.

The question related to the limitations on what can be known by the mind has a logical component related to Gödel's Incompleteness Theorem and logical paradoxes (Davis, 1965), which can also apply to limits of knowing in physics and biology (Herrnstein, 1985; Grandpierre and Kafatos, 2012; Kak, 2012; Buser et al., 2013).

In this article, we focus on the relationship between reality and our conceptions of it as mediated by consciousness. This mediation may be seen through the lenses of neuroscience and physics. We first show how the mind constructs reality; later we discuss how the nonlocality inherent in quantum theory and, therefore, in physical process, is veiled by consciousness so that it appears to be local (Kak, 2014; Kafatos and Kak, 2014). This veiling is a characteristic of consciousness as process and it mirrors the way quantum theory conceals all the details of the state of a single particle by means of the Heisenberg Uncertainty Principle. Finally, we consider how indirect evidence in terms of anomalous probabilities in certain events can be adduced as evidence in support of reality being quantum-like at a fundamental level.

Interpretations of Quantum Theory

The Copenhagen Interpretation of quantum mechanics separates the physical universe in two parts, the first part is the system being observed, and the second part is the human agent, together with his instruments. The extended agent is described in mental terms and it includes not only his apparatus but also instructions to his colleagues on how to set up the instruments and report on their observations. The Heisenberg cut (also called the von Neumann cut) is the hypothetical interface between quantum events and an observer's information, knowledge, or conscious awareness. Below the cut everything is governed by the wave function; above the cut a classical description applies.

In the materialist conception, which leaves out quantum processes since they were until recently thought not to play a role in brain processes, consciousness is an epiphenomenon with biology as ground. But, although this is the prevalent neuroscience paradigm, there is no way we can justify the view that material particles somehow acquire consciousness on account of complexity of interconnections between neurons. There is also the panpsychist position that consciousness is a characteristic of all things, but in its usual formulation as in the MWI position that will be discussed later it is merely a restatement of the materialist position so as to account for consciousness (Strawson, 2006). On the other hand is the position that consciousness is a transcendent phenomenon – since it cannot be a thing – which interpenetrates the material universe. The human brain, informed by the phenomenon of consciousness, has self-awareness to contemplate its own origins.

We do know that mental states are correlated with activity in different parts of the brain. With regard to mental states several questions may be asked:

1. Do mental states interact with the quantum wavefunction?
2. Are mental states governed by quantum laws?
3. Are there limitations to what the mind can know about reality?

There is considerable literature on research done on each of these questions.

The question of interaction between mental states and the wavefunction was addressed by the pioneers of quantum theory and answered in the affirmative in the Copenhagen Interpretation (CI) (von Neumann, 1932/1955). In CI, the wavefunction is properly understood epistemologically, that is, it represents the experimenter's knowledge of the system. Upon observation there is a change in this knowledge. It postulates the observer without explaining how the capacity of observation arises (Schwartz et al., 2005; Stapp, 2003). Operationally, it is a dualist position, where there is a fundamental split between observers and objects. There is the added variance or interpretation of CI followed by von Neumann in his discussion of the quantum cut which implies that although the separation between observers and objects has to be brought into actual experimental setups, that it is not fundamental.

In the realistic view of the wavefunction as in the Many Worlds Interpretation (MWI), there is no collapse of the wavefunction. Rather, the interaction is seen through the lens of decoherence, which occurs when states interact with the environment producing entanglement (Zurek, 2003). By the process of decoherence the system makes transition from a pure state to a mixture of states that observers end up measuring.

In the MWI interpretation, which may be called the ontic interpretation, the wavefunction has objective reality. The problem of collapse of the wavefunction is sidestepped by speaking of interaction between different subsystems. But since the entire universe is also a quantum system, the question of how this whole system splits into independent subsystems arises. Furthermore, if the wavefunction has objective reality, then consciousness must be seen to exist everywhere (Tegmark, 1998). Actually, this particular point was again implied by von Neumann's extension of CI who held that the wave function collapses upon observation, that the wave function is real, and that the world is quantum, which is decidedly contrary to MWI and Tegmark's recent interpretation but in agreement with the works of Wigner and Stapp, as well as our own. We emphasize that the MWI resolution is not helpful because it sees consciousness only as a correlate without any explanation for agency and freedom. There are other interpretations of quantum mechanics not as popular as either CI or MWI that will not be considered in this paper.

Construction Of Reality By The Mind

If we side-step for now the question of where the self is located and focus on the question of how the self relates to reality, we come up with several issues: First, we note that the mind is an active participant, together with the sensory organs, in the construction of reality (e.g. Wheeler, 1990: Kafatos and Nadeau, 2000; Gazzaniga, 1995; Chopra, 2014). An example of this is the phantom limb phenomenon in which there is a sensation of pain in a missing or amputated limb (Melzack, 1992). There are other cases where the phantom limb does not even correspond to anatomical reality. A recent research paper reported the case of a 57-year-old woman who was born with a deformed right hand consisting of only three fingers and a rudimentary thumb. After a car crash at the age of 18, the woman's deformed hand was amputated, which gave rise to feelings of a phantom hand that was experienced as having all five fingers (McGeoch and Ramachandran, 2012).

In the case of injury to the brain, the construction of reality by mind seen most clearly in terms of deficits that persist even though the related sensory information is reaching the brain. Consider agnosia, which is failure of recognition that is not due to impairment of the sensory input or a general intellectual impairment. A visual agnosic patient will be unable to tell what he is looking at, although it can be demonstrated that the patient can see the object. Prosopagnosic patients are neither blind nor intellectually impaired; they can interpret facial expressions and they can recognize their friends and relations by name or voice, yet they do not recognize specific faces, not even their own in a mirror. Electrodermal recordings show that the prosopagnosic responds to familiar faces although without awareness of this fact. It appears that the patient is subconsciously registering the significance of the faces.

In a recent study (Rezlescu et al., 2014) of prosopagnosia the authors consider the role of the face-specific and the expertise (information processing) aspects in the recognition mechanism. According to the face-specific theory, upright faces are processed by face-specific brain mechanisms, whereas the expertise hypothesis claims faces are recognized by fine-grained visual processing. The authors used greebles, which are objects designed to place face-like demands on recognition mechanisms, in their experiments. They present compelling evidence that performance with greebles does not depend on face recognition mechanisms. Two individuals with acquired prosopagnosia displayed normal greeble learning despite severely impaired face performance. This research supports the theory that face- specific mechanisms are essential for recognition of faces.

Similar counterintuitive behavior of the mind includes the following: in alexia, the subject is able to write while unable to read; in alexia combined with agraphia, the subject is unable to write or read while retaining other language faculties; in acalculia, the subject has selective difficulty in dealing with numbers.

There are anecdotal accounts of blind people who can see sometime and deaf people who can likewise hear. Some brain damaged subjects cannot consciously see an object in front of them in certain places within their field of vision, yet when asked to guess if a light had flashed in their region of blindness, the subjects guess right at a probability much above that of chance.

One may consider that the injury in the brain leading to blindsight causes the vision in the stricken field to become automatic. Then through retraining it might be possible to regain the conscious experience of the images in this field. In the holistic explanation, the conscious awareness is a correlate of the activity in a complex set of regions in the brain. No region can be considered to be producing the function by itself although damage to a specific region will lead to the loss of a corresponding function (Kak, 2000).

Split Brains

The corpus callosum connects the two hemispheres of the brain and each eye normally projects to both hemispheres. By cutting the optic-nerve crossing, the chiasm, the remaining fibers in the optic nerve transmit information to the hemisphere on the same side. Visual input to the left eye is sent only to the left hemisphere, and input to the right eye projects only to the right hemisphere.

Experiments on split-brain human patients raise questions related to the nature and the seat of consciousness. For example, a patient with left-hemisphere speech does not know what his right hemisphere has seen through the right eye. The information in the right brain is unavailable to the left brain and vice versa. The left brain responds to the stimulus reaching it whereas the right brain responds to its own input. Each half brain learns, remembers, and carries out planned activities and it is as if each half brain works and functions outside the conscious realm of the other.

Roger Sperry and his associates performed a classic experiment on cats with split brains (Sperry et al., 1956; Sperry, 1980). They showed that such cats did as well as normal cats when it came to learning the task of discriminating between a circle and a square in order to obtain a food reward, while wearing a patch on one eye. This showed that one half of the brain did as well at the task as both the halves in communication. When the patch was transferred to the other eye, the split-brain cats behaved different from the normal cats, indicating that their previous learning had not been completely transferred to the other half of the brain.

It appears that for split brains nothing is changed as far as the awareness of the patient is considered and the cognitions of the right brain were linguistically isolated all along, even before the commissurotomy was performed. The procedure only disrupts the visual and other cognitive-processing pathways.

The patients themselves seem to support this view. There seems to be no antagonism in the responses of the two hemispheres and the left hemisphere is able to fit the actions related to the information reaching the right hemisphere in a plausible theory. For example, consider the test where the word "pink" is flashed to the right hemisphere and the word "bottle" is flashed to the left. Several bottles of different colors and shapes are placed before the patient and he is asked to choose one. He immediately picks the pink bottle explaining that pink is a nice color. Although the patient is not consciously aware of the right eye having seen the word "pink" he, nevertheless, "feels" that pink is the right choice for the occasion. In this sense, this behavior is very similar to that of blindsight patients.

The brain has many modular circuits that mediate different functions. Not all of these functions are part of conscious experience. When these modules related to conscious sensations get "crosswired," this leads to synesthesia. One would expect that similar joining of other cognitions is also possible. A deliberate method of achieving such a transition from many to one is a part of some meditative traditions. It is significant that patients with disrupted brains never claim to have anything other than a unique awareness.

If shared activity was all there was to consciousness, then this would have been destroyed or multiplied by commissurotomy. Split brains should then represent two minds just as in freak births with one trunk and two heads we do have two minds. But that is never the case.

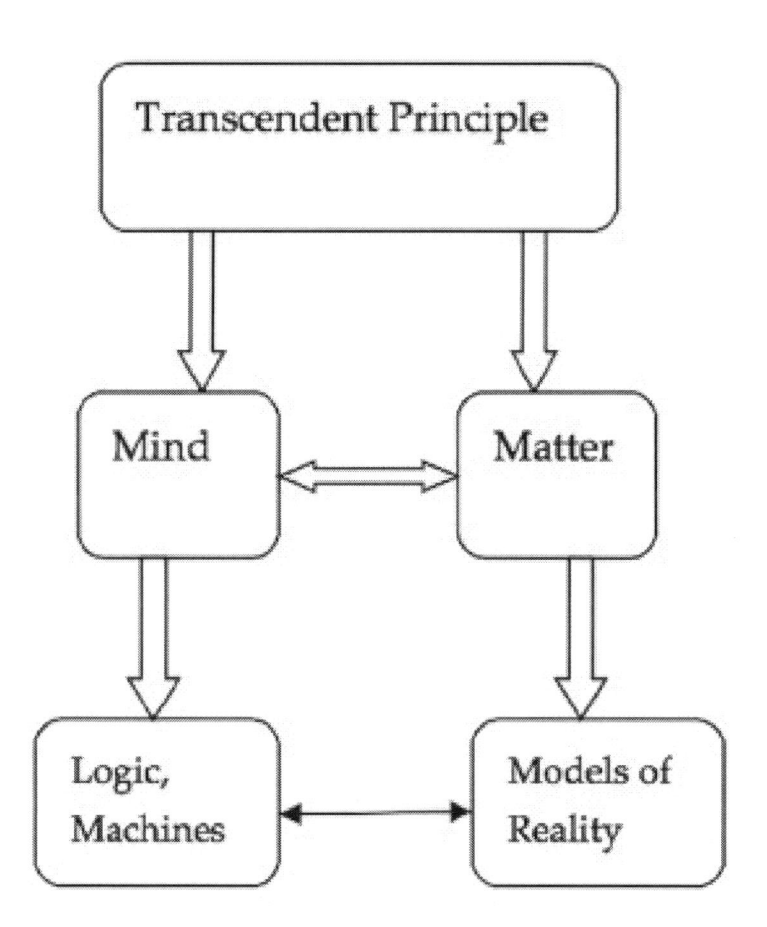

Figure 1. Universe as projection of a transcendent principle (broad arrow is projection; narrow arrow is full representation)

The experiments of Benjamin Libet showed how decisions made by a subject arise first on a subconscious level and only afterward are translated into the conscious decision (Libet, 1983). Upon a retrospective view of the event, the subject arrives at the belief that the decision occurred at the behest of his will.

Quantum Physics, **Retrocausation, PreCognition, Entanglement,**

In Libet's experiment the subject was to choose a random moment to flick the wrist while the associated activity in the motor cortex was measured. Libet found that the unconscious brain activity leading up to the conscious decision by the subject began approximately half a second before the subject consciously felt that he had taken his decision. But this is not to be taken as an example of retrocausation; rather, this represents a lag in the operation of the conscious mind in which this construction of reality by the mind occurs.

Orthodox Quantum Mechanics And Complementarity

As shown in Figure 1, we have dealt with models of reality constructed by the mind, and theories of physics, both of which are at the bottom layer of reality. Now we consider the next higher layer and see how we make observations. To record an observation in a laboratory requires a certain conception of the components of the system and a hypothesis related to the process. The observations are inferred from the readings on instruments or photographic traces. More direct observations register directly with our senses and here the intention in that the sense organ, such as the eye, focuses on a specific object to the exclusion of the remainder of the visual scene indicates that there is some kind of an interaction between consciousness and matter that leads to the observation.

Within the Copenhagen Interpretation, von Neumann provided a mathematical treatment (von Neumann, 1932; Bohr, 1934; Heisenberg, 1958) of this question by speaking of two different processes at work and doing so made one look for the interaction of consciousness with matter in the first process: *Process 1*. This is a non-causal, thermodynamically irreversible process in which the measured quantum system ends up randomly in one of the possible eigenstates (physical states) of the measuring apparatus together with the system. The probability for each eigenstate is given by the square of the coefficients c_n of the expansion of the original system state

$$|\varphi\rangle$$

$$c_n = \langle \varphi_n | \phi \rangle$$

This represents the collapse of the wavefunction.

Process 2. This is a reversible, causal process, in which the system wave function evolves deterministically. The evolution of the system is described $c_n = \langle \varphi_n | \phi \rangle$ by a unitary operator $U(t_2, t_1)$ depending on the times t_1 and t_2, so that

$$|\varphi_2\rangle = U(t_2, t_1)|\varphi_1\rangle$$

The evolution operator is derived from the Schrödinger equation

$$i\hbar \frac{d|\varphi\rangle}{dt} = H|\varphi\rangle$$

When the Hamiltonian H is time independent, U has the form:

$$U(t_2, t_1) = e^{-\frac{i}{\hbar} H(t_2 - t_1)}$$

Von Neumann was guided by the principle of psycho-physical parallelism which requires that it must be possible to describe the extra-physical process of the subjective perception as if it were in reality in the physical world. This principle is not appreciated much nowadays but it is justified since psychological states must have a corresponding physical correlate. von Neumann described the collapse of the wave function as requiring a cut between the microscopic quantum system and the observer. He said it did not matter where this cut was placed:

> The boundary between the two is arbitrary to a very large extent. ... That this boundary can be pushed arbitrarily deeply into the interior of the body of the actual observer is the content of the principle of the psycho-physical parallelism -- but this does not change the fact that in each method of description the boundary must be put somewhere, if the method is not to proceed vacuously, i.e., if a comparison with experiment is to be possible. Indeed experience only makes statements of this type: an observer has made a certain (subjective) observation; and never any like this: a physical quantity has a certain value. (von Neumann, 1932).

To the extent that the collapse provides a result in a statistical sense, there appears to be a choice made by Nature. The other choice is made by the observer whose intention sets the measurement process in motion.

These two processes are an instance of complementarity of which the wave-particle duality of the Copenhagen Interpretation is a more commonly stated example. In reality, complementarity is a general principle that characterized all experience. Although it is sometimes seen as emerging out of complexity (Theise and Kafatos, 2013), here we view it as a fundamental principle that organized reality.

The question of complementarity was matter of debate in Greek thought. Thus reality was taken as change by Heraclitus and as things and relationships by Parmenides. In Indian thought several fundamental dichotomies such as matter and consciousness, physical reality and its descriptions, and analysis and synthesis are given. Some of the complementarities that are part of the contemporary discourse are:

- Waves and particles – quantum theory

- Being and time – in philosophy (Heidegger, 1962)
- Law and freedom – in physics and psychology
- Holistic and reductionist views – in system theory
- Matter and consciousness.

If we take complementarity as the common thread in phenomena that are described at different levels, then one might suspect that quantum theory should also underlie consciousness.

Veiled Monlocality

According to quantum mechanics reality is nonlocal and objects separated in time and space can be strongly correlated. Thus for a pair of entangled particles billions of miles apart, an observation on one particle causes an instant collapse of the wavefunction of the twin particle. Entanglement also persists across time and an observation made now can change the past as in Wheeler's delayed choice experiment (Wheeler, 1990). Yet there is no way one can confirm such an entanglement for a specific pair of particles for any attempt at verification will be defeated by the collapse of the wavefunction. Probability experiments for ensembles of particles to separate classical from quantum effects do exist (Bell, 1964).

If reality is nonlocal why does it appear to our senses as local and separated? The idea of veiled nonlocality is that consciousness disguises its wholeness and nonlocality in order to produce local processes. This idea arose out of the experimental fact that no loophole free test of nonlocality has yet been found (Kak, 2014). It can be seen as quite like the Heisenberg's Uncertainty Principle which places limits on the description of the state of a specific particle.

This filtering process allows for specific observations and thoughts in a classical world of everyday experience, while keeping quantum and general relativistic processes out of sight. Another example of veiled nonlocality in gravitation is the hypothesis of cosmic censorship (Penrose, 1999), which describes the inability of distant observers to directly observe the center of a black hole, or "naked singularity." (Kafatos and Kak, 2014)

Veiled nonlocality is like a fuzzification that breaks up a whole system into several locally connected subsystems. We illustrate below in Figure 2 the general idea.

The breakup of a whole into subsystems can be shown through simple observation. The five senses cannot perceive the quantum world, and yet perception depends upon it. The quantum world is hidden from us the way the operation of the brain is hidden. If you think the word "elephant" and see an image of the animal in your mind's eye, you aren't aware of the millions of neurons firing in your brain in order to produce them. Yet those firings -- not to mention the invisible cellular operations that keep every part of your body alive

-- are the foundation of the brain's abilities.

Just as the image of an elephant is the visible end point of veiled processes, the material world is founded on a veiled reality. Moreover, to produce a single mental image, the whole brain must participate. Specific areas, mainly the visual cortex, produce mental pictures, but they are coordinated with everything else the brain does, such as sustaining the cerebral cortex, which recognizes what an image is, and maintaining a healthy body. This points to a profound link between the brain and the cosmos. The two are inseparable. In fact, in our view, complementarity assures that they appear as separate and one causing the other but in fact they are aspects of undivided wholeness brought about by consciousness, which is undivided, nonlocal and whole (see also below).

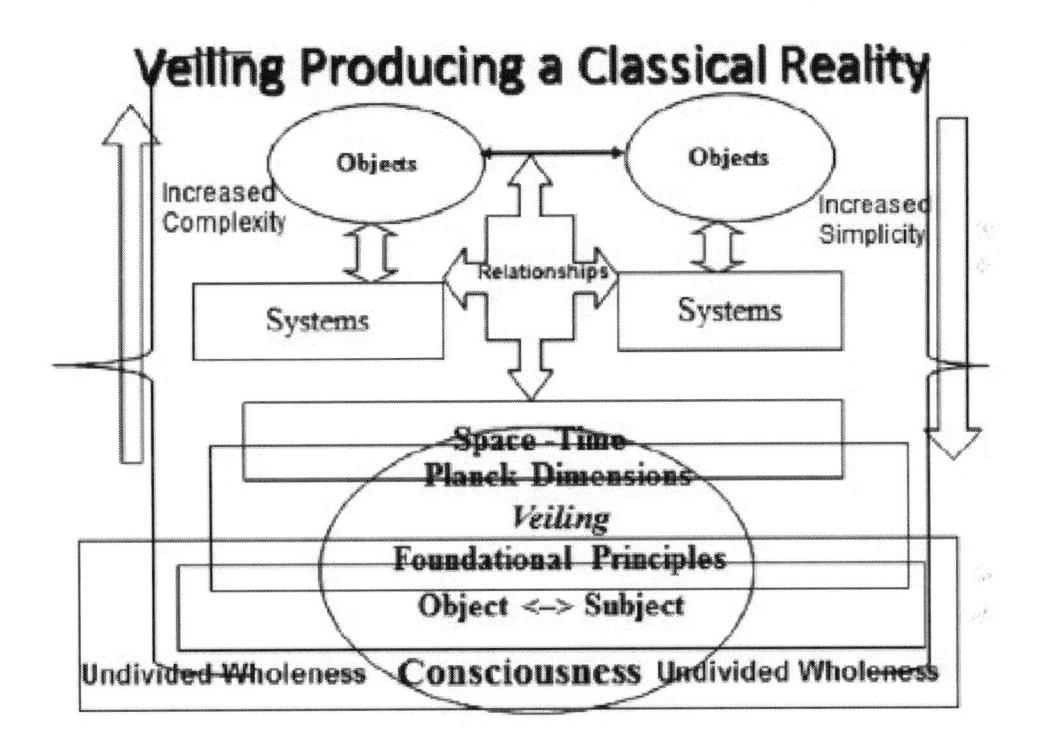

Figure 2: Breakup of the undivided wholeness of consciousness and implied quantum wholeness through the veiling process.

The veiling of reality is in consonance with the idea of the mind constructing its reality. Such a veiling even occurs in the scientific process which filters out and discards a huge portion of human experience -- almost everything one would classify as subjective. Its model is just as selective, if not more so, than the model which shapes a religious or metaphysical reality. As far as the brain is concerned, neural filtering is taking place in all models, whether they are scientific, spiritual, artistic, or psychotic. The brain is a processor of inputs, not a mirror to reality.

If our brains are constantly filtering every experience, there is no way anyone can claim to know what is "really" real. You can't step outside your brain to fathom what lies beyond it. Just as there is a horizon for the farthest objects that emit light in the cosmos, and a farthest horizon for how far back in time astronomy can probe, there is a farthest horizon for thinking. The brain operates in time and space, having linear thoughts that are the end point of a selective filtering process. So whatever is outside time and space is inconceivable, and unfiltered reality would probably blow the brain's circuits, or simply be blanked out.

Consciousness as Foundational Principle of Reality

We propose, as shown in Figure 2, that consciousness-based reality limited by a fundamental veiling provides meaning and the appearance of what we say is objective reality, systems, objects and relationships at all levels of organization. The quantum and the classical worlds aren't separated merely by a physical gap. On one side the behavior of the quantum is meaningless, random, and unpredictable. A subatomic particle has no purpose or goal. On the other side, in the classical world, it goes without saying that each of us lives our life with purpose and meaning in mind. To accept this as self-evident is crucial to getting up every morning, so arcane disputations about free will and determinism are, pragmatically speaking, not relevant to the more fundamental question: Can randomness produce meaning, and if so, how?

To lead a meaningless existence is intolerable, so it's ironic that quantum physics bases the cosmos on meaningless operations, and doubly ironic when you consider that physics itself is a meaningful activity. The phrase "participatory universe," sums up how the very process of observation changes the outcome (Wheeler, 1990). Observation not only changes the outcome in a random sense, but it can actually be used to steer to unfolding of a physical system to whichever way one chooses by the quantum Zeno effect (Misra and Sudarshan, 1974; Kak, 2007).

By definition reality is complete; therefore, whatever purpose and meaning we find in it, using limited human capacities, is a fragment of a pre-existing state, which we term the state of infinite possibilities. The fragment cannot be the whole, although in what may appear as strange, the part implies the whole (Kafatos and Nadeau, 2000; Nadeau and Kafatos, 1999). And the whole is more than the sum of the parts, because no amounts of parts, no matter how many, form the whole. This state is veiled from us, just as the existence of every possible subatomic particle is hidden from us. The concept of a field contains within it this relationship between the whole and its parts. There is no reason to exclude the field of consciousness from exhibiting the same relationship to its parts, whence the insight that there can be only one consciousness, not many (Schrödinger, 1974).

Consciousness, Neuroscience, Time Travel

In a matter–alone conception of the universe, we cannot conceive of a reality that has no objects in it, and only pure consciousness. However, such a matter alone conception goes against the very nature of a quantum universe as quantum theory contains observation through measurement. Yet without awareness, nothing can be perceived. Having placed its trust in "reality as given," science overlooks the self- evident fact that nothing can be experienced without consciousness. It is a more viable candidate for "reality as given" than the physical universe. Even if the materialist position is accepted that claims consciousness is an epiphenomenon, one cannot escape the fact that consciousness existed from the very beginning although it may have been in a latent form.

If enlightenment consists of seeing beyond the veiling that accompanies a commonsensical view of the universe, it too isn't some sort of obscure mysticism but recognition that self-awareness can know itself. The mind isn't only the thoughts and sensations constantly streaming through it. There is a silent, invisible foundation to thought and sensations. Until that background is accounted for, individual consciousness mistakes itself, and in so doing it cannot help but mistake what it observes. This is expressed in a Vedic metaphor about the wave and the ocean: A wave looks like an individual as it rises from the sea, but once it sinks back, it knows that it is ocean and nothing but ocean.

Cosmic consciousness, then, isn't just real -- it's totally necessary. It rescues physics and science in general from a dead end -- the total inability to create mind out of matter -- and gives it a fresh avenue of investigation. We exist as creatures with a foot in two worlds that are actually one, divided by appearance and reality. Consciousness as a transcending principle provides a way to bridge the two processes of quantum theory. We emphasize that this view is the only one that is ultimately self-consistent. Material views of reality ultimately run into unanswerable conundrums and inconsistencies or the need of strange views such as the MWI which is founded on the existence of real outcomes without the agency of consciousness.

Conclusion

Quantum theory has reached the point where the source of all matter and energy is a vacuum, a nothingness that contains all the possibilities of everything that has ever existed or could exist. These possibilities then emerge as probabilities before "collapsing" into localized quanta, manifesting as the particles in space and time that are the building blocks of atoms and molecules.

Where do the probabilities exist? Where is the exquisite mathematics that we have at our disposal to be found? Some sort of "real space", or material-like space? That of course makes no sense. The probability of an event (even an event like winning the lottery or flying on the day a blizzard strikes) only exists as long as there is someone to ask the question of what may happen and to measure the outcomes when they occur. So probabilities and other mathematical expressions, which are the foundation of modern quantum physics, imply the existence of observation. Countless acts of observation give substance and reality to what would otherwise be ghosts of existence. This solves the so-called "measurement problem" of quantum theory which is there if one assumes a reality independent of observation.

It is more elegant, self-consistent and far easier to accept as a working hypothesis that sentience exists as a potential at the source of creation, and the strongest evidence has already been put on the table: Everything to be observed in the universe implies consciousness. Some theorists try to rescue materialism by saying that information is encoded into all matter, but "information" is a mental concept, and without the concept, there's no information in anything, since information by definition must ultimately contain meaning (even if it is a sequence of 0s and 1s as in computer language), and only minds grasp meaning. Besides, assuming that this kind of bit information is an encoded property of matter implies hidden variables (the bits) which have been ruled out by the Bell (1964) Theorem and laboratory experiments related to it (Aspect, Dalibard and Roger, 1984). Does a tree falling in the forest make no sound if no one is around to hear it? Obviously not. The crash vibrates air molecules, but sound needs hearing in order for these vibrations to be transformed into perception.

We've proposed that consciousness creates reality and makes it knowable -- if there's another viable candidate, it must pass the acid test: Transform itself into thoughts, feelings, images, and sensations. Science isn't remotely close to turning the sugar in a sugar bowl into the music of Mozart or the plays of Shakespeare. Your brain converts blood sugar into words and music, not by some trick of the molecules in the brain, since they are in no way special or privileged. Rather, your consciousness is using the brain as a processing device, moving the molecules where they are needed in order to create the sight, sound, touch, taste, and smell of the world.

In everyday life, we get to experience the miracle of transformation that causes a three-dimensional world, completed by the fourth dimension of time, to manifest before our eyes. The great advantage of experience is that it isn't theoretical. Reality is never wrong, and all of us are embedded in reality, no matter what model we apply to explain it.

The indirect examination of consciousness through the process of veiling as sketched in this paper can be tested by means of experiments. For example, the proposed theory can be refuted if loophole-free tests to confirm nonlocality are devised. On the other hand, cognitive processes with anomalous probabilities would lend support to our thesis.

We finally note that the cut of Heisenberg/von Neumann does not exist anywhere: The observer must be one, all observers are appearances of distinct or independent entities, taking on an apparent "reality" through the veiling action. As stated above, our thesis resolves the measurement problem. In reading von

Neumann, there is a strong hint that he also held this view.

REFERENCES

Aspect, A., Dalibard, J., Roger, G. (1981). Physical Rev. Letters, 47, p. 460.

Bell, J. (1964) On the Einstein Podolsky Rosen paradox. Physics 1 195–200.

Bohr, N. (1934) Atomic theory and the description of nature. Cambridge: Cambridge University Press.

Buser, M., Kajari, E., and Schleich,, W.P. (2013). Visualization of the Gödel universe. New Journal of Physics 15, 013063.

Chopra, D. (ed.) (2014). Brain, Mind, Cosmos. Trident Media, New York.

Davis, M. (1965). The Undecidable: Basic papers on undecidable propositions, unsolvable problems and computable functions. Raven Press, New York.

Freeman, W. and Vitiello, G. (2006) Nonlinear brain dynamics as macroscopic manifestation of underlying many-body dynamics. Physics of Life Reviews 3: 93-118.

Gazzaniga, M.S. (1995) The Cognitive Neurosciences. Cambridge, MA The MIT Press. Hameroff, S. and Penrose, R. (2003) Conscious events as orchestrated space-time selections. NeuroQuantology 1: 10-35.

Heidegger, M. (1962) Being and Time, trans. by John Macquarrie & Edward Robinson. London: SCM Press.

Heisenberg, W. (1958) Physics and Philosophy: The Revolution in Modern Science, London: George Allen & Unwin.

Herrnstein, R.J. (1985). Riddles of natural categorization. Phil. Trans. R. Soc. Lond. B 308: 129-144.

Kafatos, M. and Nadeau, R. (2000) The Conscious Universe. Springer.

Kafatos, M. and Kak, S. (2014) Veiled nonlocality and cosmic censorship. arXiv:1401.2180

Kak, S. (2000) Active agents, intelligence, and quantum computing. Information Sciences 128: 1-17

Kak, S. (2007) Quantum information and entropy. International Journal of Theoretical Physics 46, 860-876.

Kak, S. (2012) Hidden order and the origin of complex structures. In Swan, L., Gordon, R., and Seckbach, J. (editors), Origin(s) of Design in Nature. Dordrecht: Springer, 643- 652.

Kak, S. (2014) From the no-signaling theorem to veiled non-locality. NeuroQuantology 12: 1- 9.

Lambert, N. et al, (2013) Quantum biology. Nature Physics 9, 10–18.

Libet, B. et al. (1983) Time of conscious intention to act in relation to onset of cerebral activity (readiness-potential) - The unconscious initiation of a freely voluntary act. Brain 106: 623–642.

McGeoch, P., and Ramachandran, V., (2012), The appearance of new phantom fingers post- amputation in a phocomelus, Neurocase, 18 (2), 95-97.

Melzack, R. (1992). Phantom limbs. Scientific American (April): 120–126.

Misra, B. and Sudarshan, E. C. G. (1977). The Zeno's paradox in quantum theory. Journal of Mathematical Physics 18, 758–763.

Nadeau, R. and Kafatos, M. (1999). The Non-local Universe: The New Physics and Maters of the Mind, Oxford University Press, Oxford.

Penrose, R. (1994) Shadows of the Mind. New York: Oxford.

Penrose, R. (1999) The question of cosmic censorship. J. Astrophys. Astr. 20: 233–248.

Roy, S. and Kafatos, M. (2004). Quantum processes and functional geometry: new perspectives In brain dynamics. FORMA, 19, 69.

Rezlescu, C., Barton, J. J. S., Pitcher, D. & Duchaine, B. (2014) Normal acquisition of expertise with greebles in two cases of acquired prosopagnosia. Proc Natl Acad. Sci. USA http://dx.doi.org/10.1073/pnas.1317125111.

Schrödinger, E. (1974). What is Life? and Mind and Matter. Cambridge University Press.

Schwartz, J.M., Stapp, H.P., Beauregard, M. (2005) Quantum physics in neuroscience and psychology. Phil. Trans. Royal Soc. B 360: 1309-1327.

Sperry, R. (1980) Mind-brain interaction: Mentalism, yes; dualism, no. Neuroscience 5 (2): 195–206

Sperry, R. W., Stamm, J. S., and Miner, N. (1956) Relearning tests for interocular transfer following division of optic chaism and corpus callosum in cats. J. Compar. Physiol. Psych. 49: 529-533.

Stapp, H.P. (2003) Mind, Matter, and Quantum Mechanics. New York: Springer-Verlag. Strawson, G. (2006) Realistic monism: Why physicalism entails panpsychism. Journal of Consciousness Studies 13: 10–11, 3–31.

Tegmark, M. (1998) The interpretation of quantum mechanics: many worlds or many words? Fortsch. Phys. 46: 855-862.

Theise, N.D. and Kafatos, M. (2013) Complementarity in biological systems: a complexity view. Complexity 18: 11-20.

Von Neumann, J. (1932/1955) Mathematical Foundations of Quantum Mechanics, translated by Robert T. Beyer, Princeton, NJ: Princeton University Press.

Wheeler, J.A. (1990). Information, physics, quantum: the search for links. In Complexity, Entropy, and the Physics of Information, W.H. Zurek (Ed.). Addison-Wesley, pp. 3- 28.

Wigner, E. (1983). "The Problem of Measurement", In: Quantum Theory and Measurement, J.A. Wheeler, & W.H. Zurek (Eds.), Princeton University Press, Princeton.

Zurek, W.H. (2003) Decoherence, einselection, and the quantum origins of the classical. Rev. Mod. Phys. 75: 715-775

IV: Brain, Mind, Cosmology, Causality, and Time

8. Space, Time and Consciousness
Chris King

Emeritus, University of Auckland

Abstract

This paper presents a potential mechanism for the conscious brain to anticipate impending opportunities and threats to survival through massively parallel weak quantum measurement (MPWQM) induced by the combined effects of edge of chaos sensitivity and phase coherence sampling of brain states. It concludes that the underpinnings of this process emerged in single-celled eucaryotes in association with (a) excitability-induced sensitivity to electro-chemical perturbations in the milieu as an anticipatory sense organ and (b) cell-to-cell signaling necessary for critical phases in the life cycle.

Keywords: space, time, consciousness, evolution, neurodynamics, chaos, quantum entanglement, weak quantum measurement

1: Introduction: Consciousness Entangled

Subjective consciousness poses the ultimate dilemma for the scientific description of reality. We still have no idea of how the brain generates it, or even how, or why, such an objectively elusive phenomenon can come about from the physiology of brain dynamics. The problem is fundamental because, from birth to death, the sum total of all our observations of the physical world, and all our notions about it, come exclusively through our subjective conscious experience. Although neuroscience has produced new techniques for visualizing brain function, from EEG and MEG to PET and fMRI scans, which show a parallel relationship between mental states and brain processing, these go no way in themselves to solving the so-called hard problem of consciousness research — how these objective physiological processes give rise to the subjective effects of conscious experience.

One key to the possible role of subjective consciousness is that it appears to be a product of coordinated brain activity involving diverse regions operating together in a coherent manner so as to anticipate environmental challenges (see section 3).

This leads to another critical question: "Why did nervous systems evolve subjective consciousness?" If nervous systems are able to fully provide adaptive solutions simply as heuristic computers, there is no role for extraneous brain functions that simply add a subjective shadow reality, with no adaptive function, and presumably a physiological cost. A digital computer is a purely functional entity, so has no role for a subjective aspect, no matter how complex it becomes.

Diverse higher animal nervous systems appear to work on a common basis of edge- of-chaos excitation (see section 5) that arose in excitable single cells before multi- celled organisms evolved (see section 2), which, in humans is accompanied by subjective consciousness (King 2008). This suggests that subjectivity is a critical survival attribute, which has been reinforced by natural selection, its key role being anticipating opportunities and threats to survival (see section 4). Strategic decision-making in the open environment is notorious for being computationally intractable because of super-exponential runaway as the number of contingencies increases (see section 4). By contrast, vertebrate brains have a common mechanism of massively parallel processing using wave phase coherence to distinguish ground noise from attended signal, accompanied by transitions at the edge of chaos (see section 5), which successfully resolves intractability in real time.

A non-computational form of space-time anticipation may aid this process (see section 6). Chaotic sensitivity and self-organized criticality combined with stochastic resonance may enable the ongoing brain state to become sensitive to quantum uncertainties through nested instabilities running from the molecular level, through cell organelles and neurons to global activations, when the global context is critically poised (see section 5). Quantum entangled phase coherence sampling accompanying the wave excitations of brain states could then provide a means for anticipation of future threats to survival through massively parallel weak quantum measurement (see section 6). We shall explore how this capacity might provide an explanation for subjective consciousness and the notion of free-will.

2: Underpinnings of Consciousness Emerged in Single Celled Eucaryotes

The neurodynamic processes underpinning subjective consciousness are evolutionarily ancient, are based on fundamental bifurcations evident in biogenesis, and originate in single-celled protista before the emergence of multi-celled animals and nervous systems (King 2002, 2011).

Excitable membranes are universal to eucaryote cells, as is the need to sense electrochemical and nutrient changes in their milieu. The sodium channel key to the axon potential, for example, arose in founding single-celled eucaryotes. Chay and Rinzells (1985) model of bursting and beating derived from the alga Nitella demonstrates the widespread nature of chaotic excitability arising before animals and plants diverged. Similar excitability has been observed through cAMP dynamics in the social amoeba Dictyostelium (Mestler 2011) and action potentials in Paramecium (Hinrichsen & Schultz 1988).

The elementary neurotransmitter types, many of which are fundamental amino acids (glutamate, glycine, GABA) or amines derived from amino acids (serotonin, dopamine, histamine, choline) have primordial relationships with the membrane, as soluble molecules with complementary charge relationships to the hydrophilic ends of the phospholipids, which later became encoded in protein receptors.

Tryptophan, the amino acid from which serotonin (5-hydroxytryptamine) is generated, plays a key role in the transfer of electric charge in the earliest forms of photosynthesis. To make serotonin from tryptophan, oxygen is needed. Thus, serotonin is made specifically in unicellular systems capable of photosynthesis and the cellular production of oxygen. Consequently serotonin is up to 100 times more plentiful in plants, and animals have ceased to synthesize tryptophan depending on plants for their supply. This relationship with light continues to this day in human use of melatonin to define the circadian cycle and serotonin in wakefulness and sleep, with light deprivation causing depression through serotonin (Azmitia 2010).

The 5-HT1a receptor is estimated to have evolved 750 million to 1 billion years ago, (Peroutka & Howell 2004, Peroutka 2005) long before the Cambrian radiation. This places the emergence of receptor proteins and their neurotransmitters as occurring before the multicellular nervous systems, as cell-to-cell signalling molecules essential for survival, reproduction and positive and negative responses to nutrition and danger. It also explains that neurotransmitters originated from direct signalling pathways between the cell membrane and gene expression in the nucleus of single cells. Key enzymes in neurotransmitter pathways may have become ubiquitous through horizontal gene transfer from bacteria placing their emergence even earlier (Iyer et al. 2004).

Receptor proteins, second signalling pathways and key neurotransmitters occur widely in single-celled protists. Both Crithidia and Tetrahymena contain norepinephrine, epinephrine, and serotonin (Blum 1969). Aggregation of slime molds such as Dictyostelium is mediated by cyclic-AMP and uses glutamate and GABA (Halloy et al. 1998, Goldbeter 2006, Taniura et al. 2006, Anjard & Loomis 2006, Brizzi & Blum 1970, Essman 1987, Takeda & Sugiyama 1993, Nomura et al. 1998). Tetrahymena pyriformis also has circadian light-related melatonin expression (Köhidai et al. 2003). Trypanosoma cruzi can be induced to differentiate by increased cAMP levels that resulted from addition of epinephrine (Gonzalez- Perdomo et al. 1988). Species of Entamoeba secrete serotonin and the neuropeptides neurotensin and substance P (McGowan et al. 1985) and release and respond to catecholamine compounds during differentiation from the trophozoite stage into the dormant or transmissible cyst stage (Eichinger et al. 2002, 2005). Plasmodium falciparum malaria replication can be blocked by 5HT1a agonists (Locher et. al 2003).

This leads to a picture where the essential physiological components of conscious brain activity arose in single-celled eucaryotes, both in intra and intercellular communication, and in the chaotic excitability of single cells in sensing and responding to their environment. These include ion channel based excitability and action potentials, neurotransmitter modulated activity based on specific receptor proteins, membrane-nucleus signalling and precursors of synaptic communication.

Edge of chaos dynamics is a natural consequence of excitability providing arbitrary sensitivity to disturbances caused by predators and prey in the active environment. It is a function critical for survival in both single-celled and multicellular organisms, providing a selective advantage for the evolution of chaotically excitable brains from chaotically excitable cells. Once in place, this form of active anticipation, if linked to the anticipatory quantum process we are going to investigate, would then lead to a continuing use of edge of chaos wave coherence processing, subsequently expanded to primitive nervous systems as multicellular organisms evolved. One can see such strategically purposive behaviour in both single celled protists such as paramecium and in active human cells such as neutrophils hunting and consuming bacteria (King 2008).

Consequently the major neuroreceptor classes have a very ancient origin, with the 5HT1 and 5HT2 families diverging before the molluscs, arthropods and vertebrates diverged, close to the level of the founding metazoa. Sponges, with only two cell types, express serotonin (Wayrer et al. 1999) and have been shown to have the critical gene networks to generate synapses, in a pre-coordinated form (Conaco et al. 2012). Coelenterates already have all the key components of serotonin pathways, involved in signalling by sensory cells and neurons, despite having only a primitive nerve network (McCauley et al. 1997, Umbriaco et al. 1990). Given its ancient origin serotonin is also found to play a key role in development and embryogenesis in Molluscs (Buznikov et al. 2001, 2003, sea urchins (Brown and Shaver) and mammals, where the expression of serotonin receptors occurs at the earliest stages, activated by circulating plasma serotonin from the mother.

The metabotropic (protein-activating) glutamate and GABA receptors likewise go back to the social amoeba Dictyostelium discoideum, where there is a family of 17 GABA receptors and a glutamate receptor involved in differentiation (Taniura et al 2006). The glutamate-binding ""fly trap"" section of both ionotropic and metabotropic glutamate receptors show homologies with the bacterial periplasmic amino-acid binding protein (Felder et al 1999, Oh et al 1994, Lampinen et al 1998). The membrane-spanning section of the iGluRs also show homology with the bacterial voltage-gated K+ channel (Chen et al 1999). These changes are already in place in the cyanobacterial ionotropic glutamate receptor. The fact that an iGluR has also been found in Arabidopsis (Turano et al 2001) shows this class entered the eucaryotes before the plants, animals and fungi diverged. Elements of the protein signalling pathways, such as protein kinase C, essential to neuronal synaptic contact originated close to the eucaryote origin (Emes et al. 2008, Ryan & Grant 2009). Likewise the Dlg family of postsynaptic scaffold proteins, which bind neurotransmitter receptors and enzymes into signaling complexes originated before the divergence of the vertebrates and arthropods (Nithianantharajah et al. 2012).

Thus we can see how the survival modalities of complex organisms have continued to be mediated by classes of neurotransmitters modulating key motivational, aversive and social dynamics, from single cells to multi-celled organisms, with ascending central nervous system complexity. There are thus strong parallels in how the key classes of neurotransmitters modulate affect in organisms as diverse as arthropods and vertebrates.

In higher animals, 5-HT continues in its role as a homeostatic regulator in adjusting the dynamic interactions of these many functions within the organism, and how the organism interacts with the outside world, elaborated in humans into a variety of functions including the sleep-wakefulness cycle, triggering the psychedelic state, depression and social delinquency (King 2012). Similarly, dopamine and nor- epinephrine pathways modulate reward and vigilance, forming a spectrum of fundamental strategic responses in humans, including motor coordination roles whose overstimulation or disruption can lead to Parkinsons, dependency and psychosis. Reports of increased social dominance in primates (Edwards and Kravitz, 1997) and improved mood and confidence in social interactions in humans after using drugs which increase serotonin levels are well documented (Kramer, 1993; Young and Leyton, 2002).

Functional studies in the honey bee and fruit fly have shown that serotonergic signaling participates in aggression, sleep, circadian rhythms, responses to visual stimuli, and associative learning (Blenau & Thamm 2011). Serotonin in lobsters regulates socially relevant behaviours such as dominance-type posture, offensive tail flicks, and escape responses (Kravitz, 2000, Sosa et al. 2004). In insects, dopamine acts instead as a punishment signal and is necessary to form aversive memories (Barron et al. 2007, Schwaerzel et al. 2003, Selcho et al. 2009). In flies dopamine modulates locomotor activity, sexual function and the response to cocaine, nicotine, and alcohol (Hearn et al. 2002). Octopamine, the arthropod analogue of norepinephrine, regulates desensitization of sensory inputs, arousal, initiation, and maintenance of various rhythmic behaviors and complex behaviors such as learning and memory, and endocrine gland activity (Farooqui 2007). Web building in spiders is likewise affected by stimulants and psychedelics (Dunn).

These neurotransmitters are thus playing a similar role in humans in modulating the excitable brain to attune it to survival objectives that these same signaling molecules had in single celled eucaryote social and reproductive behaviours.

Moreover, although most neurophysiological investigations of arthropod and mollusc neural ganglia tend to be recordings of single neuronal action potentials (e.g. Paulk & Gronenberg 2005), ""silent"" cells with graded electrical responses are also integral to the function of small neuronal circuits (Kandel 1979), and as we have already noted, chaotic excitability occurs in bursting and beating action potentials in amoebae, Paramecium and simple algae. Furthermore studies in both molluscs (Schütt & Basar 1992) and arthropods (Kirschfeld 1992) have demonstrated coherent gamma-type oscillations. These results lead us to the hypothesis that there is a common basis of attentive processing in the gamma band across wide branches of the metazoa, based on edge-of-chaos processing and wave phase coherence, despite their highly varied neuroanatomies (Basar et al. 2001).

3: Consciousness - Coordinated Activity Anticipating Future Challenges

The organization of the cerebral cortex and its underlying structures, consist of a series of microcolumns vertically spanning the three to six layers of the cortex, acting as parallel processing units for an envelope of characteristics, a hologram-like featural mathematical transform space. Typical features represented in particular cortical regions include sensory attributes such as the line orientation and binocular dominance of visual processing, tonotopic processing of sounds, somato-sensory bodily maps, and higher level features such as facial expressions and the faces of individuals, leading to the strategic executive modules of the prefrontal cortex and our life aims and thought processes.

The many-to-many nature of synaptic connections forms the basis of this abstract representation, which is also adaptive through neural plasticity. Space and time also become features in the transform mapping, so that certain e.g. parietal areas have major roles in spatial navigation while other areas, for example in the temporal lobes elicit experiences of a memory-episodic nature. The hippocampus pivotal in consolidating sequential memory also appears to function as a spatial GPS, emphasizing the mutual relation between space and time in transform space. A key role of wave-based brain processing is to harness this transform representation to predict, using experiential memory and contextual clues, the ongoing nature of opportunities and threats to survival.

Subjective consciousness involves coordinated whole-brain activity (Baars 1997, 2001), as opposed to local activations, which reach only the subconscious level, as evidenced in both experiments on conscious processing and the effects of dissociative anaesthetics (Alkire et al. 2008). Attempts to find the functional locus of subjective consciousness in brain regions have arrived at the conclusion that active conscious experiences are not generated in a specific cortical region but are a product of integrated coherent activity of global cortical dynamics (Ananthaswamy 2009, 2010). This distributed view of conscious brain activity is consistent with experimental studies in which the cortical modules we see activated in fMRI and PET scans correspond to salient features of subjective conscious experience.

This implies that the so-called Cartesian theatre of consciousness is a product of the entire active cortex and that the particular form of phase coherent, edge-of-chaos processing adopted by the mammalian brain is responsible for the manifestation of subjective experience. This allows for a theory of consciousness in which preconscious processing e.g. of sensory information can occur in specific brain areas, which then reaches the conscious level only when these enter into coherent global neuronal activity integrating the processing, as Baars global workspace theory (1997, 2001) proposes, rather than being a product of a specific region such as the supplementary motor cortex (Eccles 1982, Fried et al. 1991, Haggard 2005).

The approach is also consistent with there being broadly only one dominant stream of conscious thought and experience at a given time, as diverse forms of local processing give way to an integrated global response. A series of experiments involving perceptual masking of brief stimuli to inhibit their entry into conscious perception (Sergent et al. 2005, Sigman and Dehaene 2005, 2006, Dehaene and Changeux 2005, Del Cul et al. 2007, 2009, Gaillard et al. 2009), studies of pathological conditions such as multiple sclerosis (Reuter et al. 2009, Schnakers 2009), and brief episodes in which direct cortical electrodes are being used during operations for intractable epilepsy (Quiroga et al. 2008) have tended to confirm the overall features of Baars model (Ananthaswamy 2009, 2010). EEG studies also show that under diverse anesthetics, as consciousness fades, there is a loss of synchrony between different areas of the cortex (Alkire et al. 2008). The theory also tallies with Tononis idea of phi, a function of integrated complexity used as a measure of consciousness (Barras 2013, Pagel 2012).

The brain regions involved in our sense of self - the actor-agent behind conscious states - are specifically activated in idle periods, in the so-called default circuit, whose function appears to be adaptively envisaging future challenges. The default network (Fox 2008, Zimmer 2005) encompasses posterior-cingulate/precuneus, anterior cingulate/mesiofrontal cortex and temporo-parietal junctions, several of which have key integrating functions. The ventral medial prefrontal (Macrae et al. 2004) is implicated in processing risk and fear. It also plays a role in the inhibition of emotional responses, and decision-making. It has been shown to be active when experimental subjects are shown imagery they think apply to themselves. The precuneus (Cavanna & Trimble 2006) is involved with episodic memory, visuo- spatial processing, reflections upon self, and aspects of consciousness. The insulae are also believed to be involved in consciousness and play a role in diverse functions usually linked to emotion and the regulation of the body's homeostasis, including perception, motor control, self-awareness, cognitive functioning, and interpersonal experience. The anterior insula is activated in subjects who are shown pictures of their own faces, or who are identifying their own memories. The temporo- parietal junction is known to play a crucial role in self-other distinction and theory of mind. Studies indicate that the temporo-parietal junction has altered function during simulated out of body experiences (Ananthaswamy 2013).

Although subjective consciousness involves the entire cortex in coherent activation, brain scans highlight certain areas of pivotal importance, whose disruption can impede active consciousness. Three regions associated with global workspace have been identified as key participants in these higher integrative functions, the thalamus which is a critical set of relay centres underlying all cortical areas and possibly driving the EEG, lateral prefrontal executive function and posterior parietal spatial integration (Bor 2013). Another set of two regions, anterior cingulate and fronto- insular are highlighted in the saliency circuit (Williams 2012) in which von Economo (VEN) bipolar neurons provide fast connectivity between regions to maintain a sense of the conscious present providing a sense of immediate anticipation of the ongoing external and internal condition (saliency and interoception). These appear prominently in large brained animals, including humans, elephants and cetaceans where there is greater need to rapidly stitch together related processing areas critical to the ongoing conscious state.

Several researchers have highlighted specific aspects of consciousness in an attempt to understand how it evolved. Higher integrative processing associated with global workspace has been extended to other animals such as apes and dolphins (Wilson 2013). Another approach suggests that making integrative decisions socially would have aided better environmental decision-making concerning hard to discern situations involving the combined senses in which social discussion aids survival, such as two hunters trying to assess whether dust on the prairie suggests running from lions or hunting buffalo, or women discussing where to find hard to get herbs from the visual appearance, taste and smell of a sample (Bahrami et al. 2010).

4: Conscious Survival in the Wild

To discover what advantage subjective consciousness has over purely computational processing, we need to examine the survival situations that are pivotal to organisms in the open environment and the sorts of computational dilemmas involved in decision-making processes on which survival depends.

Many open environment problems of survival are computationally intractable and would leave a digital antelope stranded at the crossroads until pounced upon by a predator, because they involve a number n of factors, which increase super-exponentially with n. For example, in the traveling salesman problem - finding the shortest path around n cities - the calculation time grows super-exponentially with the factorial $(n-1)!/2$ - the number of possible routes which could be taken, each of which needs to me measured to find the shortest path (King 1991). There are probabilistic methods which can give a sub-optimal answer and artificial neural nets solve the problem in parallel by simulating a synaptic potential energy landscape, using thermodynamic annealing to find a local minimum not too far from the global one. Vertebrate brains appear to use edge of chaos dynamics to similar effect.

Open environment problems are intractable both because they fall into this broad class and also because they are prone to irresolvable structural instabilities, which defy a stable probabilistic outcome. Suppose a gazelle is trying to get to the waterhole along various paths. On a probability basis it is bound to choose the path, which, from its past experience, it perceives to be the least likely to have a predator, i.e. the safest. But the predator is likewise going to make a probabilistic calculation to choose the path that the prey is most likely to be on given these factors i.e. the same one. Ultimately this is an unstable problem that has no consistent computational solution.

There is a deeper issue in these types of situation. Probabilistic calculations, both in the real world and in quantum mechanics, require the context to be repeated to build up a statistical distribution. In an interference experiment we get the bands of light and dark color representing the wave amplitudes as probability distributions of photons on the photographic plate only when a significant number have passed through the apparatus in the same configuration. The same is true for estimating a probabilistically most viable route to the waterhole. But real life problems are plagued by the fact that both living organisms and evolution itself are processes in which the context is endlessly being changed by the decision-making processes. Repetition occurs only in the most abstract sense, which is one reason why massively parallel brains we have are so good at such problems.

Finally, in many real life situations, there is not one optimal outcome but a whole series of possible choices, any or all of which could lead either to death, or survival and reproduction. This is the super-abundance problem we shall investigate shortly. Despite having complex brains, even humans are very inferior computers with a digit span of only six or seven and a calculation capacity little better than a pocket calculator. We all know what we do and what conscious animals do in this situation. They look at the paths forward. If they have had a bad experience on one they will probably avoid it, but otherwise they will try to assess how risky each looks and make a decision on intuitive hunch to follow one or the other. In a sense, all their previous life experience is being summed up in their conscious awareness and their contextual memory. The critical point is that consciousness is providing something completely different from a computational algorithm, it is a form of real time anticipation of threats and survival that is sensitively dependent on environmental perturbation and attuned to be anticipatory in real time just sufficiently to jump out of the way and bolt for it and survive. Thus the key role of consciousness is to keep watch on the unfolding living environment, to be paranoid to hair-trigger sensitivity for any impending hint of a movement, or the signs, or sound of a pouncing predator — an integrated holographic form of space-time anticipation.

5: Edge of Chaos Sensitivity and Phase Coherence Processing

From the work of Walter Freeman (1991, Skarda & Freeman 1987) and others (Basar et al. 1989) it has been established that the electroencephalogram shows characteristics of chaos, including broad-spectrum frequency activity, strange attractors with low fractal dimension and transitions from high-energy chaos, to learned, or new attractors, during sensory and perceptual processing. A fundamental property of chaos is sensitivity to arbitrarily small perturbations - the butterfly effect.

Between the global, the cellular and the molecular level are a fractal cascade of central nervous processes, which, in combination, make it theoretically possible for a quantum fluctuation to become amplified into a change of global brain state. The neuron is itself a fractal with multiply branching dendrites and axonal terminals, which are essential to provide the many-to-many synaptic connections between neurons, which make adaptation and the transform representation of reality possible. Furthermore, like all tissues, biological organization is achieved through non-linear interactions which begin at the molecular level and have secondary perturbations upward in a series of fractal scale transformations through complex molecules such as enzymes, supra-molecular complexes such as ion channels and the membrane, organelles such as synaptic junctions, to neurons and then to neuronal complexes such as cortical mini-columns and finally to global brain processes.

Because neurons tend to tune to their threshold with a sigmoidal activation function, which has maximum limiting slope at threshold, they are capable of becoming critically poised at their activation threshold. It is thus possible in principle for a single ion channel, potentially triggered by only one or two neurotransmitter molecules, if suitably situated on the receptor neuron, e.g. at the cell body, where an action potential begins, to act as the trigger for activation (King 2008).

The lessons of the butterfly effect and evidence for transitions from chaos in perceptual recognition suggest that if a brain state is critically poised, the system may become sensitive to instability at the neuronal, synaptic, ion-channel, or quantum level.

A variety of lines of evidence have demonstrated that fluctuations in single cells can lead to a change of brain state. In addition to sensitive dependence in chaotic systems, stochastic resonance (Liljenström and Uno 2005), in which the presence of noise, somewhat paradoxically, leads to the capacity of ion channels to sensitively excite hippocampal cells and in turn to cause a change in global brain state. In this sense noise is equivalent to the properties of dynamical chaos, which distribute through the dynamic pseudo-randomly preventing the dynamic getting locked in a stable attractor. Such a dynamic is thus able to explore its dynamical space, just as thermodynamic annealing is used in artificial neural nets to avoid them becoming locked in sub-optimal local minima.

Chandelier cells, which are more common in humans than other mammals, such as the mouse, and were originally thought to be purely inhibitory, are axon-axonal cells, which can result in specific poly-synaptic activation of pyramidal cells. It has been discovered (Molnar 2008, Woodruff and Yuste 2008) that chandelier cells are capable of changing the patterns of excitation between pyramidal neurons that drive active output to other cortical regions and to the peripheral nervous system, in such a way that single action potentials are sufficient to recruit neuronal assemblies that are proposed to participate in cognitive processes.

6: Quantum Reality, Sentience and Intentional Will

Many scientists assume that all human activity must be a product of brain function and that any notion of conscious will acting on the physical is delusory. This flies in contradiction to our subjective assessment that we are autonomous beings with voluntary control over our fates. To claim free will is a delusion leads to a catatonic impotence of consciousness and contradicts the assumptions of legal accountability, where we assume a person of sound mind is physically responsible for the consequences of their consciously intentional actions.

Many physicists, from Arthur Eddingtons citation of the uncertainty of position of a synaptic vesicle in relation to the thickness of the membrane on, have drawn attention to the fact that the quantum universe is not deterministic in the manner of classical causality and that quantum uncertainty provides a causal loophole, which might make it possible for free will to coexist in the quantum universe.

Biology is full of phenomena at the quantum level, which are essential to biological function. Enzymes invoke quantum tunneling to enable transitions through their substrates activation energy. Protein folding is a manifestation of quantum computation intractable by classical computing. When a photosynthetic active centre absorbs a photon, the wave function of the excitation is able to perform a quantum computation, which enables the excitation to travel down the most efficient route to reach the chemical reaction site (McAlpine 2010, Hildne et al. 2013).

Quantum entanglement is believed to be behind the way some birds navigate in the magnetic field (Amit 2012, Courtland 2011). Light excites two electrons on one molecule and shunts one of them onto a second molecule. Their spins are linked through quantum entanglement. Before they relax into a decoherent state, the Earth's magnetic field can alter the relative alignment of the electrons' spins, which in turn alters the chemical properties of the molecules involved. Quantum coherence is an established technique in tissue imaging, demonstrating quantum entanglement in biological tissues at the molecular level (Samuel 2001, Warren 1998).

Although the brain needs to able to be resilient to noise in its stable functioning, in the event of a critically poised dynamic in which there is no stable determining outcome, several key processes, may make the brain state capable of being sensitive to fluctuations at the quantum level. These include chaotic sensitivity, self- organized criticality, the amplifying effects of chandelier cells and stochastic resonance (King 2008, 2012).

Consciousness, Neuroscience, Time Travel

Karl Pribram (1991), in the notion of the holographic brain, has drawn attention to the similarity between phase coherence processing of brain waves in the gamma frequency range believed to be responsible for cognitive processes and the wave amplitude basis of quantum uncertainty in reduction of the wave packet and quantum measurements based on the uncertainty relation Et h , where the relation is determined by the number of phase fronts to be counted.

This raises an interesting implication, that the evolution of nervous systems may have arrived at a neurodynamic homologous with quantum processes at the foundation of physics, suggesting that quantum entanglement in brain states could in turn be a basis for active biological anticipation of immediate threats to survival through the forms of subjective consciousness the brain generates.

In quantum mechanics, not only are all probability paths traced in the wave function, but past and future are interconnected in a time-symmetric hand-shaking relationship, so that the final states of a wave-particle or entangled ensemble, on absorption, are boundary conditions for the interaction, just as the initial states that created them are. The transactional interpretation of quantum mechanics expresses this relationship neatly in terms of offer waves from the past emitter/s and confirmation waves from the future absorbers, whose wave interference becomes the single or entangled particles passing between. When an entangled pair are created, each knows instantaneously the state of the other and if one is found to be in a given state, e.g. of polarization or spin, the other is immediately in the complementary state, no matter how far away it is in space-time. This is the spooky action at a distance, which Einstein feared because it violates local Einsteinian causality — particles not communicating faster than the speed of light.

The brain explores ongoing situations which have no deductive solution, by evoking an edge-of-chaos state which, when it transitions out of chaos, results in the aha of insight learning. The same process remains sensitively tuned for anticipating any signs of danger in the wild. This is pretty much how we do experience waking consciousness. If this process involves sensitivity to quantum indeterminacy the coherent excitations would be quantum entangled, invoking new forms of quantum computation.

However quantum entanglement cannot be used to make classical causal predictions, which would formally anticipate a future event, so the past-future hand- shaking lasts only as long as a particle or entangled ensemble persist in their wave function.

Weak quantum measurement (WQM) is one way a form of quantum anticipation could arise. Weak quantum measurement (Aharonov et al. 1988) is a process where a quantum wave function is not irreversibly collapsed by absorbing the particle but a small deformation is made in the wave function whose effects become apparent later when the particle is eventually absorbed e.g. on a photographic plate in a strong quantum measurement. Weak quantum measurement changes the wave function slightly mid-flight between emission and absorption, and hence before the particle meets the future absorber involved in eventual detection. A small change is induced in the wave function, e.g. by slightly altering its polarization along a given axis (Kocsis et al. 2011). This cannot be used to deduce the state of a given wave-particle at the time of measurement because the wave function is only slightly perturbed, and is not collapsed or absorbed, as in strong measurement, but one can build up a prediction statistically over many repeated quanta of the conditions at the point of weak measurement, once post-selection data is assembled after absorption.

This suggests (Merali 2010, Cho 2011) that, in some sense, the future is determining the present, but in a way we can discover conclusively only by many repeats. Focus on any single instance and you are left with an effect with no apparent cause, which one has to put it down to a random experimental error. This has led some physicists to suggest that free-will exists only in the freedom to choose not to make the post- selection(s) revealing the futures pull on the present. Yakir Aharonov, the co- discoverer of weak quantum measurement (Aharonov et al. 1988) sees this occurring through an advanced wave travelling backwards in time from the future absorbing states to the time of weak measurement. What God gains by playing dice with the universe, in Einsteins words, in the quantum fuzziness of uncertainty, is just what is needed, so that the future can exert an effect on the present, without ever being caught in the act of doing it in any particular instance: ""The future can only affect the present if there is room to write its influence off as a mistake""', neatly explaining why no subjective account of prescience can do so either.

Weak quantum measurements have been used to elucidate the trajectories of the wave function during its passage through a two-slit interference apparatus (Kocsis et al. 2011), to determine all aspects of the complex waveform of the wave function (Hosten 2011, Lunden et al. 2011), to make ultra sensitive measurements of small deflections (Hosten & Kwiat 2008, Dixon et al. 2008) and to demonstrate counterfactual results involving both negative and positive post-selection probabilities, which still add up to certainty, when two interference pathways overlap in a way which could result in annihilation (Lundeen & Steinberg 2009).

WQM provides a potential way that the brain might use its brain waves and phase coherence to evoke entangled (coherent) states that carry quantum encrypted information about immediate future states of experience as well as immediately past states, in an expanded envelope - the quantum present. It is this coordinated state that corresponds to subjective experience of the present moment, encoded through the parallel feature envelope of the cerebral cortex, including the areas associated with consciousness.

Effectively the brain is a massively parallel ensemble of wave excitations reverberating with one another, through couplings of varying strength in which excitations are emitted, modulated and absorbed. Interpreted in terms of quantum excitations, the ongoing conscious brain state could be a reverberating system of massively parallel weak quantum measurements (MPWQMs) of its ongoing state. This could in principle give the conscious brain a capacity to anticipate immediate future threats through the intuitive avenues of prescience, paranoia and foreboding. This suggests that the reverberating ensemble of the quantum present could provide an intuitive form of anticipation complementing computational predictions.

This would require significant differences from the post-selection paradigm of weak quantum measurement experiments, which are designed to produce a classically confirmed result from an eventual statistical distribution in the future. In the brain, consciousness being identified with the coherent excitations and hence the entangled condition could reverse the implication of backwards causality of advanced waves, with the future effectively informing the present of itself in quantum encrypted form through the space-time expansion of the quantum present. Discovering a molecular-biological basis for such an effect would pose an ultimate challenge to experimental neuroscience.

An indication of how quantum chaos might lead to complex forms of quantum entanglement can be gleaned from an ingenious experiment forming a quantum analogue of a kicked top using an ultra-cold cesium atom kicked by a laser pulse in a magnetic field: the classical dynamical space of the kicked top, showing domains of order where there is periodic motion and complementary regions of chaos where there is sensitive dependence on initial conditions. In the quantum system (middle pair), in the ordered dynamic (left), the linear entropy of the system (bottom pair) is reduced and there is no quantum entanglement between the orbital and nuclear spin of the atom. However in the chaotic dynamic (right) there is no such dip, as the orbital and nuclear spins have become entangled as a result of the chaotic perturbations of the quantum tops motion (Chaudhury et al. 2009, Steck 2009). This shows that, rather than the suppression of classical chaos seen in closed quantum systems (King 2013), reverberating chaotic quantum systems can introduce new entanglements.

The prevailing theory for loss of phase coherence and entanglement is decoherence caused by the interaction of a wave-particle with other wave particles in the environmental milieu. The coherence of the original entanglement becomes perturbed by other successive forms of entanglement, which successively reduce the coherence exponentially over time in the manner of an open system chaotic billiards. However in a closed universe, such as the global excitations of a brain state, decoherence does not necessarily approach the classical limit, but may retain encoded entangled information, just as the above example of the quantum kicked top does in a simpler atomic system, which could be referenced by the brain in the same way multiple hippocampal representations over time can, as an organism explores a changing habitat. Intriguingly, continued weak quantum measurement, rather than provoking decoherence tends to preserve entanglement because the ordered nature of the weak quantum measurements reduces the disordered nature of the environment (Hosten 2011, Lundeen et al. 2011). Massively parallel weak quantum measurements in the brain might thus function to maintain the ongoing entanglement.

7: Unraveling the Readiness Potential

Many aspects of brain function display dynamic features, which show the brain is focused on attempting to anticipate ongoing events. When a cat is dropped into unfamiliar territory, the pyramidal cells in its hippocampus become desynchronized and hunt chaotically, in what is called the orienting reaction, until the animal discovers where it is, or gains familiarity with its environment, when phase synchronization ensues (Coleman & Lindsley 1975). In a similar manner, the EEG will show a desynchronized pattern when a subject is listening for a sound which is irregularly spaced, but will fall into a synchronized pattern if the subject can confidently anticipate when the next sound is going to occur. Greater capacity to shift synchronization rather than it remaining locked has been associated with higher IQ (Jung-Beeman 2008).

This kind of processing is consistent with a computational process involving transitions from chaos to order. The chaotic regime acts both to provide sensitive dependence on any changes in boundary conditions such as sensory or cognitive inputs, at the same time as preventing the dynamical system getting caught in a suboptimal attractor, by providing sufficient energy to cause the process to fully explore the space of dynamical solutions. Artificial neural net annealing and quantum annealing both follow similar paradigms using random fluctuations and uncertainty to achieve a similar global optimization. Such a dynamic also allows for ordered deductive computation, but enables the system to evolve chaotically when the ordered process fails to arrive at a computational solution. In combination with quantum entanglement and massively parallel weak quantum measurement, as we have seen, this process may enable the ongoing conscious state to be anticipative.

However, a historical experiment suggested that, far from anticipating reality in real time, conscious awareness of a decision might actually lag behind unconscious brain processing which is already leading to the decision, although being placed by subjective experience at the time the conscious decision was made. In 1983, neuroscientist Benjamin Libet and co-workers (1983, 1989) asked volunteers wearing scalp electrodes to flex a finger or wrist. When they did, the movements were preceded by a dip in the signals being recorded, called the "readiness potential". Libet interpreted this RP as the brain preparing for movement. Crucially, the RP came a few tenths of a second before the volunteers said they had decided to move. Libet concluded that unconscious neural processes determine our actions before we are ever aware of making a decision. Since then, others have quoted the experiment as evidence that free will is an illusion - a conclusion that was always controversial, particularly as there is no proof the RP represents a decision to move.

With contemporary brain scanning technology, Soon et al. (2008) were able to predict with 60% accuracy whether subjects would press a button with their left or right hand up to 10 seconds before the subject became aware of having made that choice. This doesn't of itself negate conscious willing because these prefrontal and parietal patterns of activation merely indicate a process is in play, which may become consciously invoked at the time of the decision, and clearly many subjects (40% of trials) were in fact making a contrary decision. Neuroscientist John-Dylan Haynes, who led the study, notes: "I wouldn't interpret these early signals as an 'unconscious decision', I would think of it more like an unconscious bias of a later decision" (Williams 2012).

The assumption that Libets RP is a subconscious decision has been undermined by subsequent studies. Instead of letting volunteers decide when to move, Trevena and Miller (2010) asked them to wait for an audio tone before deciding whether to tap a key. If Libet's interpretation were correct, the RP should be greater after the tone when a person chose to tap the key. While there was an RP before volunteers made their decision to move, the signal was the same whether or not they elected to tap. Miller concludes that the RP may merely be a sign that the brain is paying attention and does not indicate that a decision has been made. They also failed to find evidence of subconscious decision-making in a second experiment. This time they asked volunteers to press a key after the tone, but to decide on the spot whether to use their left or right hand. As movement in the right limb is related to the brain signals in the left hemisphere and vice versa, they reasoned that if an unconscious process is driving this decision, where it occurs in the brain should depend on which hand is chosen, but they found no such correlation.

Schurger and colleagues (2012) have elucidated an explanation. Previous studies have shown that, when we have to make a decision based on sensory input, assemblies of neurons start accumulating evidence in favor of the various possible outcomes. A decision is triggered when the evidence favoring one particular outcome becomes strong enough to tip its associated assembly of neurons across a threshold. The team hypothesized that a similar process happens in the brain during the Libet experiment. They reasoned that movement is triggered when this neural noise generated by random or chaotic activity accumulates and crosses a threshold. The team repeated Libet's experiment, but this time if, while waiting to act spontaneously, the volunteers heard a click they had to act immediately. The researchers predicted that the fastest response to the click would be seen in those in whom the accumulation of neural noise had neared the threshold - something that would show up in their EEG as a readiness potential. In those with slower responses to the click, the readiness potential was indeed absent in the EEG recordings. "We argue that what looks like a pre-conscious decision process may not in fact reflect a decision at all. It only looks that way because of the nature of spontaneous brain activity." Both these newer studies thus cast serious doubt on Libets claim that a conscious decision is made after the brain has already put the decision in motion, leaving open the possibility that conscious decisions are actually made in real time.

Some aspects of our conscious experience of the world make it possible for the brain to sometimes construct a present that has never actually occurred. In the "flash-lag" illusion, a screen displays a rotating disc with an arrow on it, pointing outwards. Next to the disc is a spot of light that is programmed to flash at the exact moment the spinning arrow passes it. Instead, to our experience, the flash lags behind, apparently occurring after the arrow has passed (Westerhoff 2013). One explanation is that our brain extrapolates into the future, making up for visual processing time by predicting where the arrow will be, however, rather than extrapolating into the future, our brain is actually interpolating events in the past, assembling a story of what happened retrospectively, as was shown by a subtle variant of the illusion (Eagleman and Sejnowski 2000). If the brain were predicting the spinning arrow's trajectory, people would see the lag even if the arrow stopped at the exact moment it was pointing at the spot. But in this case the lag does not occur. If the arrow begins stationary and moves in either direction immediately after the flash, the movement is perceived before the flash. How can the brain predict the direction of movement if it doesn't start until after the flash? The perception of what is happening at the moment of the flash is determined by what happens to the disc after it. This seems paradoxical, but other tests have confirmed that what is perceived to have occurred at a certain time can be influenced by what happens later. This again does not show that the brain is unable to anticipate reality because it applies only to very short time interval spatial reconstructions by the brain, which would naturally be more accurate by retrospective interpolation.

8: Prescience - Three Personal Experiences

To fathom situations where real time anticipation may have occurred without any prevailing causal implication leading up to it, we need to turn to rare instances of prescience with no reasonable prior cause. These kinds of events tend to be rare, apocryphal and lack independent corroboration, like stories of telepathic connection or the sense of foreboding that a relative has died, which later receives confirmation. Paradoxically some of the most outstanding examples can come from alleged precognitive dreaming rather than the waking state, which tends to be more circumscribed by commonsense everyday affairs.

As a student, I picked up and read ""An Experiment with Time"" by J W Dunne (1929), which outlined double blind experiments in which the dream diaries led to as many accounts linking to future events in the peoples lives as they did to past experiences. A few weeks later I had a horrific double nightmare that I was being agonizingly stung. In the dream it was a spider which I couldnt remove because it would leave poison fangs inside me (as a bee or wasp does) and in the second dream it had returned to sting me again when I was distracted, as one often is in dreams. At eight in the morning as my wife woke, I recounted the nightmares in detail to her, before falling asleep again. About an hour later I was stung wide-awake by a wasp that had flown in the window, which my wife had opened after getting up. Suddenly the dream I had reported to my wife had become a reality. A skeptic might try to interpret this as a coincidence - merely an application of Bayes theorem of conditional probabilities - but the complete absence of any such dream before, drove home to me that dreaming, and by implication waking experiences too, have properties violating classical causality. The fact that it closely followed on reading the book gave this prescience an added dimension, capped by the fact that the scientist providing an introduction to the work was none other than Arthur Eddington, who had suggested quantum uncertainty of the synaptic vesicle as a basis for free will.

This raises a series of questions about coincidence and Carl Jungs (1952) notion of synchronicity, the idea that seemingly unrelated events and experiences may be caught up in a deeper correspondence, as reminiscent of quantum entanglement as phase coherence in brain processing appears to be. Many peoples personal accounts attest to a currency of such prescience.

A month before the twin towers fell in New York I wrote a song and posted the lyrics online. They contained several prescient lines, one invoking jihad: ""When it comes to the final struggle, jihad of the biosphere, theres only one true rogue nation - the great American shaitan"". The lyrics continue with a lament for the dark canyons of lower Manhattan among the fallen towers: ""walking in the twilight, down in the valley of shadows"", and then the plane ""Well fly so high well pass right to the other side and never fall in flames again"". Then I watched live in prescient horror as one of the two planes struck the tower and passed right through, coming out in a burst of flames on the other side. The lyrics continue with the genocide - ""when will you comprehend the damage you have wrought in your indiscretion, can we undo the death trance you have set in motion?"" The last line closed with ""Can we bear it all again? It thus presciently echoed the Mayor of New Yorks own words on TV ""This …… will be more than any of us can bear"". I was singing about a mass extinction of life, but why the Islamic jihad, meeting Icarus descending?

All shamanic practitioners have to answer a question of coincidence. When curing a sick person, it is not to explain why the person has contracted tuberculosis or leprosy i.e. that the respective bacteria were infectious, or their immune system was weakened, but why this person caught this sickness at this particular time.

In the process of writing this paper, I awoke from a dream in which I was gazing at a pregnant woman, touching her on the shoulders, absorbed in the glowing beauty and fecundity of her pregnant state. Then when I sat down to look at the news next morning, I found myself watching this time-lapse video clip taken of a womans 40 week pregnancy by photographer, Nicole Gourley.

9: Anticipating the Multiverse

The central enigma of quantum reality is the causality-violating reduction of the wave packet. We see this in Schrödingers cat paradox a cat set to be killed by a radioactive scintillation breaking a cyanide flask. In quantum reality the cat is both alive and dead with differing probabilities, but in our subjective experience, when we open the box the cat is either alive, or dead, with certainty. Reduction also occurs when a wave is absorbed as a particle in an interference experiment. Quantum mechanics appears to preserve all the conceivable outcomes in parallel superposition. Not only is Schrödingers cat both alive and dead, but Napoleon has both won and lost the battle of Waterloo. Many of these strategic outcomes, indeed all accidents of history, depend on uncertainties that go, in principle, right down to the quantum level.

There is continuing debate among physicists about how and where in the causal chain, reduction of the wave packet actually occurs. While decoherence theories suggest this may occur simply through interaction of single or entangled states with other particles, the wave function of the entire universe is in effect one single multi- particle entangled state and so the whole notion of a single line of history unfolding seems to be something only our conscious awareness is able to determine.

Several of the founding quantum physicists adhered to this view. John von Neumann suggested that quantum observation is the action of a conscious mind and that everything in the universe that is subject to the laws of quantum physics creates one vast quantum superposition. But the conscious mind is different, being able to select out one of the quantum possibilities on offer, making it real - to that mind. Max Planck, the founder of quantum theory, said in 1931, "I regard consciousness as fundamental. I regard matter as derivative from consciousness." Werner Heisenberg also maintained that wave function collapse - the destruction of quantum superposition - occurs when the result of a measurement is registered in the mind of an observer. In Henry Stapps words we are "participating observers" whose minds cause the collapse of superpositions. ""Before human consciousness appeared, there existed a multiverse of potential universes. The emergence of a conscious mind in one of these potential universes, ours, gives it a special status: reality"" (Brooks 2012). This is effectively a complement to the anthropic principle of physical cosmology in which conscious observers are selective boundary conditions on the laws of nature in the universe (Barrow and Tipler 1988).

Thus another idea of the role of subjective consciousness is that it is a way the universe can solve the super-abundance of multiverses to bring about a natural universe in which some things do happen and other things dont. One of the most central experiences of our transient mortal lives is historicity — that there is a line of actual history, in which each of us, however small and insignificant our lives, are participating in bringing the world into actual being, albeit sometimes rather diabolically in times of exploitation, but with some reflection on our own transience, perhaps reaching towards a more enlightened existence, in which the passage of the generations is able to reach towards a state where the universe comes to consciously understand itself ever more deeply and completely.

The idea that consciousness collapses the wave functions of the universe leads to some counter-intuitive conclusions, because it implies that the consciousness itself is inducing the historical collapse that is in turn causing my brain to have a memory of this process.

On the other hand, the notion of the brain using entanglement provides a paradigm for resolving many of the contradictory situations that arise when classical causality is applied to anticipatory processes. A premonition being either a cause of a future event or caused by it leads to contradiction, which is resolved in the space-time hand-shaking of the entanglement.

The process goes like this: Memory systems are used to form a model of the quantum collapsed history already experienced, which is sequentially stored in the hippocampus and then semantically re-encoded into the cortical feature envelope so that it can be interrogated from any contingent perspective. The conscious cerebral cortex contains a dynamical system of entangled states, which together envelop a space-time region extending a limited distance into both the past and future - the quantum-delocalized present. The cortical envelope thus maintains a state of context-modulated sensitively-dependent dynamic excitation which generates our conscious sense of the present moment by encoding the immediate past and future together in a wave function representation entangled in the global coherent dynamic.

The quantum present would extend over the entangled life-time of the coherent excitations, incorporating quantum-encrypted information about the immediate past and future of the organism into the current state of subjective experience. The quantum present provides the loophole in classical causality that permits intentional will, or free-will to alter brain states and hence physical states in the world through behavior, as an effect of the entanglement. An external observer will simply see a brain process sensitively dependent on quantum uncertainty.

It may also be possible for the brain to encode entangled states in a more permanent form. Highly active brain states have been shown in fMRI studies to elicit changes in cerebral activation lasting over 24 hours (Heaven 2013, Harmelech et al. 2013). Long-term potentiation and memory processes are in principle permanent and may involve epigenetic changes (Levenson & Sweatt 2005).

REFERENCES

Aharonov Y., Albert D.Z., Vaidman L. (1988) How the Result of a Measurement of a Component of the Spin of a Spin-2 Particle Can Turn Out to be 100 Physical review letters 60, 1351.

Alkire M, Hudetz A., Tononi G. (2008) Consciousness and Anesthesia Science 322/5903 876-880. Amit G. (2012) 'Eye bath' to thank for quantum vision in birds New Scientist http://www.newscientist.com/article/dn22199-eye-bath-to-thank-for-quantum-vision-in-birds.html

Ananthaswamy A. (2009) 'Consciousness signature' discovered spanning the brain New Scientist 17 March. http://www.newscientist.com/article/dn16775-consciousness-signature-discovered- spanning-the-brain.html

Ananthaswamy A. (2009) Whole brain is in the grip of consciousness New Scientist 18 March. http://www.newscientist.com/article/mg20127004.300-whole-brain-is-in-the-grip-of- consciousness.html

Ananthaswamy A. (2010) Firing on all neurons: Where consciousness comes from New Scientist 22 March. http://www.newscientist.com/article/mg20527520.400-firing-on-all-neurons-where- consciousness-comes- from.html

Ananthaswamy A. (2013) The self: Trick yourself into an out-of-body experience New Scientist 20 Feb.

Anjard C., Loomis W. (2006) GABA induces terminal differentiation of Dictyostelium through a GABAB receptor Development 133 2253-2261 doi:10.1242/dev.02399

Azmitia E. C. (2010) Evolution of Serotonin: Sunlight to Suicide Muller C & Jacobs B (Eds.) Handbook of Behavioral Neurobiology of Serotonin Elsevier ISBN 978-0-12-374634-4 DOI: 10.1016/B978-0-12-374634-4.00034-4

Baars, B. J. (1997) In the Theatre of Consciousness: Global Workspace Theory, A Rigorous Scientific Theory of Consciousness. Journal of Consciousness Studies, 4/4 292-309 8.

Baars, B. J. (2001) In the Theater of Consciousness Oxford University Press US.

Bahrami B., Olsen K., Latham P., Roepstorff A., Rees G., Frith C. (2010) Optimally Interacting Minds Science 329/5995 1081-5.

Barras C (2013) Mind maths: The sum of consciousness New Scientist 6 February. http://www.newscientist.com/article/mg21729032.400-mind-maths-the-sum-of-consciousness.html

Barron, A. B., Maleszka, R., Vander Meer, R. K., & Robinson, G. E. (2007). Octopamine modulates honey bee dance behavior. Proceedings of the National Academy of Sciences, 104(5), 1703- 1707. DOI:10.1073/pnas.0610506104. PMC 1779631.

Barrow J, Tipler F (1988) The Anthropic Cosmological Principle, Oxf. Univ. Pr., Oxford.

Basar E., Basar-Eroglu J., Röschke J., Schütt A., (1989) The EEG is a quasi-deterministic signal anticipating sensory-cognitive tasks, in Basar E., Bullock T.H. eds. Brain Dynamics Springer- Verlag, 43-71.

Basar E, Basar-Erogluc C, Karakas S, Schürmann M (2001) Gamma, alpha, delta, and theta oscillations govern cognitive processes International Journal of Psychophysiology 39 241-8. Blenau W & Thamm M (2011) Distribution of serotonin (5-HT) and its receptors in the insect brain with focus on the mushroom bodies. Lessons from Drosophila melanogaster and Apis mellifera Arthropod Structure & Development 40 381-394.

Bor D (2013) Consciousness: Watching your mind in action New Scientist 20 May.

Brooks M (2012) Reality: How does consciousness fit in? New Scientist 3 October.

Brizzi G & Blum J (1970) Effect of growth conditions on serotonin content of Tetrahymena pyriformis Journal of Eukaryotic Microbiology (J. Protozool.) 17/4 553-555.

Buznikov, G. A., Lambert, W. H., & Lauder, J. M. (2001). Serotonin and serotonin-like substances as regulators of early embryogenesis and morphogenesis. Cell and tissue research, 305(2), 177-186.

DOI10.1007/s004410100408

Buznikov, G. A., Lambert, W. H., & Lauder, J. M. (2001). Serotonin and serotonin-like substances as regulators of early embryogenesis and morphogenesis. Cell and tissue research, 305(2), 177-186. Cavanna A., Trimble M. (2006) The precuneus: a review of its functional anatomy and behavioural correlates Brain 129 564-583 doi:10.1093/brain/awl004.

Chaudhury S, Smith A, Anderson B, Ghose S, Jessen P (2009) Quantum signatures of chaos in a kicked top Nature 461 768-771.

Chay T.R., Rinzel J. (1985), Bursting, beating and chaos in an excitable membrane model, Biophys. J. 47, 357-366.

Chen, G. Q., Cui, C., Mayer, M. L., & Gouaux, E. (1999). Functional characterization of a potassium- selective prokaryotic glutamate receptor. Nature, 402(6763), 817-821

Cho A. (2011) Furtive Approach Rolls Back the Limits of Quantum Uncertainty Science 333 690-3. Coleman, J. R., Lindsley, D. B. (1975). Hippocampal electrical correlates of free behavior and behavior induced by stimulation of two hypothalamic-hippocampal systems in the cat. Experimental neurology, 49(2), 506-528.

Conaco, C., Bassett, D. S., Zhou, H., Arcila, M. L., Degnan, S. M., Degnan, B. M., & Kosik, K. S. (2012). Functionalization of a protosynaptic gene expression network. Proceedings of the National Academy of Sciences, 109(Supplement 1), 10612-10618. doi/10.1073/pnas.1201890109.

Courtland R. (2011) Quantum states last longer in birds' eyes New Scientist http://www.newscientist.com/article/mg20927963.000-quantum-states-last-longer-in-birds-eyes.html

Deacon T. (2012) Consciousness is a matter of constraint 30 November.

Del Cul A., Baillet S., Dehaene S. (2007) Brain dynamics underlying the nonlinear threshold for access to consciousness. PLoS Biol 5(10) e260. doi:10.1371/journal.pbio.0050260

Del Cul A., Dehaene S., Reyes P., Bravo E., Slachevsky A. (2009) Causal role of prefrontal cortex in the threshold for access to consciousness Brain 132 2531——2540.

Dehaene S., Changeux J.P. (2005) Ongoing spontaneous activity controls access to consciousness: A neuronal model for inattentional blindness. PLoS Biol 3(5) e141.

Dixon, P. B., Starling, D. J., Jordan, A. N., & Howell, J. C. (2009) Ultrasensitive beam deflection measurement via interferometric weak value amplification. Physical review letters, 102(17), 173601.

Dunn T. Spiders on drugs http://www.trinity.edu/jdunn/spiderdrugs.htm Dunne J. W. (1929) An Experiment with Time A & C Black, London.

Eagleman, D. M., Sejnowski, T. J. (2000) Motion integration and postdiction in visual awareness. Science. 287/5460 2036-8.

Eagleman, D.M., Sejnowski, T.J. (2000) Flash Lag Effect: Differential latency, not postdiction: Response. Science. 290/5494 1051a

Eccles J.C. (1982) The Initiation of Voluntary Movements by the Supplementary Motor Area Arch Psychiatr Nervenkr 231 423-441.

Eichinger, D., Coppi, A., Frederick, J., & Merali, S. (2002). Catecholamines in Entamoebae: Recent (re) discoveries. Journal of biosciences, 27(6), 589-593.

Eichinger, L., Pachebat, J. A., Glöckner, G., Rajandream, M. A., Sucgang, R., Berriman, M., ... & Hauser, H. (2005). The genome of the social amoeba Dictyostelium discoideum. Nature, 435(7038), 43-57.

Edwards, D.H., Kravitz, E.A. (1997) Serotonin, social status and aggression Curr. Opin. Neurobiol. 7 812-819.

Emes, R. D., Pocklington, A. J., Anderson, C. N., Bayes, A., Collins, M. O., Vickers, C. A., ... & Grant, S. G. (2008). Evolutionary expansion and anatomical specialization of synapse proteome complexity. Nature neuroscience, 11(7), 799-806.

Essman E (1987) The serotonergic system in Tetrahymena pyriformis International Journal of Clinical & Laboratory Research 17/1 77-82.

Farooqui T. (2007) Octopamine-Mediated Neuromodulation of Insect Senses Neurochem Res 32 1511-1529 DOI10.1007/s11064-007-9344-7

Felder C. et al. (1999) The venus flytrap of periplasmic binding proteins: an ancient protein module present in multiple drug receptors AAPS Pharm Sci. 1/2 E2.

Fox D. (2008) The secret life of the brain New Scientist 5 Nov.

Freeman W. (1991) The physiology of perception. Sci. Am. 264 Feb 35-41.

Fried I., Katz A., McCarthy G., Sass K., Williamson P., Spencer S. (1991) Functional Organization of Human Supplementary Motor Cortex Studied by Electrical Stimulation The Journal of Neuroscience, 1(11) 3656-3666.

Gaillard R., Dehaene S., Adam C., Clemenceau S., Hasboun D. (2009) Converging intracranial markers of conscious access. PLoS Biol 7(3) e1000061. doi:10.1371/journal.pbio.1000061

Goldbeter A (2006) Oscillations and waves of cyclic AMP in Dictyostelium: A prototype for spatio- temporal organization and pulsatile intercellular communication Bull Math Biol 68 1095-1109. Gonzales-Perdomo, M., Romero, P., & Goldenberg, S. (1988). Cyclic AMP and adenylate cyclase activators stimulate Trypanosoma cruzi differentiation. Experimental parasitology, 66(2), 205-212.

Haggard P. (2005) Conscious intention and motor cognition TRENDS in Cognitive Sciences 9/6 290- 295.

Halloy, J., Lauzeral, J., & Goldbeter, A. (1998). Modeling oscillations and waves of cAMP in Dictyostelium discoideum cells. Biophysical chemistry, 72(1), 9-19.

Harmelech, T., Preminger, S., Wertman, E., Malach, R. (2013). The Day-After Effect: Long Term, Hebbian-Like Restructuring of Resting-State fMRI Patterns Induced by a Single Epoch of Cortical Activation. The Journal of Neuroscience, 33(22), 9488-9497.

Hearn, M. G., Ren, Y., McBride, E. W., Reveillaud, I., Beinborn, M., & Kopin, A. S. (2002). A Drosophila dopamine 2-like receptor: Molecular characterization and identification of multiple alternatively spliced variants. Proceedings of the National Academy of Sciences, 99(22), 14554-9.

Heaven D. (2013) Echoes in the brain open a window on yesterday New Scientist 27 Jun.

Hildner, R., Brinks, D., Nieder, J. B., Cogdell, R. J., & van Hulst, N. F. (2013). Quantum Coherent Energy Transfer over Varying Pathways in Single Light-Harvesting Complexes. Science, 340(6139), 1448-1451. doi: 10.1126/science.1235820

Hinrichsen, R. D., Schultz, J. E. (1988) Paramecium: a model system for the study of excitable cells. Trends in neurosciences, 11(1), 27-32.

Hosten O. (2011) How to catch a wave Nature 474 170-1.

Hosten O., Kwiat P. (2008) Observation of the Spin Hall Effect of Light via Weak Measurements Science 319 787-790.

Iyer, L. M., Aravind, L., Coon, S. L., Klein, D. C., & Koonin, E. V. (2004). Evolution of cell—cell signaling in animals: did late horizontal gene transfer from bacteria have a role?. TRENDS in Genetics, 20(7), 292-299.

Jung, Carl G. (1993) [1952]. Synchronicity: An Acausal Connecting Principle. Bollingen, Switzerland: Bollingen Foundation. ISBN 978-0-691-01794-5.

Jung-Beeman, Mark (2008) The Eureka Hunt New Yorker July 28 84/22 40.

King C. C. (1991) Fractal and Chaotic Dynamics in Nervous Systems 1991 Progress in Neurobiology, 36 279-308.

King C. C. (2002) Biocosmology http://www.dhushara.com/book/bchtm/biocos.htm

King C. C. (2008) The Central Enigma of Consciousness Nature Precedings 5 November 2008 Journal of Consciousness Exploration and Research 2(1) 2011 http://www.dhushara.com/enigma/enigma.htm

King C. C. (2011) The Tree of Life: Tangled Roots and Sexy Shoots, DNA Decipher Journal 1(1) 73- 109.

King C. C. (2012) Entheogens, the Conscious Brain and Existential Reality http://www.dhushara.com/psyconcs/psychconsc8.htm

King C. C. (2013) Exploring Quantum, Classical and Semi-Classical Chaos in the Stadium Billiard Quanta 3/1 16-31 DOI: 10.12743/quanta.v3i1.23

Kirschfeld, K., (1992). Oscillations in the insect brain: do they correspond to the cortical gamma waves of vertebrates? Proc. Natl. Acad. Sci. USA 89, 4764-8.

Köhidai, L., Vakkuri, O., Keresztesi, M., Leppaluoto, J., & Csaba, G. (2003). Induction of melatonin synthesis in Tetrahymena pyriformis by hormonal imprinting-A unicellular" factory" of the indoleamine. Cellular And Molecular Biology-Paris-Wegmann-, 49(4), 521-524.

Kocsis, S., Braverman, B., Ravets, S., Stevens, M. J., Mirin, R. P., Shalm, L. K., & Steinberg, A. M. (2011). Observing the average trajectories of single photons in a two-slit interferometer. Science, 332(6034), 1170-1173.

Kramer, P. (1993) Listening to Prozac Penguin, New York.

Kravitz E.A. (2000) Serotonin and aggression: insights gained from a lobster model system and speculations on the role of amine neurons in a complex behavior J. Comp. Physiol. 186 221-238.

Lampinen, M., Pentikäinen, O., Johnson, M. S., & Keinänen, K. (1998). AMPA receptors and bacterial periplasmic amino acidbinding proteins share the ionic mechanism of ligand recognition. The EMBO journal, 17(16), 4704-4711.

Lampinen M. et al. (1998) AMPA receptors and bacterial periplasmic amino acid-binding proteins share the ionic mechanism of ligand recognition EMBO 17/16 4704—4711.

Levenson J., Sweatt D. (2005) Epigenetic Mechanisms In Memory Formation Nature Reviews Neuroscience 6 108.

Libet, B., Gleason, C. A., Wright, E. W., & Pearl, D. K. (1983). Time of conscious intention to act in relation to onset of cerebral activity (readiness-potential) the unconscious initiation of a freely voluntary act. Brain, 106(3), 623-642

Libet B. (1989) The timing of a subjective experience Behavioral Brain Sciences 12 183-5. (See also Libet, B. et al. (1985) Behav. Brain Sci. 8, 529-566.)

Liebeskind, B. J., Hillis, D. M., & Zakon, H. H. (2011). Evolution of sodium channels predates the origin of nervous systems in animals. Proceedings of the National Academy of Sciences, 108(22), 9154-9159. doi/10.1073/pnas.1106363108

Liljenström Hans, Svedin Uno (2005) Micro-Meso-Macro: Addressing Complex Systems Couplings Imperial College Press.

Locher, C. P., Ruben, P. C., Gut, J., & Rosenthal, P. J. (2003). 5HT1A serotonin receptor agonists inhibit Plasmodium falciparum by blocking a membrane channel. Antimicrobial agents and chemotherapy, 47(12), 3806-3809.

Lundeen J. S., Steinberg A. M. (2009) Experimental Joint Weak Measurement on a Photon Pair as a Probe of Hardys Paradox Physical review letters 102 020404.

Lundeen J.S., Sutherland B., Patel A., Stewart C, Bamber C. (2011) Direct measurement of the quantum wavefunction Nature 474 188-191.

Macrae C., Heatherton T., Kelley W. (2004) A Self Less Ordinary: The Medial Prefrontal Cortex and You.in Cognitive Neurosciences III. Ed Michael S. Gazzaniga. MIT Press. http://dartmouth.edu/~thlab/pubs/04_Macrae_etal_CogNeuroIII.pdf

McAlpine K. (2010) Nature's hot green quantum computers revealed New Scientist 3 February, http://www.newscientist.com/article/mg20527464.000-natures-hot-green-quantum-computers- revealed.html

McCauley, D.W. (1997) Serotonin plays an early role in the metamorphosis of the hydrozoan Phialidium gregarium Dev. Biol. 190 229-240.

McGowan , K., Guerina, V., Wicks, J., & Donowitz, M. (1985). Secretory hormones of Entamoeba histolytica. In Ciba Foundation Symposium 112-Microbial Toxins and Diarrhoeal Disease (pp. 139- 154). John Wiley & Sons, Ltd.

Merali Z. (2010) Back From the Future Discover Magazine August 26.

Mestler, T. (2011). Excitable Signal Relay and Emergent Behavior in the Social Amoeba Dictyostelium discoideum. Princeton University.

Murch, K. W., Weber, S. J., Macklin, C. & Siddiqi, I. (2013) Observing single quantum trajectories of a superconducting quantum bit Nature 502, 211—214.

Molnár, G., Oláh, S., Komlósi, G., Füle, M., Szabadics, J., Varga, C., ... & Tamás, G. (2008). Complex events initiated by individual spikes in the human cerebral cortex. PLoS biology, 6(9), e222.

Nithianantharajah, J., Komiyama, N. H., McKechanie, A., Johnstone, M., Blackwood, D. H., St Clair, D., ... & Grant, S. G. (2012). Synaptic scaffold evolution generated components of vertebrate cognitive complexity. Nature neuroscience. doi:10.1038/nn.3276.

Nomura, T., Tazawa, M., Ohtsuki, M., Sumi-Ichinose, C., Hagino, Y., Ota, A., & Nagatsu, T. (1998). Enzymes related to catecholamine biosynthesis in Tetrahymena pyriformis. Presence of GTP cyclohydrolase I. Comparative

Biochemistry and Physiology Part B: Biochemistry and Molecular Biology, 120(4), 753-760.

Oh, B. H., Kang, C. H., De Bondt, H., Kim, S. H., Nikaido, K., Joshi, A. K., & Ames, G. F. (1994). The bacterial periplasmic histidine-binding protein. structure/function analysis of the ligand-binding site and comparison with related proteins. Journal of Biological Chemistry, 269(6), 4135-4143. Pagel M. (2012) How to measure consciousness New Scientist 1 August http://www.newscientist.com/article/mg21528762.000-how-to-measure-consciousness.html

Paulk, A. C., Gronenberg, W. (2008). Higher order visual input to the mushroom bodies in the bee, Bombus impatiens. Arthropod structure & development, 37(6), 443-458.

Peroutka S. (1995) Serotonin receptor subtypes: Their evolution and clinical significance CNS Drugs 4 suppl 18-28.

Peroutka, S., Howell, T. (1994) The molecular evolution of G-protein-coupled receptors: focus on 5- hydroxytryptamine receptors Neuropharmacology 33 319-324.

Pribram, K. H. (1991). Brain and perception: Holonomy and structure in figural processing. Psychology Press.

Quiroga R., Mukamel R., Isham E., Malach R., Fried I. (2008) Human single-neuron responses at the threshold of conscious recognition PNAS 105/9 3599-3604.

Reuter, F., Del Cul, A., Malikova, I., Naccache, L., Confort-Gouny, S., Cohen, L., ... & Audoin, B. (2009). White matter damage impairs access to consciousness in multiple sclerosis. Neuroimage, 44(2), 590-599.

Ryan T., Grant S. (2009) The origin and evolution of synapses Nature Reviews Neuroscience 10 701- 12.

Samuel E. (2001) Seeing the seeds of cancer New Scientist 24 Mar 42-45. 137

Schnakers C. (2009) Detecting consciousness in a total locked-in syndrome: An active event-related paradigm Neurocase 15/4 271-7.

Schurger A., Sitt J., Dehaene S. (2012) An accumulator model for spontaneous neural activity prior to self-initiated movement PNAS DOI: 10.1073.pnas.1210467109

Schütt, A., Basar, E., (1992). The effects of acetylcholine, dopamine and noradrenaline on the visceral ganglion of Helix Pomatia II: Stimulus evoked field potentials. Compar. Biochem. Physiol. 102C, 169-176.

Schwaerzel, M., Monastirioti, M., Scholz, H., Friggi-Grelin, F., Birman, S., & Heisenberg, M. (2003).

Dopamine and octopamine differentiate between aversive and appetitive olfactory memories in Drosophila. The Journal of neuroscience, 23(33), 10495-10502.

Selcho, M., Pauls, D., Han, K. A., Stocker, R. F., & Thum, A. S. (2009). The role of dopamine in Drosophila larval classical olfactory conditioning. PLoS One, 4(6), e5897.

Sergent C., Baillet S., Dehaene S. (2005) Timing of the brain events underlying access to consciousness during the attentional blink Nature Neuroscience 8/10 1391-1400.

Sheldrake R. (1981) A New Science of Life: the hypothesis of formative causation, Los Angeles, CA: J.P. Tarcher, 1981 ISBN 978-1-84831-042-1.

Sigman M., Dehaene S. (2005) Parsing a cognitive task: A characterization of the minds bottleneck. PLoS Biol 3(2) e37.

Sigman M., Dehaene S. (2006) Dynamics of the central bottleneck: Dual-task and task uncertainty. PLoS Biol 4(7) e220. DOI: 10.1371/journal.pbio.0040220

Skarda C., Freeman W., (1987) How brains make chaos in order to make sense of the world Behavioral and Brain Sciences 10 161-195.

Soon C., Brass M., Heinze H., Haynes J. (2008) Unconscious determinants of free decisions in the human brain. Nature Neuroscience 11(5): 543—545. doi:10.1038/nn.2112.

Sosa, M. A., Spitzer, N., Edwards, D. H., & Baro, D. J. (2004). A crustacean serotonin receptor: cloning and distribution in the thoracic ganglia of crayfish and freshwater prawn. Journal of Comparative Neurology, 473(4), 526-537.

Steck D. (2009) Passage through chaos Nature 461 736-7.

Takeda N., Sugiyama K. (1993) Metabolism of biogenic monoamines in the ciliated protozoan, Tetrahymena pyriformis Comparative biochemistry and physiology 106/1 63-70.

Taniura, H., Sanada, N., Kuramoto, N., & Yoneda, Y. (2006). A metabotropic glutamate receptor family gene in Dictyostelium discoideum. Journal of Biological Chemistry, 281(18), 12336-12343. Trevena J., Miller J. (2010) Brain preparation before a voluntary action: Evidence against unconscious movement initiation Consciousness and Cognition, 19/1, 447-456 DOI: 10.1016/j.concog.2009.08.006

Turano, F. J., Panta, G. R., Allard, M. W., & van Berkum, P. (2001). The putative glutamate receptors from plants are related to two superfamilies of animal neurotransmitter receptors via distinct evolutionary mechanisms. Molecular biology and evolution, 18(7), 1417-1420.

Umbriaco, D., Anctil, M., & Descarries, L. (1990). Serotoninimmunoreactive neurons in the cnidarian Renilla koellikeri. Journal of Comparative Neurology, 291(2), 167-178.

Warren W. (1998) MR Imaging contrast enhancement based on intermolecular zero quantum coherences Science 281 247.

Westerhoff J (2013) The self: You think you live in the present? New Scientist 20 February. Weyrer, S., Rutzler, K., & Rieger, R. (1999). Serotonin in Porifera? Evidence from developing Tedania ignis, the Caribbean fire sponge

(Demospongiae). Memoirs-Queensland Museum, 44, 659-666

Williams C. (2012) Are these the brain cells that give us consciousness? New Scientist 20 Jul. Wilson C. (2013) Consciousness: Why aren't we all zombies? New Scientist 15 May.

Woodruff A ., Yuste R. 2008 Of Mice and Men, and Chandeliers PLOS Biology 6/9 243.

Young, S.N., Leyton, M. (2002) The role of serotonin in human mood and social interaction. Insight from altered tryptophan levels Pharmacol. Biochem. Behav. 71 857-865.

Zimmer C. (2005) The Neurobiology of the Self Scientific American Nov 93.

9. Many Mansions: Special Relativity, Higher-Dimensional Space, Neuroscience, Consciousness and Time

John Smythies

Director, Integrative Neuroscience Program, Center for Brain and Cognition, University of California, San Diego

Abstract:
This paper first reviews what is known about the neural correlates of consciousness (NCCs), both at the nerve network and cellular levels, that lays stress the importance of small differences in the dynamic microanatomy of wiring patterns orchestrated by an epigenetic code. However, information about NCCs themselves does not throw light on the logically different problem of how NCCs are related to the phenomenal experiences they induce. To tackle that problem I suggest it is necessary to reformulate our basic concepts of space, time and matter, as well as replacing the psychoneural Identity Theory (IT), whose defects I outline. In particular neuroscience needs to be based on special relativity (SR) in place of the Newtonian cosmology that forms its present framework. This replaces the concept of neurons as 3D entities, that generate physicochemical events in a separate time, with the concept that neurons are 4D structures, whose world lines are extended in the 4D block universe of SR. This requires, as Broad noted, an ontologically independent conscious observer located in a space of its own (phenomenal space). The hypothesis of material dualism suggests that physical space-time (4D) and phenomenal space (3 spatial dimensions) plus 1 dimension of real time—t2, are cross-sections of a common higher dimensional space that are in relative motion in t2 along the time axis of the block universe. This movement generates the 'now' and the passage of the time that we experience. The contents of phenomenal space are our sensations, images and thoughts all causally related to (but not identical with) particular brain events. This hypothesis has implications for what has been called "the idea of another world".

Keywords: Neural correlates of consciousness, epigenetic code, identity theory, special relativity, time, material dualism, Plato's cave, out-of-the-body experiences, near-death experiences, body-image, Hindu psychology, astral body

Introduction

Whereas our knowledge of the neural correlates of consciousness (NCCs) is steadily increasing under the impetus provided by advanced scanning and other neuroscientific techniques (Fingelkurts et al. 2013), the same cannot be said about what this correlation amounts to. A review of the current literature reveals a sorry state of entangled confusion in the very basic concepts with which this attempt is framed, relating in particular to our ideas about space, time, matter and phenomenal consciousness. The dominant hypothesis today is the Identity Theory (IT) which states that consciousness, however defined, must be identical with certain events in particular regions in the brain: but concrete details of how this is done have singularly failed to materialize. The purpose of this essay is to argue that, to order to clarify and solve this problem, it will be necessary to reformulate our basic concepts of space, time and matter. As a preliminary I will present a short account of what is currently known about NCCs.

Neural Correlates Of Consciousness

This information can be presented at two scales—at the macroscopic level and the cellular level. In the former case the leading hypothesis is the Global Workspace theory (Bartolomei and Naccache 2011). This proposes that conscious processing results from coherent neuronal activity between widely distributed brain regions, with fronto-parietal associative cortices as key elements. The main activity involved is the synchronization of neuronal oscillations, in particular in two synchronized networks—the retrolandic (cognitive network) and the frontal (executive control network) (Leon-Carrion et al. 2012). These authors suggest that the executive control network could facilitate the synchronization and coherence of large populations of distant cortical neurons using high frequency oscillations on a precise temporal scale. They suggest that the synchrony between anterior and retrolandic regions is essential to awareness. Other aspects of the role of integrated networks in NCCs are presented by Achard et al. (2012), Demertzi et al. 2013, Orpwood et al. 2013, Lewis et al. (2012) and Smythies et al. (2014).

At the neuronal level some interesting information has been obtained. Evidence from de- and re-afferentation and sensory pathway rerouting experiments show that the modality of a sensory neuron (i.e. whether it is an auditory or a visual neuron) is determined, not by where it is located, but by where it's afferent inflow comes from (Smythies and Edelstein 2013). For example, in blind patients skilled in Braille, axons from the hand region of the somatosensory cortex grow and take over adjacent, and now inactive, neurons in the visual cortex. If these ex-visual neurons are now activated by transcranial magnetic stimulation, the patient feels a touch on the finger. This change is orchestrated by the epigenetic code, whereby the afferent axon transfers instructional epigenetic material (transcription factor proteins and a variety of RNAs including microRNAs) to the post-synaptic neuron, which is thereby restructured (Smythies and Edelstein 2013). The differences between the different sensory cortices (visual, auditory and somatosensory) consist of subtle differences in the microanatomical wiring patterns (brought about in part by the epigenetic code) between these brain areas (Linden and Schriener 2003). Yet the results of their activation results in the enormously qualitatively different type of sensation experienced.

This raises another question. What is it about a neuron's activity that determines what kind of conscious sensation results—is it the pattern of action potentials? Or are sub-threshold dendritic potentials involved? Or it the total electrical field? There is some experimental evidence that throws some light upon this question. If we stimulate the retina with a flashing (stroboscopic) light at a frequency of between 6-18 Herz with both eyes open, the subject will see a series of simple, regular geometrical flickering patterns with such forms as parallel lines, grids, checkerboards, spirals, concentric circles and mazes (Smythies 1959a,b; 1960). However if we stimulate only one retina, the results is quite different. In this case the same geometrical patterns (called the bright phase) appear but these are soon replaced with another quite different series of pattern (called the dark phase). These are non-flickering oily swirls like oil on water or boiling lava, usually in two colors with green and red predominant. These two types of pattern then alternate in retinal rivalry. There is evidence that the dark phase patterns arise in the cortical neurons belonging to the closed eye (Brown and Gebhart 1948). The stimulation may come from the adjacent active neurons belonging to the open eye via direct current carried by the dense interneuronal network provided by the interlinked dendrites of GABAergic interneurons (Fukada et al. 2006). Thus the digital code carried by the open-eye cortical neurons appears to result in the perception of digital flickering geometrical patterns— whereas the analog code carried by the GABAergic network appears to result in the perception of analog oily swirls. However, none of this evidence is directly relevant to the question of what is the relation between these brain events and the events that take place on the stage of consciousness. To answer this we will have to take a look at our fundamental ideas about space, time and matter—starting with space and time.

Special Relativity

One foundational problem here is that, whereas Special Relatively (SR) is accepted in physics, the disciplines of biology and psychology are mired in Newtonian cosmology. Biology still deals with a Newtonian cosmos in which 3D spatially extended organisms move, evolve and behave in a separate time. In contrast, in the Minkowskian block universe of SR, an organism is a stationary 4D material object extended in a 4D space- time. The Earth, for instance is not a globular 3D object rotating in a spiral pathway around another 3D globular object—the sun: rather it is a stationary 4D hyperhelix that is wound around the stationary 4D sun. In Newtonian cosmology matter exists only at the 'now' of time, whereas in SR cosmology matter exists from the beginning to the end of time. Furthermore in SR cosmology the 'now' and the 'passage' of time are not supplied by the physical universe but are subjective 'illusions'. Newtonian cosmology and SR cosmology involve quite different accounts of Darwinian evolution. In the former organisms are born, grow and die over time. In the SR cosmology the evolution of organisms expresses the fact that the stationary world lines of the atoms that constitute them possess a more complicated structure if we examine them 'later' as compared with 'earlier' locations in the block universe. Darwinian evolution is a dynamic process in Newtonian cosmology, but it is a matter of 4D structure in SR cosmology. The appearance of dynamic changes in the physical world experienced by conscious observers is an illusion generated by the movement of the observing consciousness along the time-like dimension of the 4D block universe from the place labeled 'past' towards the place labeled 'future'. Current biology and neuroscience have not accommodated this fact and it is time to do so.

As I am not a physicist it will be fitting to list a series of statements presenting this case from a number of leading physicists and cosmologists Louis de Broglie was one of the founders of quantum theory. He put it thus: "Each observer, as his time passes, discovers, so to speak, new slices of space-time which appear to him as successive aspects of the material world, though in reality the ensemble of events constituting space-time exist prior to his knowledge of them... the aggregate of past, present and future phenomena are in some sense given a priori." (De Broglie, 1959)."

Russell Stannard, Emeritus Professor of Physics, at the Open University makes the static sculpture of the SR universe plain:

"Physics itself recognizes no special moment called 'now' — the moment that acts as the focus of 'becoming' and divides the 'past' from the 'future'. In four-dimensional space-time nothing changes, there is no flow of time, everything simply is...It is only in consciousness that we come across the particular time known as 'now'... It is only in the context of mental time that it makes sense to say that all of physical space-time is. One might even go so far as to say that it is unfortunate that such dissimilar entities as physical time and mental time should carry the same name!" (Stannard, 1987).

Penrose himself (1994) says that in the universe described by special relativity *"...particles do not even move, being represented by "static" curves drawn in space-time". Thus what we perceive as moving 3D objects are really successive cross sections of immobile 4D objects past which our field of observation is sweeping."*

The list continues:

—Quine (1982): *"A drastic departure from English is required in the matter of time. The view to adopt is the Minkowskian one, which sees time as a fourth dimension on a par with the three dimensions of space."*

—Lloyd (1978): *"For the Quinean, what differences we see between past, present and future pertain to our limited mode of access to reality."*

—Heller (1984): *"...a physical object is not an enduring hunk of matter but an enduring spatio-temporal hunk of matter."*

—Eddington (1920): *"Events do not happen: they are just there, and we come across them ... [as] the observer on his voyage of exploration."*

—Weyl (1922): *"The objective world simply is, it does not happen. Only to the gaze of my consciousness crawling upward along the life-line [world line] of my body does a section of this world come to life as a fleeting image."*

—Werth (1978) makes the important point that this new formulation applies to somatic sensation as well as to vision: *"Our apparent body ['body image' is the neurological name for this] at each instant is simply a 'slice' of our four-dimensional body. That is the experiencing subject sequentially 'intersects' his four-dimensional body and 'projects' the sequence of three-dimensional intersections upon the 'screen'of his consciousness: his body appears to him as being ever changing though in physical reality it is a static and immutable four-dimensional object."*

—Lastly Broad (1953): *"...if we assume one additional spatial dimension beside the three we can observe, and if we suppose that our field of observation at any onemoment is confined to the content of a {3,4}-fold which moves uniformly at right angles to itself along a straight line in the {3,4}-fold, then there is no need to assume any other motion in the universe. This one uniform rectilinear motion of the observer's field of observation, together with the purely geometrical properties of the stationary material threads in the four-fold, will account for all the various observed motions (various in both magnitude and direction) of the material particles which are the appearances of these threads in the successive fields of observation."*

Broad also points out that his assumption requires two 'times'. Time 1 has become fused with space into space-time. But a real time — t2 — is still required in which the 'observer's field of observation' moves through space-time. However, these statements raise a problem. De Broglie speaks of 'each observer', Lloyd of 'our limited mode of access to reality', Eddington of 'the observer on his voyage of exploration', Broad of 'the observer's field of observation'. In these instances the terms 'observer' and 'our' cannot refer, as is usual, to the physical body of the scientist, for this is composed of the world lines of the atoms that make up the physical body and that belongs to the immobile block universe. So what is the "observer", "field of observation", and "gaze of my consciousness" that travels from the past into the future marking the fleeting 'now' of time as it does so? Before we can tackle this question we will need to look at another source of confusion underlying this whole subject. This is the almost universal confusion between physical space and phenomenal space that has its roots in the attempt to believe in two mutually incompatible theories of perception—namely Naïve Realism (NR) and the Representative Theory (RT).

Scientific theories emerge from a background of "common sense" ideas about the world. Early cosmological theories reflected the "obvious" fact that the earth is flat and that the heavens arch above it as a great crystal dome. It took centuries of observation and challenging of dogmas to realize that this model is mistaken.

Likewise, most people today, including many philosophers, believe that NR gives an accurate account of how we perceive the world. That is that the colored objects that fill our visual field really are physical objects, or at least the surfaces of physical objects. In contrast the current scientific theory of (visual) perception holds that these colored objects, that we experience, are the products of a long and complex mechanical process which involves light rays landing on the retina and setting off neuronal reactions that spread to the visual cortex where the percepts that we experience are manufactured. Thus our visual system works like television and not like the simple telescope proposed by NR. The contents of our visual fields that we experience are constructs of the visual nervous system and are not direct views of external objects. However, unfortunately most scientists who may adhere to RT in their laboratories, slip back unwittingly into NR as soon as they get home. A similar state of affairs holds in the other spatially organized senses such as somatic sensation. Neurologists have known for more than a century that the body we experience is not the physical body itself but is an image of the body (the "body image") constructed by representative mechanisms of perception. As the neurologist Paul Schilder (1942, 1950) said "...the empirical method leads immediately to a deep insight that even our own body is beyond our immediate reach, that even our own body justifies Prospero's words "We are such stuff as dreams are made on and our little life is rounded by a sleep."

If we reject NR, and accept RT fully, then we can recognize how the problem of the relationship between phenomenal space and physical space arises. Physical space is that in which physical objects are located and extended. Phenomenal space is that in which our spatially organized sensations are located and extended. Under NR these two spaces can be topologically and geometrically the same. Phenomenal objects and events and physical objects and events can be located in the same location in space where they appear to be to the naïve observer. Whereas, under RT one of either of two other systems is necessary. The first is that phenomenal objects and physical objects can be in the same space but must be in different parts of it. Objects and events in phenomenal space can be located in the physical brain that is located in the physical world (as in the Identity Theory IT)—but they cannot be located in the external physical world where they appear to the naïve observer to be. The second possibility is that phenomenal space and physical space are different spaces as expressed in the cosmological theory known as material dualism, which will be the subject of the next section.

Material Dualism

Material dualism is based on the premise that phenomenal space is a real space, with which we have direct experience, and which possesses topological and geometrical properties in which the events we experience are located (French 1987). Wright (1983) proposes that there is a primary 'phenomenal field' that actually exists in which sensations and after-images are located. Fitzgerald

Consciousness, Neuroscience, Time Travel

(1978) says "None [visual sensations] are located out in physical space: all are in a visual phenomenal space with causal relations with the observer's brain in that the brain's doings produce the sense-data in this space, and indeed the space itself." The neurologist Jason Brown (1991) gets it right I think: "Space itself is an object: volumetric, egocentric, and part of the mind of the observer...Mind is positioned in a space of its own making...We wonder about the limits of the universe but never ask what is beyond the space of a dream."

Phenomenal space may be defined in the following manner. If you obtain a spatially extended visual after-image (e.g. by looking at a square green illuminated surface and then closing your eyes) you can observe that the boundary of that after-image forms a closed Jordan curve that uniquely divides phenomenal space into one inside and one outside. This is an undeniable ontological and topological property of the after-image located in your own phenomenal space (visual field). The image is constructed by causal relations with specific NCCs in your visual cortex.

The fact that the after-image cannot be identical with these NCCs, and so that IT is false, can be shown by the following argument. Leibniz's Law of the Identity of Indiscernibles states that, for any entity A to be identical with an entity B, they must have all properties in common. For example the entities Monte Cervino and the Matterhorn have all their properties in common. 'Monte Cervinio' and 'Matterhorn' are just different names for the same entity. Whereas the after-image described above is square, red and its boundary forms a Jordan curve—whereas their causal NCCs have the shape of a distributed net, the color grey (according to Poirot) and their boundary does not form a Jordan curve. The after-image can be read by the conscious observer to have the same information as the NCCs (i.e. that there is a square red entity out there) but only if that observer already knows that green lights produce red after-images. A similar situation holds for a TV set. The events portrayed have one format inside the set and quite a different format on the TV screen. The events inside the set are not identical with the events on the screen.

It might be argued that this argument fails to recognize the difference between an objective color (physical reflectance and adsorption spectra) and subjective color. In reply I would argue that the word 'color' here is being used to two different senses. The nature of a phenomenal color is simply experienced as a raw fact not amenable to further analysis, whereas the nature of an 'objective' color is discovered by scientific measurements. The 'inner nature' of an objective color—like the 'inner nature' of matter— is not something that we can discover.

With these considerations in mind, it is possible to ask what are the topological and geometrical relations between physical space (ignoring time for the moment) and phenomenal space. There are four, and only four, possible answers.

—PheS is topologically inside all of phyS within the range of vision (as expressed in NR).

—PheS is topologically inside only that portion of phyS that is inside the brain (as expressed in RT and IT).

—PheS and phyS are two different cross-sections of a common higher dimensional space (as suggested by C.D. Broad (1923), Bernard Carr (2008) and myself (Smythies 1994).

—PheS and phyS are wholly different spaces and bear no spatial relations with each other: only causal relations link their contents (physical and phenomenal events respectively (as suggested by Price 1953).

Since we know that NR is false, and the IT fails the Leibniz test, our choice is limited to the other two hypotheses. Of these the fourth can be treated as a minor variant of the third, so I will next discuss the Broad-Carr-Smythies hypothesis in some detail.

Dualistic concepts of mind reach far back into human history, but, in most, the autonomous mind was thought of in terms of non-material spirits in the manner later crystalized by Descartes. However, ideas that the mind might incorporate material elements were developed by the Hindu philosophers of the classical era. They suggested that the mind was material like the body, but of a form of matter so diaphanous as to be undetectable by ordinary instruments.

Joseph Priestly (1777) was the first in the West to take up this topic:

"But how anything could have extension, and yet be immaterial, without coinciding with our idea of mere empty space, I know not. I am therefore bound to conclude, that the sentient principle in man, containing ideas which certainly have parts [is] not the simple, indivisible, and immaterial substance that some have imagined it to be; but something that has real extension and therefore may have the other properties of matter."

The Cambridge philosopher C.D. Broad took the next, and very significant, step in 1923 when he wrote:

"For reasons already stated, it is impossible that sensa should literally occupy places in scientific space, though it may not, of course, be impossible to construct a space-like whole of more than three dimensions, in which sensa of all kinds, and scientific objects literally have places. If so, I suppose, that scientific space would be one kind of section of such a quasi-space, and e.g. a visual field would be another kind of section of the same quasi-space (pp. 392-393)."

Further details of this new theory were supplied by Smythies (1956) who provided links with both neurology and introspectionist psychology. The concept that phenomenal space and physical space are ontologically different spaces has also been expressed briefly by others: Ayer (1940) "...John Stuart Mill and Berkeley fail to distinguish properly between physical space and sensible space." — Russell (1948) *"...the space in which the physical table is located must be different from the space we know by experience"* —and Moore (1971) *""...it seems to me just possible that the two sensations in question [those belonging to two coins] though not circular in my private space may yet be circular in physical space"*. Unfortunately this work was almost ignored in the 1950s and 1960s owing to the rise of linguistic philosophy. Bernard Carr (2008) was the first physicist to enter this field when he published his theory that phenomenal space and physical space are both cross-sections of a higher-dimensional space. He writes: *"My proposal is that mental and physical space can be integrated into a communal space which is higher dimensional, in the sense that it has more than the three dimensions perceived by our physical sensors. This involves what I call a "Universal Structure."* (see Smythies 1994 pp. 149-150 for details).

This hypothesis can be illustrated by the following introspective experiments.

Sit down in a comfortable chair in a dark room, close your eyes and observe what you experience. First direct your attention to the dark visual field in front of you. Ask yourself what this darkness consists of. Is it just nothing? However, that is not right as evidenced by the difference between retinal blindness and cortical blindness. In the former patients experience a black field whereas in the latter they experience nothing at all.

When you open your eyes your visual field becomes filled with colored mobile shapes i.e. your visual sensations. When you close your eyes again these vanish to be replaced by the black uniform field. So what lies behind this field? IT would say that other neurons lie behind it. This hypothesis entails that the black field itself consists of activity in NCCs and what lies behind them must be activity in pre-NCCs.

A similar situation holds in the case of somatic sensations. Sitting in your chair examine your bodily sensations. These will consist of a variety of feelings of touch, pressure, stretch, itch, proprioception, tingling and others that make up your body image. IT holds that these are composed of NCCs and what lies immediately outside them is activity in pre-NCCs—because your body image that you experience is identical, according to IT, not with your actual physical body but with Penfield's homunculus located in your somatosensory cortex. In contrast, the theory of material dualism holds that all the contents of your phenomenal consciousness, including your phenomenal visual field and your body-image, lie outside the brain altogether and are located in a space of their own.

We can illustrate this by considering how IT and this theory deal with the image of Plato's cave (Smythies 1994). In this model, prisoners have always been strapped to posts and cannot move their heads. Behind them a great fire burns. Statues of objects are carried behind them so that their shadows fall on the wall of the cave in front of the prisoners. Would not the prisoners, Plato asks, then consider that these shadows represent reality? In IT Plato's cave is the brain and the shadows represent NCCs in the visual cortex. In material dualism the cave is an extra part of the human organism (the 'consciousness module') located outside the brain in a (higher-dimensional) space of its own that has an inner screen on which the shadows (our actual visual sensations) are projected. For a modern version of Plato's cave we can use television. A mad scientist takes an infant and straps a TV set over its eyes with an external camera so that the developing child will see on its TV screen whatever the camera is pointing at. Would not this child grow up believing that the TV images, that are all she ever sees, were the objects televised themselves that she can feel by touch (Wright 1983).

To account for the 'now' and the passage of time in this new theory we must add a detail. As Broad (1953) noted in the earlier quotation, the theory needs two times. Time in the sense of past, present and future becomes the fourth dimension of space, as in SR. The 'now' and the passage of time requires a second real time $t2$ in which the time- traveling 'field of observation' described above travels from the past into the future. Geometrically this can be expressed in the following way. Any n-dimensional structure (say a cube) can be cut by two (n-1) D structures (in this case planes). The resulting planes can either be parallel or they can intersect. The 4D cross-section A of Carr's 5D Universal Structure that contains phenomenal space-time and the cross-section B that is physical space-time are different cross-sections of the Universal Structure. If these cross- sections are not parallel but intersect then they (and their contents) can be in relative motion to each other along the time axis of B in $t2$. This entails that the 'now' of time is wherever A (that carries phenomenal consciousness) has reached in its travel along B. The existence of this second time might find confirmation in some Near Death Experience reports which mention that during the experience the notions of time or of duration may disappear leading to notions of "no time" or "eternal present" along with the feeling of a 'second sort of time" (Jourdan 2011).

The proposed causal interactions between events in NCCs and events in phenomenal consciousness can be represented by higher-dimensional vectors. Causal interactions between events wholly inside A, or wholly inside B, can be represented by vectors (straight lines) that are geometrically wholly within A or wholly within B. Interactions between events in A and events in B can be represented by vectors that originate in one of these and terminate in the other, crossing the dimensional interface between these two spaces as they do so. A dimensional interface is any surface that an n- dimensional space presents to an n+1 dimensional space of which it is a cross section— for example the surface of a plane that is a cross-section of a cube.

Material dualism also has implications for "the idea of another world" (Price 1953). If phenomenal events are ontologically separate from physical events and exist in a space (or space-time) of their own, they could continue to exist in some form after the wreck of the brain. This hypothesis is congruent with the proposal that the physical world exists as a means of communication between Selves that lie outside it in a phenomenal world of their own. This supports the notion that the events reported by observers during out-of-the-body experiences (OBEs) and near-death experiences (NDEs) suggest what these post-mortem experiences may be like (Jourdan 2010). However, the theory suggests one basic change. Since it holds that the Self and its phenomenal consciousness were never geometrically in the physical body, the term out-of-the-body needs changing. During our present life the Self is located in the body-image, not the body, under any valid theory. To return to Plato's cave: the picture drawn by Plato refers to events during our present life. After the death of the brain the observers may be freed from their chains and can leave the cave (or sensorium) and are able to explore the beautiful countryside around it. Hindu psychology has long held that, in addition to our physical body, we also have an astral body that could survive the death of the physical body. Material dualism supports this idea but with an important difference. The Hindu theory is based on the idea that we experience the physical body: and the astral body only hovers in the wings as it were. However, the truth may be is that which we always experience—i.e. the body image— metamorphoses under different circumstances into the astral body during OBEs, NDEs and after death itself.

Conclusion

The theory of material dualism represents a paradigmic change in our basic concepts of space, time and matter.

—The real world, according to Newton, consists of 3D physical objects that exist in a separate time. Matter only exists at the 'now' of time.

—According to Special Relativity the real world consists a 4D block of matter in which t1 time and physical space are fused into space-time. Matter exists throughout space-time. The 'now' of time and the passage of time are held to be 'subjective illusions'.

—In material dualism the real world consists of a 5D spatiotemporal system (involving t1) of which the Einsteinian physical world and the phenomenal world are located in two different cross-sections. These both exist in real time t2. Thus the world contains two different kinds of matter i.e. physical matter (atoms and fields: brains and planets) and phenomenal matter (sensations, images and thoughts). The two interact via causal relations between events in the NCCs in the brain and events in phenomenal consciousness.

REFERENCES

Achard S, Delon-Martin C, Vértes PE, Renard F, Schenck M, Schneider F, Heinrich C, Kremer S, Bullmore ET. (2012) Hubs of brain functional networks are radically reorganized in comatose patients. Proc Natl Acad Sci U S A. 2012 Dec 11;109(50):20608-13. doi: 10.1073/pnas.1208933109. Epub 2012 Nov 26.

Ayer, A.J. (1940) comments I P. Laslett (ed.) The Physical Basis of mind.. Blackwell., Oxford, UK.

Bartolomei F1, Naccache L. (2011) The global workspace (GW) theory of consciousness and epilepsy. Behav Neurol. 2011;24(1):67-74. doi: 10.3233/

BEN-2011-0313.

Broad, C.D. (1923), Scientific Thought (London: Routledge & Kegan Paul).

Broad, C.D. (1953) Religion, Philosophy and Psychical Research .(London: Routledge & Kegan Paul).

Brown and Gebhardt JW. (1948) Visual field articulation in the absence of spatial stimulus gradients. J. Exp. Psychol. 38, 188-200.

Brown, J. W. The Life of the Mind. London, Erlbaum, 1988.

Carr, B. (2008) Worlds apart? Proceedings of the Society for Psychical Research, 59, 1- 96.

de Broglie, L. (1959), 'A general survey of the scientific work of Albert Einstein', in Albert EinsteinPhilosopher-Scientist, ed. P.A. Schlipp (New York: Harper & Row), pp. 107–28.

Demertzi A, Soddu A, Laureys S. (2013) Consciousness supporting networks. Curr Opin Neurobiol. 2013 Apr;23(2):239-44. doi: 10.1016/j.conb.2012.12.003. Epub 2012 Dec 27. Review.

Eddington, A.S. (1920), Space, Time and Gravitation (Cambridge: Cambridge University Press).

Fingelkurts, A.A., Fingelkurts, A.A., Bagnato, S., Boccagni, C., and Galardi, G. (2013) Dissociation of vegetative and minimally conscious patients based on brain operational architectonics: factor of etiology. Clinical Electroencephalography and Neuroscience, 2013 May 10. [Epub ahead of print].

Fitzgerald, P. (1978) Review of C.W.K. Mundle Perception, Facts and Theories, in Philosophy of Science, 45, 165-169.

French R.E. (1987) The Geometry of Vision and the Mind Body Problem. New York: Lang.

Fukada H, Kosaka T, Singer W, Gakuske RA. (2006) Gap junctions among dendrites of cortical GABAergic neurons establish a dense and widespread intercolumnar network. J.Neurosci. 26, 8589-8604. doi: 10.1523/JNEUROSCI.4076-05.2006.

Heller, M. (1984), 'Temporal parts of four dimensional objects', Philosophical Studies, 46, pp. 323–34.

Jourdan J-P. (2010) Deadline. Paris: Les Tres Orangers.

Jourdan J-P. (2011) Near Death Experiences and the 5th dimensional spatio-temporal perspective. J. Cosmol. 14, 4743-4762. http://journalofcosmology.com/Consciousness152.html

Leibniz G. W. (1981) New Essays on Human Understanding. (P. Remnant and J. Bennett Tr.). Cambridge University Press. 1981.

Leon-Carrion J, Leon-Dominguez U, Pollonini L, Wu MH, Frye RE, Dominguez- Morales MR, Zouridakis G. (2012) Synchronization between the anterior and posterior cortex determines consciousness level in patients with traumatic brain injury (TBI). Brain Res. 2012 Oct 2;1476:22-30. doi: 10.1016/j.brainres.2012.03.055. Epub 2012 Mar 29.

Lewis LD1, Weiner VS, Mukamel EA, Donoghue JA, Eskandar EN, Madsen JR, Anderson WS, Hochberg LR, Cash SS, Brown EN, Purdon PL. (2012) Rapid fragmentation of neuronal networks at the onset of propofol-induced unconsciousness. Proc Natl Acad Sci U S A. 2012 Dec 4;109(49):E3377-86. doi: 10.1073/pnas.1210907109. Epub 2012 Nov 5.

Lloyd, G. (1978), 'Time and existence', Philosophy, 53, pp. 215–28.

Moore, G.E. (1971) Philosophical Studies. Harcourt and Brace, New York, US.

Orpwood R. (2013) Qualia could arise from information processing in local cortical networks. Front Psychol. 2013 Mar 14;4:121. doi: 10.3389/fpsyg.2013.00121. eCollection 2013.

Penrose, R. (1994), Shadows of the Mind: A Search for the Missing Science of Consciousness (Oxford: Oxford University Press).

Price, H.H. (1953) Survival and the idea of another world. Proc.Soc. Psychical Res. 50, 1-25.

Priestly, J. (1777) Disquisitions relating to Matter and Spirit. Johnson, London, UK.

Russell, B. (1948) Human Knowledge. Its Scope and Limits. Allen and Unwin, London, UK (pp. 45 & 582-593).

Quine, W.V. (1982), Methods of Logic (4th edn.) (Cambridge, MA: Harvard University Press).

Schilder, P. Mind. Perception and Thought in their Constructive Aspects. New York. Columbia University Press. 1942.

Schilder, P. The Image and Appearance of the Human Body. New York. International Universities Press, 1950.

Smythies J. (1959a) The stroboscopic patterns. Part I. The dark phase. Brit.J.Psychol. 50, 106-116.

Smythies J. (1939b) The stroboscopic patterns. Part II. The phenomenology of the bright phase and after-images. Brit.J.Psychol. 50, 305-324. DOI: 1e.1111/j.2044- 8295.1959.tb00710.x

Smythies J. (1960) The stroboscopic patterns: part III. Further experiments and discussion, Brit.J.Psychol, 247-255. DOI: 10.1111/j.2044-8295.1960.tb00747.x

Smythies, J. The Walls of Plato's Cave. Aldershot, Averbury Press, 1994.

Smythies J, Edelstein L. (2013) Transsynaptic modality codes in the brain: possible involvement of synchronized spike timing, microRNAs, exosomes and epigenetic processes. Frontiers in Intregrative Neuroscience. 2012;6:126. doi: 10.3389/fnint.2012.00126. Epub 2013 Jan 4.

Stannard, R. (1987), 'Making sense of God's time', The Times, August 22nd.

Werth, L.F. (1978), 'Normalizing the paranormal', American Philosophical Quarterly, 15, pp. 47–56.

Weyl, H. (1922), Space-Time-Matter (London: Constable).

Wright (1983) Inspecting images. Philosophy. 58. 57-72.

10. Brain, Consciousness, and Causality
Andrea Nani[1], Andrea E. Cavanna[2,3]

[1]School of Psychology, University of Turin, Italy.
[2]Dept. of Neuropsychiatry, BSMHFT, University of Birmingham, UK.
[3]Dept. of Neuropsychiatry, Institute of Neurology, University College London

Abstract

Consciousness seems to be a fundamental ingredient of human life: our common sense tells us that without it we would not behave in the same way. However since the end of the XIX century among some philosophers and scientists have become increasingly familiar with a counterintuitive position on the place of consciousness in nature, known as epiphenomenalism. Epiphenomenalism excludes from scientific accounts of human behavior any appeal to conscious processes occurring in the brain. Its main claim is that conscious experience is an epiphenomenon of brain activity, without causal powers in terms of volition and action. This paper examines the issue whether consciousness can be regarded as a mere epiphenomenon from both the theoretical and empirical perspective. The epiphenomenalist theory is analyzed with reference to the work of leading neuroscientist Gerald Edelman and neurological syndromes defined by key alterations in conscious domains. It is argued that conscious states are likely to play essential causal roles in the scientific account of how the brain brings about the voluntary actions that contribute to form our deepest personal identities.

Key Words: Causality, Consciousness, Edelman, Epiphenomenalism, Neurology

1. Introduction. The Temptation of Epiphenomenalism

One of the most enduring and intriguing questions for both philosophical and scientific researchers is whether we have conscious minds capable to control and produce the motivations for all our actions. In the light of our common experience, an affirmative answer to that question would be but a platitude. Indeed, brains seem to be capable to create a great variety of mental events. Love, hate, sadness, joy, sorrow, pleasure, shame, grief, delight, and resentment are only a few of the many different psychological states composing our rich mental lives. However, on more accurate reflection the solution to the problem of the nature of consciousness would not appear as evident as it may seem at first sight.

There are, in fact, philosophers and neuroscientists who firmly believe that mental properties, particularly the conscious ones, are wholly epiphenomenal with respect to brain processes (Edelman 2004, 2007; Fuster 2003). According to epiphenomenalism, mental properties are superfluous by-products of the function of our cerebral mechanism, just like the images which are reflected by mirrors are not made by glass, and the shadows which objects cast on the ground are not parts of those objects. This view is not new: in a famous conference held at the British Association for the Advancement of Science, Thomas Henry Huxley compared consciousness to the steam whistle of a locomotive (Huxley 1884). In contrast with Descartes, Huxley did not consider animals as unconscious machines, but was very perplexed with regard to the exact function of consciousness and hypothesized that conscious states played no role in behavioral mechanisms (Huxley 1874). Just as the steam whistle of a locomotive did not influence the work of the locomotive's motor, he thought, so animal consciousness could neither cause nor modify animal behavior. Being also a strenuous advocate of Darwin's theory of natural selection, Huxley assumed for reasons of biological continuity that there are no differences between animal and human consciousness (Huxley 1884).

We can affirm that the modern shape of the problem of epiphenomenalism was set with Huxley, even though he never used this word in his writings. In effect, the Modern Age (post-Cartesian) thinkers did not tend to contrast sharply the concepts of mind and body. Ancient Greeks had a much broader idea of mind, closely linked with bodily functions (Bremmer 1983). Mind (i.e. the soul) was considered the principle of life capable to animate the body in order for it to perform its basic biological processes, such as breathing, digestion, procreation, growth, motion, and, for humans, also other sophisticated processes of life, such as thinking, perceiving, imagining, and reasoning. Of course every school of ancient Greek philosophy had its own concept of "soul". For instance, Aristotle thought of the soul as the body's system of active abilities to accomplish the vital functions that organisms naturally perform, e.g. nutrition, movement or thought (Nussbaum and Rorty 1992). On the other hand, Stoic philosophy of mind conceived the soul itself as a corporeal entity (Inwood 2003). This position was similar to that of Epicurus, who taught the soul to be a kind of body composed of atomic particles (Kerferd 1971). Perhaps Plato and the Pythagoreans held the closest concept of soul to the Cartesian view. They maintained it to be as something incorporeal and able to exist independently of the body (Lorenz 2008; Huffman 2009). Nevertheless, it is only after Descartes' philosophy that the debate of mental causation was to be set down in its modern form. An important echo of this debate is to be found in the discussion upon the automatism of behavior raised by Huxley in the second half of the nineteenth century.

Positions similar to those embraced by Huxley are still held by some contemporary philosophers and neuroscientists. Epiphenomenalists would be willing to explain the origin of consciousness in the same way as we can physically explain how mirrors produce reflected images or bodies cast their shades on the ground. According to this explanatory model, all our psychological states should, in theory, be accounted for entirely in terms of scientific vocabularies which contain no mental concepts.

A similar approach, at least in relation to its practical consequences for psychological research, was maintained by Burrhus Skinner with his theory of "radical behaviorism". Skinner supported the view that mental terms could be completely paraphrased in behavioral terms, or eliminated from explanatory discourse altogether (Skinner 1974). Therefore, all accounts of human behavior would have to be given in neutral and objective terms, such as stimulus, response, conditioning, reinforcement, and so on. This position has some analogies with epiphenomenalism in that it considers consciousness a nonphysical entity which has nothing to do with behavior.

Such epiphenomenalist lines of reasoning imply that a rigorous discourse on human actions should deny the reality of consciousness, and thereby of all mental states correlated with this phenomenon. In fact, when human beings express propositions about conscious states, they actually intend to speak of other things, specifically of certain brain physical states which are to be the unique causes of all their bodily dispositions. In William James' words, "consciousness ... would appear to be related to the mechanism of [the] body simply as a collateral product of its working, and to be completely without any power of modifying that working" (James 1890). In a sense, arguing that conscious mental phenomena are causally real would be like arguing that the black spots which are on the tails of some fishes are real eyes capable of seeing and not just evolutionary tricks for misleading predators.

The solution offered by epiphenomenalism to the problem of the nature of consciousness can be roughly summarized in the assertion that such a problem does not exist because we have no consciousness, but only the illusion of having it. Although epiphenomenalism could be somehow attractive, it does not present a satisfactory solution to the problem of what consciousness really is. The present paper will show that the epiphenomenalist perspective does not offer a consistent account of certain neuropsychological phenomena which seem to be intrinsically subjective.

We will start by examining the arguments that drive some philosophers and neuroscientists to regard consciousness as an empty concept involving outdated thinking.

2. Causal Links

Epiphenomenalism upholds that the conscious mind is not part of the physical world. This implies that, given the physical causal closure of the universe, conscious mental events cannot interact with the physical reality in any way. If we think of ourselves as consciously acting agents, the epiphenomenalist claim sounds counterintuitive: is it plausible that our conscious will cannot influence the physical world?

In order to answer this question we first need a better understanding of exactly what it means to claim that conscious mental events cannot influence physical events in any way. This doctrine implies that the only possible type of causation we have to deal with is the so-called bottom-up causality, i.e. the causation that goes from the physical level to the conscious one. According to this concept of causality, the irresistible desire for an apple pie cannot actually cause the act of eating a slice of pie, since the account of our behavior is to be determined only at the physical level, where physical entities move other physical entities. However, if all causation processes belong to the physical level, how could our ontological catalogue list phenomena which are not included in causal accounts of behavioral expressions given in physical terms only? A reasonable philosophical precept (epitomized by Occam's razor) warns us that ontology should not be expanded without necessity. Moreover, there is the problem of defining the nature of those phenomena. If these entities were completely different from physical processes, then conscious mental events would necessarily belong to a distinct ontological domain, but it is easy to conclude that in such a case epiphenomenalism would be just like a spurious kind of dualism. On the other hand, if conscious mental events were physical phenomena of a very special nature, how could we distinguish the physical events capable to cause other physical events from the ones which have no causal power at all?

We cannot actually make a distinction of this kind by means of the third person vocabulary that scientists generally use to depict their objective vision of the world. Therefore, those who trust epiphenomenalism would have to include in their ontological catalogue states, events, and processes susceptible to be described exclusively in terms of the first person perspective, and to identify these states, events, and processes with non-causal physical phenomena. In addition, this sort of phenomena would have to be put together with all the other states, events, and processes liable both to be described in terms of the third person perspective and to be identified with causal physical phenomena. However, it is by no means clear why some processes – whose nature is basically physical – would have to be described exclusively in terms of the first person perspective rather than third person perspective.

In addition to these conceptual difficulties, a further grave quandary is whether we maintain the division between causal and non-causal physical events. In other words, taken for granted such a division, what in this case would the principle of the physical causal closure precisely mean? The principle of the physical causal closure states that if a physical event has a cause that occurs at t, it has a physical cause that occurs at t. Jaegwon Kim (2005) correctly observes that the "physical causal closure does not by itself exclude nonphysical causes, or causal explanations, of physical events." For instance, there could be a nonphysical causal explanation of a physical event being the first ring of a chain of other numerous physical events. According to Kim (2005), in order to rule out this kind of explanation, we need an exclusion principle such as the following: if an event e has a sufficient cause c at t, no event at t distinct from c can be a cause of e.

Following the principle of exclusion, the sufficient cause c of e at time t may be either physical or mental. However, this instance is ruled out if the principle of exclusion is linked to the principle of the physical causal closure and the event e is identified with an event whose nature is purely physical. As a result, the two principles joined together hold that only physical events can cause other physical events.

It is important to highlight that these two principles are not stricto sensu in contrast with the folk psychology view that there is a conscious mind in every human being. Moreover, these assumptions are consistent with the hypothesis of psycho-physical parallelism, according to which there could be a distinct domain of specific conscious mental events coming to occur whenever other particular physical events come to occur. Still it is unconditionally denied that conscious mental phenomena can causally interact with physical phenomena. In fact, in order to be closed, any causal explanation is to be expressed as a chain of purely physical events. This leads us back to our previous question: if we accept that both epiphenomenal events and non-causal physical entities can by no means be part of scientific accounts given in causal physical terms, then why should we expand our ontology without necessity by including allegedly redundant conscious phenomena?

3. An Argument from Quantum Physics?

An interesting argument against the plausibility of epiphenomenalism can be derived from a specific interpretation of the theory of quantum mechanics. It is well-known that the act of measurement is a crucial aspect from the perspective of quantum theory. The implications and the account of this process has been the subject of controversy for more than seven decades and the debate does not seem to be closed yet.

The fundamental point of the argument from quantum physics holds that the observer's consciousness plays a key role in the collapse of the wave function to a certain state, described by a second-order differential equation by Erwin Schrödinger. The root of this idea can be found in the so-called Copenhagen interpretation of quantum mechanics. Although the Copenhagen interpretation is not a homogenous view (Howard 2004), Heisenberg (1955) appears to be the one who first coined the term and developed the underlying philosophy as a unitary interpretation (Heisenberg 1958). On the other hand, Niels Bohr – the Danish physicist commonly regarded as the father of the Copenhagen interpretation – never seems to have emphasized or privileged the role of the observer in the wave packet collapse (Howard 1994). Bohr argued for an interpretation of complementarity with regard to the wave-particle duality which is incompatible with Heisenberg's interpretation of wave function collapse (Gomatam 2007). Bohr's view regarding the wave function was more moderate than Heisenberg's and based on epistemological concerns, rather than ontological commitments. When Bohr referred to the subjective character of quantum phenomena he was not referring to the conscious intervention of the observer in the process of measurement, but to the context-dependent status of all physical observations (Murdoch 1987; Faye 1991). However the drastic theoretical move – outlined in Heisenberg's writings – that quantum measurement is to be understood by involving the observer's act in addition to a physical process, has become the core of the so-called Copenhagen interpretation.

This idea was further developed by other physicists. Von Neumann (1932) postulated an ad hoc intervention of an observing system in order to account for the collapse or reduction of the wave function through measurement. Quite cautiously, he never claimed that the observing system had to be conscious (von Neumann 1932). In contrast to von Neumann's view, London and Bauer (1932) attributed to the oberver's consciousness only the key role in understanding the process of quantum measurement. Such a proposal was later expanded on by the physicist Eugene Wigner, according to whom "it was not possible to formulate the laws (of quantum theory) in a fully consistent way without reference to consciousness" (Wigner 1967).

The proposal to consider consciousness as causally involved in physical state reductions was further developed following Heisenberg, von Neumann, and Wigner (Stapp 1993, 1999, 2006; Schwartz et al. 2005). Undoubtedly, this approach challenges the epiphenomenal position. In fact, how is it possible for consciousness to be non-causal if it can bring about the collapse of the wave packet? If those who champion the Copenhagen interpretation are right, then epiphenomenalism should be completely refuted. On the one hand, this approach gives a fundamental causal power to consciousness in understanding the universe; on the other hand, it has the serious shortcoming of putting consciousness outside the physical world. In fact, if consciousness really causes the collapse of the wave function, then it must be a process that is not be describable by Schrödinger equation, because otherwise it would be caught in an infinite regress. Based on these arguments, consciousness should be a nonphysical entity. Therefore, if both the principle of the physical causal closure and the principle of exclusion discussed in the previous section are true, the so-called Copenhagen interpretation of quantum mechanics does not provide a strong argument for refuting epiphenomenalism.

4. Consciousness and Causality

In view of the foregoing reflections, we would have to be very reluctant to claim the distinction between causal and non-causal physical events or between causally efficacious physical and causally efficacious nonphysical events, although these distinction are of course logically possible. Accordingly, both philosophers and neuroscientists well-disposed to epiphenomenalism have to maintain that the nature of conscious mental processes cannot be physical, since, by definition, a state/event/process exists in physical terms only if it has the property of influencing and exerting causal effects over other physical entities.

An interesting type of neuroscientific approach is exemplified by the epiphenomenalist position held by Gerald Edelman, a leading neuroscientist who has given remarkable contributions to the study of mind and consciousness and their place in nature (Edelman 1989, 1993, 2007). According to Edelman's approach ("Neural Darwinism"), different configurations or patterns of neurons compete with each other to gain constancy and stability within the brain. The Neural Darwinism approach holds that groups of neurons and the neural patterns and configurations which nerve cells form ("neural networks") are subject to natural selection, just like biological species are evolutionarily selected by the environment. Specifically, Edelman's theory of "neuronal group selection" postulates that anatomical connectivity in the brain occurs via selective mechanochemical events that take place epigenetically during development. This process creates a structurally diverse primary repertoire by differential reproduction. A second selective process occurs during postnatal behavioral experience through epigenetic modifications in the strength of synaptic connections between neuronal groups, thus creating a diverse secondary repertoire by differential amplification.

In Edelman's view, human consciousness depends on and arises from the uniquely complex physiology of the human brain. He advanced a theory of how the brain generates different levels of consciousness through multiple parallel re-entrant connections between individual cells and between larger neuronal groups, in which he endorsed an epiphenomenalist position with regard to consciousness, which is central to the scope of this paper (Edelman 2004, 2007).

It has been argued that the question whether consciousness can have a causal role in determining behavior and other mental states should find an answer supported by both conceptual and empirical considerations. (Flanagan 1992; Heil and Mele 1993; Searle 2004). Contrary to this view, Edelman seems to give pre-eminence to theoretical arguments over empirical results. In fact his thesis – that consciousness is not causal – is almost exclusively based upon the following theoretical argument:

This account [which is that conscious processes arise from enormous numbers of re-entrant interactions between different areas of the brain] implies that the fundamental neural activity of the reentrant dynamic core converts the signals from the world and the brain into a "phenomenal transform" – into what it is like to be that conscious animal, to have its qualia. The existence of such a transform (our experience of qualia) reflects the ability to make high-order distinctions or discriminations that would not be possible without the neural activity of the core.

Our thesis has been that the phenomenal transform, the set of discriminations, is entailed by that neural activity. It is not caused by that activity but it is, rather, a simultaneous property of that activity. (Edelman, 2004).

Edelman's idea is that some cerebral processes entail certain phenomenal transforms, which are the contents of our conscious mental states. Following him, we can call the cerebral processes C' and the phenomenal transforms C. We can put both C' and C in a row and index them to indicate their successive states in time: C'0–C0; C'1–C1; C'2–C2; C'3–C3; and so forth. It is crucially important to highlight how in that view only the underlying cerebral processes are endowed with causal powers, whereas the phenomenal processes entailed by those brain states are not. However the relationship between the cerebral processes and the phenomenal transforms is considered by Edelman as necessary. This necessary correlation appears to be of a metaphysical kind, i.e. a correlation that holds in every possible world. Therefore, Edelman's position does not appear to be consistent with the philosophical "zombie argument", which assumes the existence of an individual capable to behave just in the same way as conscious human beings do, but in the absence of any subjective conscious experience (in Edelman's words, an individual having C' but not C). In fact, Edelman claims that "The argument we are making here implies, however, that if C' did not entail C, it could not have identical effects" (Edelman 2004; the emphasis is not ours). Consequently, an individual lacking C (phenomenal consciousness) cannot show the same behavior of an individual who has C. Indeed, specific activities of the nervous system necessarily give rise to particular conscious sensations, which in turn cannot exist without a specific underlying activity of the brain. As a result, the zombie hypothesis should be utterly inconsistent (Jackson 1982).

Edelman's theory is central to the discourse on epiphenomenalism advanced so far. In fact, the phenomenal properties which he refers to are necessarily implied by the underlying neural activity of the brain. Strictly speaking, those phenomenal properties are not redundant but absolutely non-causal, even though they have a sort of physical nature. In addition, those properties have to be seen as by-products, since they cannot play any specific role in our scientific account of natural phenomena. Therefore, neither the principle of physical causal closure nor the principle of physical exclusion seem to be violated by Edelman's perspective.

In our view, the arguments put forward by Edelman in order to demonstrate the epiphenomenalist nature of consciousness raise a number of issues. What is most unclear is the very nature of the necessary relationship between the causal physical events (i.e., the cerebral processes or C') and the non-causal physical events (i.e., the mental processes or C). If we accept that this relationship is necessary, there is no reason to assume an ontology in which conscious mental processes and physical processes within the nervous system are distinct. If a certain property is necessarily implied by certain physical processes (in such a way that the latter could not bring about the same effect without the former, as Edelman claims), then either that very property and those physical processes are different aspects of the same entity, or that very property is part of the co-occurring physical processes. From the logical point of view, an effect cannot find its cause in one event which is the result of the sum of a causal physical state and a non-causal physical state, since the non-causal physical process cannot play any causal role at all (Heil and Mele 1993). In fact, what Edelman believes to be non-causal, "the phenomenal transform", must be provided with causal powers.

In this sense, Edelman's theory with regard to the epiphenomenalist nature of consciousness appears to be anomalous. A "true" epiphenomenalist would, in fact, plausibly think of the causal relationship between physical and conscious states as contingent rather than necessary. Thus, epiphenomenalists should be willing to accept the zombie argument, since it is logically possible to accept that, if conscious mental processes are contingent, there could be possible worlds in which they do not bring about any behavioral effect.

In addition to these theoretical considerations, empirical data can raise other reservations with regard to epiphenomenalism. For example, if we agree on depriving mental entities of their causal role, we would encounter difficulties accounting for a host of well described neurological conditions. These include, but are not limited to, blindsight (cortical blindness with preserved ability to locate objects), unilateral neglect syndrome (loss of ability to detect information coming from the left side of the body), allochiria (experience of a sensory stimulation at the contralateral side to the applied stimulus), anosognosia (denial of gross neurological deficit), prosopognosia (inability to recognize familiar faces), and somatoparaphrenia (a condition in which patients deny ownership of a limb or an entire side of their body). Arguably, these neurological disorders can be explained at least to some extent in terms of a dysfunction in the causal role played by consciousness in dealing with perceptive or proprioceptive information.

The understanding of somatoparaphrenia is an exemplar case, based on the concept of verbal manifestations commonly referred to as propositional attitudes in the tradition of analytic philosophy (Bisiach and Geminiani 1991). Propositional attitudes are all the expressions whose contents consist of subjective beliefs, desires, intentions, fears, etc. In case of a patient showing somatoparaphrenic symptoms, it is plausible to suppose that a dysfunction in the conscious processing of proprioceptive information about the patient's limb (for instance, the left leg), results in the patient holding the belief that the leg does not belong to his body. Patients with somatoparaphrenia will therefore verbally deny that they own that leg, and some of them have in fact been reported trying to reject the limb that they perceive as alien (Critchley 1974).

Somatoparaphrenia has been described, with a few exceptions, in patients suffered from right parietal (or parieto-occipital) lobe injury – and almost invariably concerns the left side of the body. This condition is usually associated with motor and somatosensory deficits, and with the syndrome of unilateral spatial neglect. In a study on 79 acute stroke right-brain-damaged patients (Baier and Karnath 2008), 12 patients showed anosognosia for hemiplegia. Eleven out of these 12 patients exhibited somatoparaphrenic symptoms, and 6 among them displayed the strong belief that their limbs belonged to another person. In other cases, body parts can be just felt by the patient as separated from the body (Starkstein et al. 1990). More complex symptoms have been described: for instance, a patient can refer to the affected limb as "a make-believe leg" (Levine et al. 1991), or as "a baby in bed" (Richardson 1992).

The spectrum of somatoparaphrenic symptoms is wide, however a distinction can be drawn between misidentifications that can be corrected by patients when the error is pointed out by the examiner, and delusions that stubbornly resist to the examiner's demonstration (Feinberg et al. 2005). It is referred to fully-fledged somatoparaphrenia only in the second category of symptoms.

In sum, somatoparaphrenic phenomena do not imply a mental illness and can be characterized as follows (Vallar and Ronchi 2009):
> the feeling of estrangeness and/or separation of the affected body parts;
> delusional beliefs of disownership of the affected body parts;
> delusional beliefs that the affected body parts belong to another person;
> complex delusional misidentifications of the affected body parts;
> associated disorders, such as supernumerary limbs, personification, and misoplegia (hatred for the affected limbs).

Overall, available data suggest that patients showing somatoparaphrenic symptoms suffer from impairments in the higher-level processes concerned with body awareness and ownership. Therefore it seems reasonable to hypothesize that if certain behaviors do not occur without specific conscious sensations accompanied by the beliefs which refer to them (i.e. mental representations of the body), then those specific conscious sensations and their consequent beliefs must play an important causal role in the process that produces this kind of behavior.

5. Conclusion

In specific scientific contexts, it seems mandatory to apply concepts which carry a commitment for a causal role for consciousness. For the sake of the unity of science, it seems justified to take the physical causal closure for granted; on the other hand, it does not seem as well justified to take for granted the clear-cut distinction traced by epiphenomenalism between the physical world and the conscious mind. However, neither the common version of epiphenomenalism (in which conscious states are contingent), nor its variant proposed by Edelman (in which consciousness and the physical world are necessarily intertwined), appear to be a valid theory to explain the nature of consciousness. The theoretical and empirical arguments advanced in this article show that it is likely for consciousness to play fundamental roles in the genesis of behavior. Undoubtedly much more is to be done, especially on the side of the empirical research, in order to unravel the actual brain mechanisms of conscious causal processes.

References

Baier B., Karnath H.O. (2008), Tight link between our sense of limb ownership and self-awareness of actions. Stroke: 486-488.

Bisiach E., Geminiani G. (1991), Anosognosia related to hemiplegia and hemianopia. Prigatano G.P., Schacter D.L. (eds.), Awareness of deficit after brain injury, Oxford: Oxford University Press.

Bremmer, J. (1983), The Early Greek Concept of the Soul, Princeton: Princeton University Press.

Critchley M. (1974), Misoplegia or hatred of hemiplegia. Mt.Sinai Journal of Medicine 41: 82-87.

Edelman, G.M. (1989), The Remembered Present: A Biological Theory of Consciousness, New York: Basic Books.

Edelman, G. M. (1993), Neural Darwinism: The Theory of Neuronal Group Selection. Neuron 10: 115-125.

Edelman, G.M. (2004), Wider than the Sky. The Phenomenal Gift of Consciousness, Yale University Press.

Edelman, G.M. (2007), Second Nature, Brain Science and Human Knowledge, Yale University Press.

Faye, J. (1991), Niels Bohr: His Heritage and Legacy. An Antirealist View of Quantum Mechanics, Dordrecht: Kluwer Academic Publisher.

Feinberg, T.E., DeLuca, J., Giacinto, J.T., Roane, D.M, Solms, M. (2005), Right-hemisphere pathology and the self. Feinberg, T.E., Keenan, J.P. (eds.) The lost self. Pathologies of the brain and identity, Oxford: Oxford University Press, pp. 10-130.

Flanagan, O. (1992), Consciousness reconsidered, MIT Press, Cambridge Mass. Fuster, J. M. (2003), Cortex and Mind, Oxford University Press.

Gomatam, R. (2007), Niels Bohr's Interpretation and the Copenhagen Interpretation — Are the two incompatible? Philosophy of Science, 74, December issue.

Heil, J. and Mele, A. eds. (1993), Mental Causation, Oxford University Press.

Heisenberg, W. (1955), The Development of the Interpretation of the Quantum Theory. In W. Pauli (ed.), Niels Bohr and the Development of Physics, 35, London: Pergamon.

Heisenberg, W. (1958), Physics and Philosophy: The Revolution in Modern Science, London: Goerge Allen & Unwin.

Howard, D. (1994), What Makes a Classical Concept Classical? Toward a Reconstruction of Niels Bohr's Philosophy of Physics. In Faye, J., and Folse, H., eds., Niels Bohr and Contemporary Philosophy. Series: Boston Studies in the

Philosophy of Science, vol. 158. Dordrecht: Kluwer Academic Publisher.

Howard, D. (2004), Who Invented the "Copenhagen Interpretation?" A Study in Mythology. Philosophy of Science 71, pp. 669-682.

Huffman, C. A. (2009), The Pythagorean Conception of the Soul from Pythagoras to Philolaus. Body and Soul in Ancient Philosophy, D. Frede and B. Reis (eds.), Berlin: Walter de Gruyter.

Huxley, T.H. (1874), On the Hypothesis That Animals Are Automata, and Its Hystory. Fortnightly Review 95: 555-580. Reprinted in Collected Essays, London: Macmillan, 1893.

Huxley, T.H. (1884), Animal Automatism, and Other Essays, Humboldt Library of Popular Science Literature, New York: I. Fitzgerald.

Inwood, B. (2003), The Cambridge Companion to the Stoics, Cambridge: Cambridge University Press.

Jackson, F. (1982), Epiphenomenal qualia. Philosophical Quarterly32: 127-136.

James, W. (1890), The Principles of Psychology, reprint. New York: Dover, 1950.

Kerferd, G., 1971, Epicurus' doctrine of the soul. Phronesis 16: 80–96.

Kim, J. (2005), Physicalism, or Something Near Enough, Princeton University Press, Princeton.

Levine, D.N., Calvanio, R., Rinn, W.E. (1991), The pathogenesis of anosognosia for hemiplegia. Neurology 41: 1770-1781.

London, F., and Bauer, E. (1939), La théorie de l'observation en mécanique quantique. Hermann, Paris. English translation: The theory of observation in quantum mechanics. In Wheeler J.A., and Zurek, W.H., eds., (1983), Quantum Theory and Measurement, Princeton University Press, Princeton.

Lorenz, H. (2008), Plato on the Soul. The Oxford Handbook of Plato, G. Fine (ed.), Oxford: Oxford University Press.

Murdoch, D. (1987), Niels Bohr's Philosophy of Physics, Cambridge: Cambridge University Press.

Neumann, J. von (1932), Die mathematischenGrundlagen der Quantenmechanik, Springer, Berlin. Reprinted in English (1955), Mathematical Foundations of Quantum Mechanics. Princeton University Press, Princeton.

Nussbaum, M. C. & Rorty, A. O., eds., (1992) Essays on Aristotle's De Anima, Oxford: Clarendon Press.

Richardson, J.K. (1992), Psychotic behavior after right hemispheric cerebrovascular accident: a case report. Arch. Phys. Med. Rehabil. 73: 381-384.

Robinson, H. M. (1978), Mind and Body in Aristotle. The Classical Quarterly 28: 105-124.

Schwartz, J.M., Stapp, H.P., and Beauregard, M. (2005), Quantum theory in neuroscience and psychology: a neurophysical model of mind/brain interaction.

Philosophical Transactions of the Royal Society B 360, 1309-1327.

Searle, J. R. (2004), Mind, a brief introduction, Oxford University Press.

Skinner, B. (1974), About Behaviorism, New York: Vintage.

Stapp, H.P. (1993), A quantum theory of the mind-brain interface. Mind, Matter, and Quantum Mechanics, Springer, Berlin, pp. 145-172.

Stapp, H.P. (1999), Attention, intention, and will in quantum physics. Journal of Consciousness Studies 6(8/9), pp. 143-164.

Stapp, H.P. (2006), Clarifications and Specifications In Conversation with Harald Atmanspacher. Journal of Consciousness Studies 13(9), pp. 67-85.

Starkstein, S.E., Berthier, M.L., Fedoroff, P., Price, T.R., Robinson, R.G. (1990), Anosognosia and major depression in 2 patients with cerebrovascular lesions. Neurology 40: 1380-1382.

Vallar, G., Ronchi, R. (2009), Somatoparaphrenia: a body delusion. A review of the neuropsychological literature. Experimental Brain Research 192: 533-551.

Wigner, E.P. (1967), Symmetries and Reflections, Indiana University Press, Bloomington.

11. Time, Altered States of Consciousness, And Neuroscience

Shaun Gallagher[1] and Juan C. González[2]

[1]Philosophy, University of Memphis (USA),
[1]School of Humanities, University of Hertfordshire (UK),
[1]Faculty of Law, Humanities and the Arts, University of Wollongong (AU),
[2]Philosophy and Cognitive Science, Universidad Autónoma del Estado de Morelos, Cuernavaca (MX)

Abstract

Philosophers like Kant and the phenomenologists argue that time is a fundamental determinant of human cognition and that any dysfunction in this determinant will lead to a loss of meaningful experience. The phenomenology and science of psychotic delusions and drug-induced experience supports this idea. At the same time some philosophers and scientists have argued that such radical changes in temporal experience can be revelatory of different ways of thinking about reality and of different meaning structures. In this paper we review the relevant philosophical arguments and scientific evidence for these positions, and examine three issues. The first is whether science is empirically investigating its own rationalist foundations when it studies altered states of consciousness. The second concerns whether experiences of delusions and hallucinations challenge our default conceptions of reality. And third, can altered states of consciousness provide a form of knowledge or put us in a position to acquire knowledge? We argue that whether changes in temporal experience reveal different and productive provinces of meaning or not, they do reveal the limits of our normally ordered meaning, the limits of science and of what we consider a rational world.

Keywords: Delusion, hallucination, phenomenology, neuroscience, reality

1. Temporal Order and Time Perception

Philosophers and scientists have argued and produced good evidence to show that temporality, a broad term meant to include experiential aspects of temporal order and time perception, is basic to a coherent experience of meaning. That is, absent a well-ordered temporal structure, experience would be chaotic and meaningless, or, at best, radically different from what it is when well-ordered temporal processes are operating properly, in line with the naturally evolved way that our human neurobiological systems work. This has been made clear in philosophers such as Kant, the phenomenologists, and in the majority of neuroscientific studies of changes in temporal experience in, e.g., schizophrenic psychosis and drug-induced states.

In contrast to this, some philosophers and scientists have argued that such modulations in temporal experience, under practices of drug-induced changes, can be revelatory of different ways of thinking about reality and of different meaning structures. That is, such experiences are not disorganized but reorganized in such a way as to produce alternative meaning and knowledge.

In this paper we examine the contrast of these two general perspectives and argue that whether changes in temporal experience reveal different and productive provinces of meaning or not, they do reveal the limits of our normally ordered meaning, the limits of science and of what we consider a rational world.

2. Philosophy: Kant and the Phenomenologists

As is well known, Kant (1965), in his Critique of Pure Reason, developed a transcendental approach, that is, an approach that purported to demonstrate the conditions of possibility for any rational experience. Basic to his project is the idea that our consciousness is ordered by two basic forms of intuition: time and space. We can think of his term 'intuition' as signifying any presentational mode of experience: perception, memory, imagination, etc. The form or structuring performance of any perception, for example, is governed by temporality and spatiality. Anything that we perceive falls into a spatial-temporal order precisely because the mind structures it that way a priori, that is, independently of what we experience. For Kant, time is more basic than space because time, but not space, applies to the mind itself. Our consciousness itself is set out in a temporal order, but, as Descartes maintained, it is not extended in space. Contemporary debates in cognitive science about embodied and distributed cognition may challenge this (see Gallagher, 2005), but we leave that aside for our purposes.

In his transcendental deduction Kant then demonstrated that there are a set of categories or pure concepts of the understanding that apply to objects of intuition in general. His table of categories comprises 12 concepts that are grouped in 4 sets: the categories of quantity: Unity, Plurality, and Totality; the categories of quality: Reality, Negation, and Limitation; the categories of relation: Inherence and Subsistence, Causality and Dependence, and Community; and the categories of modality: Possibility-Impossibility, Existence-Non-existence, Necessity-Contingency. In a nutshell, these categories presumably frame our experience of the world a priori, in such a way that when we come into cognitive contact with the world, what we perceive is not only structured spatiotemporally (given the transcendental forms of intuition), but also with a default 'format' that makes our experience of the world intelligible. Indeed, these categories allow the mind to process the objects of experience in a structured or orderly fashion and eventually make sense of them.

The idea that time is so basic to experience suggests that any disorder affecting the temporal organization of experience would also affect the categories that structure what we perceive. This is easiest to see with respect to the categories of unity and causality. Unity or identity can be thought of as inextricably linked to temporality: a unit is a persistent, sustained and/or self-contained entity persisting over time. If temporal persistence or duration were broken, then our experience would be of things lacking unity. Causality involves at a minimum an ordering of one event (the cause) happening prior to another (the effect). Alter the temporal sequencing function of experience in a way that would disallow the perception of sequence, then the category of causality would fail to work. More generally, disorders in temporal experience would result in a breakdown of the rational appearance of reality.

Phenomenologists, like Husserl (1991), Heidegger (1968) and Merleau-Ponty (1962), also place great emphasis on the importance of temporality for an ordered experience. Husserl's analysis of time-consciousness is perhaps the most influential of these accounts. According to Husserl, who was influenced by William James' (1890) empiricist idea of the stream of consciousness, each moment of one's experience is present for only an instant, and then slips away into the past, even if the object of consciousness remains present and unchanging. It is also the case that some objects of perception, for example, melodies, are constantly changing. In that case we have the successive flow of consciousness and the succession of the temporal object. Yet, in some way the flowing retreat of consciousness is able to maintain an orderly sense of the melody as it expires in time. The attempt to explain how this is possible is at the root of Husserl's account.

According to his analysis, consciousness always includes a narrowly directed intentional grasp (the primary impression) of the present moment of whatever is being experienced, for example, the current note in a melody, or the current phase of any enduring object. But the primary impression never happens in isolation; consciousness includes two other structural aspects: the retentional aspect, which provides us with a consciousness of the just-elapsed phase of the enduring object, thereby providing past-directed temporal context; and the protentional aspect, which in a more-or-less indefinite way anticipates something which is about to be experienced, thereby providing a future-oriented temporal context.

Figure 1 summarizes Husserl's model. The horizontal line ABC represents a temporal object such as a melody of several notes. The vertical lines represent abstract momentary phases of an enduring act of consciousness.

Each phase has a threefold structure consisting of:

• *primal impression* (pi), which allows for the consciousness of an object (a musical note, for example) simultaneous with the current phase of consciousness;

• *retention* (r), which retains previous phases of consciousness and their intentional content;

• *protention* (p), which anticipates experience which is just about to happen.

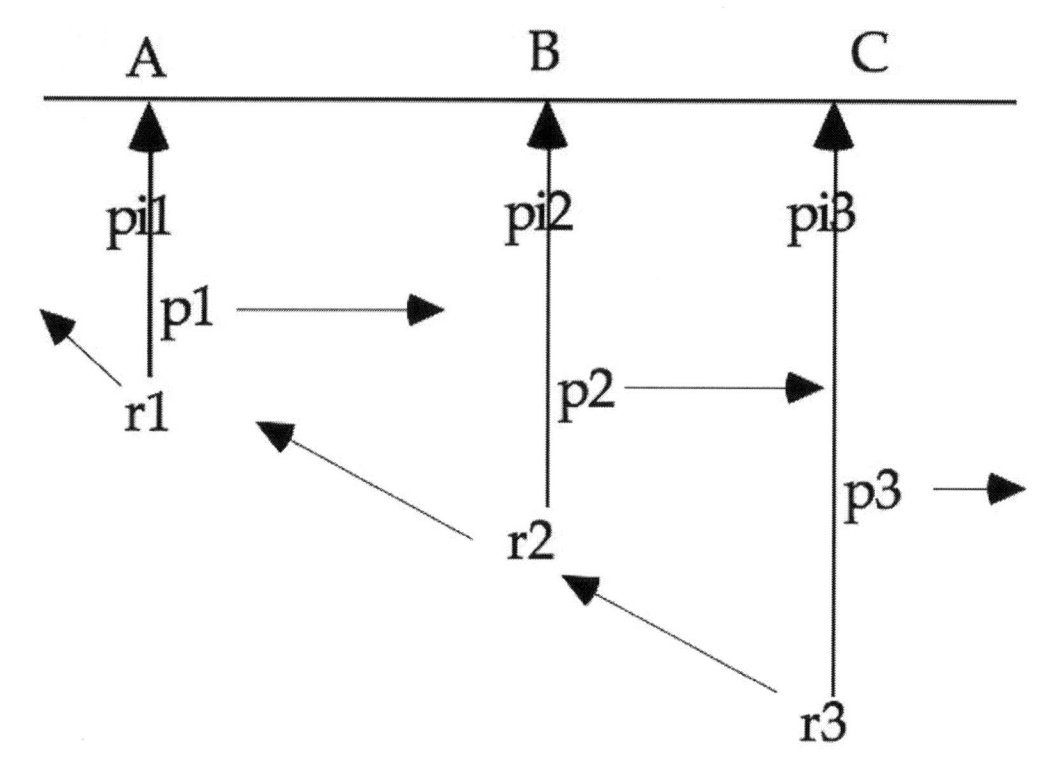

Figure 1: Husserl's model of time-consciousness (from Gallagher 1998; 2012)

Although the specific experiential contents of this structure from moment to moment progressively change, at any given moment this threefold retention-primal-impression-protention structure is present (synchronically) as a unified whole.

If, for example, we are listening to music, the retentional aspect of consciousness keeps the intentional sense of the previous notes or measures available even after they are no longer audible. On Husserl's account, consciousness retains the sense of what has just been experienced, not by retaining the event itself, but by its tacit awareness of the just-past phase of consciousness. Furthermore, as I listen I have some anticipatory sense of where the melody is going, or at the very least, that the melody is heading toward some indeterminate conclusion. The protentional aspect of the act of consciousness also allows for the experience of surprise or disappointment. If the melody is cut off prematurely, I experience a sense of incompleteness, precisely because consciousness involves an anticipation of what the imminent course of experience will provide, even if this remains relatively indeterminate.

The primal impression-retention-protention structure is the temporal organizing principle of consciousness that allows us to experience the world in a meaningful way, where the past, which is no longer there, is kept in mind, and the future, which is not yet there, is anticipated. In viewing a horse race, for example, our perception is not restricted to a current but durationless snapshot of the horses' movements – if it were, we would sense no movement at all. Perceptually, it is not as if the horses suddenly appear out of nowhere in each new moment. Rather than perceiving their present position and then adding to that the recollection of where they were a moment ago, we actually perceive the horses moving on the track as they are racing. That is, we do not engage in an act of comparative remembering in order to establish the temporal context of their current position. And yet all the previous parts of the race do not remain perceptually present in the same way as the horses' current positions. If that were the case, they would perceptually fill the entire space they had just traversed. The past movements do not remain visually present in some vague ghostly manner. Retention does not keep a set of fading images in consciousness. Rather, because of the way that our consciousness is structured, at any moment what we perceive is embedded in a temporal horizon. What we see is part of or a continuation of, or a contrasting change from what went before, and what went before is still intentionally retained so that the current moment is experienced as part of the whole movement. Consciousness retains the just past, not with the significance of something present, but with the meaning or significance of having just happened.

Other aspects of Husserl's analysis are important for solving problems that relate to the unity of consciousness, personal identity and action. For our purposes, however, it is sufficient to emphasize that in the absence of this kind of temporal structure, experience would tend to be chaotic. We live in a coherent and meaningful world precisely because we are able to navigate through a stream of experience without getting lost – and our other abilities, like our ability to move through space, or to find our way into relationships in the social world, fully depend on our temporal navigations (Gallagher and Zahavi, 2012).

3. Neuroscience, Pharmacology and Phenomenology

One can get a glimpse of what would happen if the temporal structures of experience were disrupted, by considering a case of motion agnosia. Neuronal structures in the medial temporal cortex are specialized for visual detection of motion. If this part of the cortex, and possibly other areas, including posterior temporal cortex abutting the occipital lobe, are damaged, by stroke for example, visual perception of form and color may be preserved, but perception of motion is disrupted. A person who suffers from motion agnosia experiences the world as seemingly without motion, frozen in place, for several seconds. Things then suddenly rearrange themselves in new positions. Imagine crossing a busy street where you see cars and know they are moving, perhaps by auditory clues, but simply can't see their movement (see Schenk and Zihl, 1997; Zihl et al. 1983).

Such a disruption plays havoc with the person's abilities to make sense out of the world and to act in it. Temporal continuity seems absolutely essential for making sense out of our everyday experience. This is not to deny that there is discontinuity too. We can shift rapidly from one activity to another, we can move from one situation into a completely different one, and we can even experience disruptions and breakdowns where the flow of information or activity becomes quite confused. Ultimately, if we are to restore sense to these experiences, we have to take them up into a more cohesive temporal framework.

Despite difficulties with the perception of motion, a person with motion agnosia can still function because the greater part of her non-visual experience retains some anchor in a cohesive temporal structure. Consider, however, an even more profound disruption of temporal experience. What if our ongoing experience lacked temporal coherence entirely? What if, for example, I was unable to keep the just previous moment of experience in mind long enough to write it down, or was unable to anticipate events in the next second? Would my experience make any sense at all? Husserl is motivated to ask about this possibility and admits that the temporalizing function of consciousness could fail to hold:

> Could it not be that, from one temporal moment on, all harmonious fulfillment would cease and the series of appearances would run into one another in such a way that no posited unity could ultimately be maintained... Could it not happen that... the entire stream of appearance dissolves into a mere tumult of meaningless sensations. Thus we arrive at the possibility of a phenomenological maelstrom ... it would be a maelstrom so meaningless that there would be no I and no *thou, as well as no physical world—in short, no reality* (1998, pp. 249-250).

We may approach an understanding of such radical breakdowns in temporal and semantic organization by looking at certain psychopathologies and drug-induced experiences. For a very long time we've known that schizophrenia and other psychological disorders can involve changes in temporal experience and time perception. In schizophrenia, for example, difficulties in temporally indexing events, distortions in time estimation and temporal velocity, the curtailment of coherent future time perspective, and a confusion between past, present, and future are frequent occurrences (Braus, 2002; Davalos et al. 2002; 2003; Elvevag et al. 2003; Tenckhoff et al. 2002; Volz et al. 2001).

It is also well known that chemical changes in the brain caused by the ingestion of drugs can alter the experience of time. Cannabis intoxication, for example, results in increased heart rate and abnormal EEG rhythms, disturbances in perceptual experience, and distortions in time estimation and time perspective, including flashbacks (Bech, Rafaelsen & Rafaelsen, 1973; Jones, 1978; Lieving 2006; Matthew et al. 1998; Tinklenberg et al. 1972). Opiates may produce an experiential expansion of time (Friedman, 1990; Mansky, 1978) or alteration in timing, which is a pervasive factor in motor control, perception, language production, and so forth (Meck 1996).

Psychoactive or psychedelic drugs, such as LSD, mescaline, and psilocybin, produce physiological changes with phenomenological effects and hallucinations in which time seems to either speed up or slow down. Given the complexity and inadequacy of many empirical and conceptual analyses of hallucinations (Dokic & Martin, 2012; González, 2010a,b; González & Dokic, 2009; Lehmann & González, 2009), we offer the following working definition. By 'hallucination' we understand a conscious mental state whose phenomenal and semantic content is imposed on the agent's psyche by endogenous causes, with clear and simultaneous impact on routine patterns of perception, belief and behavior of the agent. Hallucinations are hence deviant (i.e., statistically-abnormal and novel) cognitive phenomena. Subjects may also suffer from disordered temporal experience, abnormalities in time estimation and temporal orientation (Barr et al. 1972; Heimann, 1994; Hollister, 1978; Ornstein, 1969). Objective measures of these disturbances in subjective time sense suggest deficits in working memory and subjective changes in conscious states that include depersonalization and derealization (Wittmann et al. 2007).

The phenomenology of intoxication with mind-altering substances and, more specifically, with the so-called hallucinogens (mescaline, DMT, psilocybin, salvinorin A, LSD, bufotenin, etc.) — whether in traditional indigenous rituals or in isolated individuals — is truly fascinating and, as just indicated, involves changes in the experience of time and perceptual reality. Indeed, these substances:

> produce deep changes in the sphere of experience, in perception of reality, in space and time, and in consciousness of self. Depersonalization may occur. Without loss of consciousness, the subject enters a dream world that often appears more real than the normal world.... The psychic changes and unusual states of consciousness induced by hallucinogens are so far removed from similarity with ordinary life that it is scarcely possible to describe them in the language of daily living. A person under the effects of a hallucinogen forsakes his familiar world and operates under other standards, in strange dimensions and in a different time. (Schultes, Hofmann & Rätsch, 2001, p. 14).

Thus, under the influence of hallucinogens, our temporal consciousness may be profoundly modified, without necessarily losing our lucidity or cognitive contact with the world. Thus María Sabina (1888-1985), a well-known Mazatec healer and shaman, indicated that "[t]he more you go inside the world of Teonanácatl [the psilocybin mushrooms], the more things are seen. And you also see our past and our future, which are there together as a single thing already achieved, already happened" (Cited in Schultes, Hofmann & Rätsch, 2001, p. 1).

Although the "pharmacological basis of the experience of time and temporal processing is only vaguely understood" (Wittmann et al. 2007, p. 61), neuroscientific research into the pharmacological mechanism of hallucinogens and dissociative anesthetics (such as ketamine) that cause time disturbances shows an imbalance among serotonin, glutamate and dopamine neurotransmitter systems critical to psychotic symptom formation (Vollenweider, 1998).

Specifically, changes in time perception and motor timing have been linked to the dopamine system (Meck, 1996; O'Boyle et al. 1996; Rammsayer, 1999; Reuter et al. 2005), and to fronto-striatal circuits purportedly involved in timing and the coordination of temporal sequence (Meck and Benson 2002). Pharmacological manipulations of the serotonin system also affect duration discrimination abilities in humans (Rammsayer, 1989), and impairments in related frontal cortical regions, closely linked with working-memory function (or in phenomenological terms, with the temporal structure of retention-primal impression-protention) correlate significantly with altered time sense and depersonalization (Wittmann et al. 2007).

Altered time sense, in either schizophrenic psychosis or drug experiences, is not an isolated effect. Much more comes along with it, including such experiences as depersonalization and derealization which can involve dramatic changes to self and to our sense of reality. In some cases such changes, experienced, for example, as an "oceanic boundlessness," can be associated with positive mood including experiences of the sublime or heightened feelings of happiness; in other cases they can be accompanied by anxiety and paranoia, thought disorder, a disintegration of self, loss of autonomy, and feelings of being endangered, including changes in or loss of identity, lack of differentiation between ego and non-ego with respect to thought process, affective states, and body experiences, as well as deficits in self-determination in acting, thinking, feeling, and perceiving. Likewise these experiences can be accompanied by auditory and visual illusions, hallucinations, synaesthesia, and changes in perceptual meaning. These are common features of both drug and non-drug induced altered states of consciousness (see Dittrich, 1998; Vollenweider & Geyer, 2001). Thus, for example, "administration of either a serotonergic hallucinogen (e.g., psilocybin) or an NMDA [N-methyl-D-aspartate] antagonist (e.g., ketamine) elicited a psychosis-like syndrome in normal volunteers that was characterized by ego disturbances, illusions and hallucinations, thought disorders, paranoid ideations, and changes in mood and affect" (Vollenweider & Geyer, 2001, p. 497). In other words, the coherent structures of experience and reality themselves can be at stake and can be shaken right down to their most basic temporal order. It should be noted, however, that many —if not all— of these subjective effects (whether positive or negative) are the result of a complex interaction between the drug or the material cause and what is known as the "set and setting" of the altered state (i.e., the internal and external conditions in which the experience takes place). The pharmacological agent or material cause by itself is insufficient to explain, understand or predict the specific alteration of experience. Notwithstanding this, these episodes are usually accompanied by a disruption of our normal sense of reality and time experience, for better or for worse.

4. Knowledge and Reality

There are a number of philosophical questions raised by this juxtaposition of philosophical and scientific discussions of delusions and hallucinations in respect to temporality. Here we'll identify three interconnected issues and try to show how they challenge our received conceptions of knowledge and reality. The first concerns science; the second involves the concept of reality, and the third involves knowledge of a different sort.

Both Kant and Husserl were concerned about the foundations of science. That is, their attempts to work out a transcendental deduction of categories and to do a precise phenomenology, respectively, were motivated by a desire to provide epistemological and phenomenological guarantees that science can deliver (true and reliable) knowledge. Thus Kant's transcendental deduction was an attempt to provide science with the rational categories it needed to do its work – categories such as causality, unity, etc., and to give them epistemological rigor in the wake of a Humean analysis that would have made science just one more imaginative story. Husserl wanted to make firm our conception of consciousness since it is only by way of consciousness that we can do science. We need to have a clear understanding (phenomenology) of consciousness to make sure we know its limitations and the possible distortions it may introduce into the scientific enterprise.

Their analyses suggest that if the categories are not working properly, or if consciousness is not functioning properly, then it is really impossible to do science. And at the very base of our mental enterprises time or the temporal structure of consciousness has to be intact for the rest to function properly. Accordingly, only when temporality holds fast can one conduct science, including the neuroscience that examines the underlying mechanisms that explain what goes wrong when, in cases of psychotic delusion or drug-induced states, the temporal structure of experience, and with it, all other aspects of rational reality get seriously modified. No less than the Kantian analysis, the scientific study of abnormal mental phenomena is a study of rationality, or the limits of rationality – an empirical study, however, rather than a transcendental study. Indeed, science tends to pay no attention to Kant or to a worry about its own foundations. And yet it seems to be digging into its foundations when it investigates delusion and hallucination. To see why this is philosophically dicey, let's look at the second question.

It is commonplace to suggest that people who experience various delusions or hallucinations, in cases of psychotic disorders or drug-induced experience, enter into a different world or an alternative reality. Whether this is taken in a metaphorical or literal sense, the fact is that people who are in these abnormal states usually behave "as if" they were dealing with an alternative reality or world, sometimes radically different from the commonsensical one. Moreover, when telling about their experience, these people usually cast their descriptions with lucidity in perceptual and other cognitive terms. This seems to entitle one to speak of alternative, "non-ordinary realities" that one can enter or enact through certain altered states of consciousness (Duits, 1994; Grof, 2006; Huxley, 1990; Michaux, 1972). It's difficult for science to comprehend how this would be possible, since science is heavily invested in explaining the one objective world defined by physics and the established laws of nature – the default world that is also accepted without question by common sense. Philosophers sometimes talk about parallel universes or possible worlds when they do metaphysics, but this seems more of a logical exercise than anything that would lead us to believe that there are such universes. Moreover, a science of delusions and hallucinations is always a science based on the categories of non-delusional, non-hallucinated, default, objective reality, whether defined by Kant, or by physics and the established laws of nature.

William James (1890, II, pp. 291-306), however, suggested that in some sense we all live in a plurality of sub-universes, an idea developed further by the phenomenologist Alfred Schutz (1974) under the heading of 'multiple realities'. According to this idea, the experiencing subject does not live in the one unified world of meaning defined objectively by science or common sense, but in multiple realities or "finite provinces of meaning." As both James and Schutz suggest, there is a "paramount" or default reality – the reality of shared everyday life that we normally engage in, established in orderly time. This is the world where we work, earn our salary, socialize, enjoy family life, engage in scientific pursuits, and so forth. Phenomenologists call it the 'lifeworld'. But there are also multiple other realities that take us away from everyday reality. If, for example, I read a novel, or go to the theater or the cinema, or play a video game, I spend a couple of hours (measured on the clocks of paramount reality) escaping into a different sort of reality, a different time (possibly an imagined past or future) or a different place which opens up in the pages, on the stage, or on the screen.

The various "realities" (and the diverse sets of rules that govern them) are not necessarily commensurable – they have their own rationalities. Indeed, my actions or virtual actions are likely different in these different realities. Such changes in my behavior can be existential, involving a transformation of background familiarity and of the sense of reality. Usually, however, I can make sense of my behavior in these alternative realities and can understand what I'm doing from the perspective of everyday reality. There are normally clear transitions as I move from one to another. In everyday reality I can go to the theater; once in the theater I enter a different reality. At some level, when I enter into a virtual world, such as a video game, I keep account of the fact that from the perspective of everyday reality, I am playing a game. I can distance myself from the various roles that I might play or fantasize about. If I am truly engaged or fully immersed in one of these alternative realities, however, it is not simply that I adopt an alternative set of tractable beliefs or values. Rather, I may enter into it, so to speak, body and soul, or to some varying lesser degree. Indeed, the alternative reality has a certain "presence" and salience that makes it more than an intellectual exercise. I am in-the-world of the play, the film, the game, etc.; I can get excited and emotional, or remain cool under pressure; I may adopt a certain physical posture, I may act virtually, and I may engage in such action explicitly or implicitly. Sometimes, as I come back out of such realities, everyday reality can seem oddly unreal in relation to what I have been doing.

Both James and Schutz suggest that we can apply this idea to experiences of delusion and hallucination (Gallagher, 2010). It seems quite possible that one can enter into a delusional or hallucinatory reality just as one can enter into a dream reality, or a fictional reality, or a virtual reality. Like other alternative realities, some delusional realities are ones that are more or less cut off from one's everyday reality; ones that are incommensurable with the normal rules of reason that govern one's everyday default lifeworld, but ones that may have their own rules, and ones that offer a different set of affordances for action. As Louis Sass (1992, p. 109) puts it, the delusional subject "inhabits a world radically alien to that of common sense." For the schizophrenic, "the world can no longer be taken for granted" (Merleau-Ponty, 1962, p. 335).

The neuropsychatrist J. Allan Hobson (2005) has defended the idea we can identify (psychedelic) hallucinations with dream-like states that have as little or as much reality as a dream does. Some researchers claim that in certain hallucinatory states, the feeling they get from their hallucinations is not only that they are real, but that they are "more real than real." For example, Benny Shanon has stated that:

> With Ayahuasca, people often feel that what they see is 'more real than real'. From a theoretical point of view this is especially significant. Often in the philosophical and psychological literature, non-ordinary notions of reality are associated with a diminishing sense of reality. Thus, many have argued that what distinguishes true perception, on the one hand, and memories, imagination, and dreams, on the other, is that the former is clearer, more distinct, and more coherent than the latter...[think of Locke and Hume]. In contrast, the Ayahuasca experience presents cases of an enhanced sense of reality. This indicates that the reality judgment is actually a parameter that can take different values. Dreams and imagination usually decrease the value assigned to this parameter; the Ayahuasca experience may increase it. (Shanon, 2002, p. 265)

Our awareness can be immersed in a very realistic atmosphere during the Ayahuasca experience, to the point of taking the hallucinated objects for real, without doubts or skeptical attitudes.

> Normal human beings do not doubt that what they perceive is real ... the act ... of perception forces itself upon us and the moment we open our sensory faculties we are immersed in the world of our perception without any choice or room for doubt, like fish in water.... Experientially, what is felt with

powerful Ayahuasca visions is very similar" (Ibid.)

Shanon notes, however:
> Real though the Ayahuasca visions may be deemed to be, they are not usually confused with the normal perceptions of the ordinary world. Rather, the feeling is that what is seen in the visions pertains to other, separate realms. Hence, in general, at least a degree of an awareness of the real world is maintained" (Ibid.)

5. Restructured Reality

To think of delusions or hallucinations in terms of multiple realities is not to rule out a different way of putting it (one that we prefer), namely, that delusions and hallucinations involve a radical or structural (or existential) change in the way that the subject relates to everyday reality. John Kafka (1989), for example, although using the term 'multiple realities' in the context of clinical psychoanalysis, conceives of them as different organizations of everyday reality, or what he calls different 'reality organizations'. On this view there is one reality, independent of our cognitive states. When we engage or interact with reality, however, we cannot separate our cognitive frame of reference (specific to our species, ecological niche, ontogenetic development and even perhaps our current situation) from it. The realist idea is that there is one reality, which is cognitively accessed from different perspectives, although no single agent, experience, or system of beliefs can pretend to exhaust all possible descriptions of reality. What hallucinations and altered states of consciousness reveal are novel ways to experience and interact with reality – novel reality organizations. On a neo-Kantian perspective, the novelty arises through the modification of the internal relations that hold between the categories of understanding and our cognitive matrix (González, in press).

In Capgras Delusion, for example, the patient typically takes a familiar person, for example, a spouse, to be an imposter. Although rare, a subject may consider some physical objects, such as a set of tools to have been replaced by exact doubles (Ellis, 1996). One might say that the tools remain part of default reality, but that the patient's relation to the tools has been radically reconfigured (see, e.g., Dreyfus, 1987; also Kafka, 1989). This may in some cases be a more apt description than saying that the person has entered into a delusional reality where her tools have been replaced by replicas. The difference in the way of putting this, however, may reflect how comprehensive or pervasive, or "firmly sustained" (DSM IV) the delusion is. For example, the subject may still use the tools as tools rather than treating the tools as instruments with magical powers. In extreme cases, however, the delusion rather than the default reality organization becomes paramount.

As we've seen, such alterations in conscious states frequently involve different experiences of time. On one reading, this could simply be a feature of the alternative reality organization – the structural change in reality comes with some kind of change in the way that time behaves. Some philosophers and psychiatrists, however, especially those in the Kantian and phenomenological traditions (e.g., Minkowski, 1970), have suggested that the change in time experience is more basic than that, and specifically that it is the "generative disorder," which means that the change in temporality drives other aspects of the experiences.

Again, we can put this in Kantian terms and think of our everyday, default conceptual structure in terms of a grid or format that is superposed on the world, through which we cognize, represent and evaluate it. In this perspective, we may think of alternative states as different "epistemic positions" that allow for alternative forms of cognitive understanding involving modifications to the categorial grid. It could then be the case that during a delusional or hallucinatory experience the habitual relation between the world, our consciousness and the overarching conceptual structure is modified in a way that the internal-external, self/non-self relations are altered. And this could be the result of shifting categories or transformations in time-consciousness.

Indeed, it could be the case that during a hallucinatory episode the categories of quality — the category of reality in particular— are preserved or even enhanced in a way that the contents of awareness are vividly felt as real, while other categories may radically change, affecting the experience and drastically modifying it, resulting in a delusional or hallucinatory otherworldly or bizarre character. Within this system, the status of reality goes hand-in-hand with the exercise of the conceptual-categorial structure so that what we experience in delusions or hallucinations is neither more nor less real than what we experience in our everyday, paramount reality – just different. Delusions and hallucinations would therefore involve a consciousness that is either structured by an anomalous categorial grid or one that, in itself, no longer matches, partially or totally, the regular connections within the default or everyday categorial grid. To each possible disjunctive reality organization there would correspond different "epistemic positions" that allow us to do justice, in positive terms, to different types of genuine experience.

6. Science and Alternative Forms of Knowledge

It's at this point that our initial question about science links up with this idea of alternative epistemic positions or sets of modulated categories. To put the issue in its most radical form, the question is why should we think that science (or rationality) must be strictly tied to the default or everyday reality organization? Would it be possible to think that there are multiple sciences – different forms of knowledge that are relative to the different forms of reality that map onto alternative rationalities revealed by delusional or hallucinatory (or indeed, some other kinds of) experiences. The situation is analogous to thinking about geometry at the time of Kant; the idea that there could be non-Euclidian geometries was not even on the horizon. Kant wrote: "The apodeictic certainty of all geometrical propositions and the possibility of their *a priori* construction is grounded in this a priori necessity of space.... Geometry is a science which determines the properties of space synthetically, and yet a priori," (1965, pp. B/39-40), where geometry means Euclidian geometry. Likewise, Kant thought that time, as a form of intuition, was a priori and universal. But if delusions and hallucinations suggest that there are non-Kantian forms of time (analogous to non-Euclidian forms of space), we might think that there could be a science, or multiple sciences, dedicated to the alternative reality organizations that open up around these alternative temporal axes. Such alternative organizations are unhinged only in relation to Kantian time.

This motivates our third question. Might it be possible to gain a different form of knowledge from these alternative states of consciousness? It is quite easy to give an affirmative answer when we think of the alternative "realities" of the novel, the film, the theatrical play, the dream, etc. But what about delusions and hallucinations? One of us (SG) was once tracked down by a former student who was keen to convince his former professor that he had a new explanation of the universe. It was clear from his demeanor that he was suffering from a schizophrenic delusion, and I listened to his explanation with complete concern for his well being, but also with complete skepticism about his new explanation. From the perspective of my default epistemic position, and from the perspective of science, his explanation made no sense, and I was completely (and many would say justifiably) skeptical. It's not definite, however, that there was no value in his thinking.

Whether the hallucinatory experience provides (or can provide) knowledge or, at least, whether it puts (or can put) us in the position to acquire knowledge, we remain agnostic, torn between skeptical arguments of scientific and epistemological character (e.g., the orthodox definition of knowledge) and anthropological accounts, and (in the case of one author, JG) personal experiences supporting the claim that, at least in a pragmatic sense, hallucinations can dispense knowledge. The problem for establishing a definitive answer to this question (and, therefore, for escaping agnosticism on this point) may reside in the conception of knowledge at stake – what we understand by 'knowledge'. There are certainly several types of knowledge, and a proper answer to the above question may crucially hinge on the type of knowledge we have in mind and on the method of justification that we consider adequate for validating it.

A further question, however, is whether a kind of knowledge derived from hallucinations or delusions – a kind of knowledge tied to a non-Kantian temporal framework or to a consciousness with an anomalous temporal structure – would have any relevance to (or be able to explain) our default, everyday experience, or whether it would apply narrowly only to the hallucinatory or delusional experience. If, as we suspect, it may be the latter, then why should we not turn this question around and ask why we should expect that science, as we know and practice it – call it default science (which, at least on Kantian and phenomenological grounds, is tied to our default everyday rationality) – would have any relevance to (or be able to explain, or explain away) those alternative states of consciousness?

Of course, as one of the reviewers suggested, someone might say that science, operating in the default everyday mode, is able to give an account of the delusional or the hallucinatory state, whereas from within the latter states it may not be possible to account for everyday default rationality. This epistemic asymmetry is similar to waking state in contrast to dream state: it is from the standpoint of the waking or lucid state that we know about and can theorize about the dream state, whereas, within the dream state we do not know about the waking state. One might claim that this asymmetry offers an epistemic privilege to sober, rational wakefulness and thus to science and philosophy. On the one hand, this claim for epistemic superiority of waking/lucid states over the delusional/hallucinatory ones, however, appeals to rational/scientific standards, and in this sense may beg the question. On the other hand, we agree that we can best discriminate delusional and hallucinatory states only in comparison to default everyday baseline states that are reliable and have epistemic value. In commenting on Donald Davidson's stance towards classical skepticism, Richard Rorty (2003) makes a similar claim:

> Davidson's point is that retail skepticism makes sense, but wholesale skepticism does not. We have to know a great deal about what is real before we can call something an illusion, just as we have to have a great many true beliefs before we can have any false ones. The proper reply to the suggestion that [something] might be illusory is this: Illusory by comparison to what? (Rorty 2003)

Finally, however, with respect to the question of epistemic value, it's important to note that hallucinations, as we have defined them, in contrast to delusional states, are cognitive states that allow us to experience reality in an uncommon, alternative fashion, attending to aspects of it not usually processed by our default cognitive powers. In this case, then, a different epistemological picture emerges (González, 2010a). An alternative knowledge is possible either because our default epistemic categories are modified, or because our normal justification procedures are changed, and this itself could have an impact on our habitual everyday practices or behavioral patterns, and perhaps a practical effect on our lives, our relations to others and to the world. The history of mystical experience, the existence of shamanic cultures, and perhaps even poetry and art, testify to this kind of knowledge, a knowledge that is only partially commensurable with scientific rationality. Here we can picture two sets of Kantian categories, partially shifted in respect to each other, in temporal, spatial and qualitative features, constituting complementary points of view or different ways to experience reality, and yet both with real consequences on our everyday practices.

Acknowledgments SG's research is supported by the Humboldt Foundation's Anneliese Maier Research Fellowship (2012-2017).

REFERENCES

Barr, H. L. and Langs, R. (1972). LSD: Personality and Experience. Wiley-Interscience, New York.

Bech, P., Rafaelsen, L., and Rafaelsen, O. (1973). Cannabis and alcohol: Effects on estimation of time and distance. Psychopharmacologia (Berlin), 32, 373-381.

Braus, D. F. (2002). Temporal perception and organisation, neuronal synchronization and schizophrenia. Fortschr Neurol Psychiatr, 70, 591–600.

Davalos, D., Kisley, M. and Ross, R. (2002). Deficits in auditory and visual temporal perception in schizophrenia. Cog Neuropsychiat, 7, 273–282.

Davalos, D., Kisley, M. and Ross, R. (2003). Effects of interval duration on temporal processing in schizophrenia. Brain Cogn, 52, 295–301.

Dittrich, A. (1998). The standardized psychometric assessment of altered states of consciousness (ASCs) in humans. Pharmacopsychiatry, 31, 80–84.

Duits, Ch. (1994). Le Pays de l'éclairement. Le bois d'Orion. L'Isle-sur-la-Sorgue, France.

Dokic, J. and Martin, J-R. (2012). Disjunctivism, hallucinations, and metacognition. Wiley Interdisciplinary Reviews: Cognitive Science, 3 (5), 533–543.

Dreyfus, H. (1987). Alternative philosophical conceptualizations of psychopathology. In Durfee, H. A., Rodier, D. F. T. (Eds.), Phenomenology and Beyond: The Self and its Language. Kluwer Academic, Dordrecht, pp. 41-50.

Ellis, H.D. (1996). Delusional misidentification of inanimate objects: A literature review and neuropsychological analysis of cognitive deficits in two cases. Cognit. Neuropsychiatry, 1, 27–40.

Elvevag, B., McCormack, T., Gilbert, A., Brown, G. D., Weinberger, D. R. and Goldberg, T. E. (2003). Duration judgements in patients with schizophrenia. Psychol Med, 33, 1249–1261.

Friedman, W. (1990). About Time: Inventing the Fourth Dimension. MIT Press, Cambridge, US.

Gallagher, S. (2005). How the Body Shapes the Mind. Oxford University Press, Oxford, UK.

Gallagher. (1998). The Inordinance of Time. Northwestern University Press, Evanston.

Gallagher, S. (2009). Delusional realities. In Bortolotti, L., Broome, M. (Eds.), Psychiatry as Cognitive Neuroscience. Oxford: Oxford University Press, Oxford, pp. 245-66.

Gallagher, S. (2012). Phenomenology. Palgrave-Macmillan, London.

Gallagher, S. and D. Zahavi. (2012). The Phenomenological Mind, 2nd ed. Routledge, London.

González, J.C. (2010a). Du concept "hallucinogène" au concept "lucidogène" (aller-retour). In Baud, S., Ghasarian, C. (Eds.), Des Plantes Psychotropes (195-232). Ed. Imago, Paris.

González, J. C. (2010b). On pink elephants, floating daggers and other philosophical myths. Phenomenology and the Cognitive Sciences, 9, 193–211.

González, J. C. (in press). Traditional Shamanism as embodied expertise on sense and non-sense. In Cappuccio, M., Froese, T. (Eds.), Enactive Cognition at the Edge of Sense-Making. Palgrave Macmillan, London.

González, J.C. and Dokic, J. (2009). Hallucination: Vrai ou faux? Cerveau & Psychologie, 31, 49-51.

Grof, S. (2006). When the Impossible Happens. Adventures in Non-Ordinary Realities. Sounds True, Inc. Canada.

Heidegger, M. (1968). Being and Time. Trans. J. Macquarrie and E. Robinson. Harper and Row, New York.

Heimann, H. (1994). Experience of time and space in model psychoses. In Pletscher, A., Ladewig, D. (Eds.), 50 Years of LSD. Current Status and Perspectives of Hallucinogens. Parthenon Publishing Group, New York.

Hobson, J. A. (2005). Sleep is of the brain, by the brain and for the brain. Nature, 437 (7063), 1254-1256.

Hollister, L. E. (1978). Psychotomimetic drugs in man. In Iverson, L., Iverson, S., Snyder, S. (Eds.), Handbook of Psychopharmacology Vol. 11. Plenum Press, New York, pp. 389-424.

Husserl, E. (1991). On the Phenomenology of the Consciousness of Internal Time (1893-1917). Trans. J. B. Brough. Collected Works: Volume 4. Kluwer Academic Publishers, The Hague.

Husserl, E. (1998). Thing and Space: Lectures of 1907. Trans. and ed. by R. Rojceicz. Collected Works: Volume 7. Kluwer Academic Publishers, Dordrecht.

Huxley, A. (1990). The Doors of Perception & Heaven and Hell. Harper & Row, New York.

James, W. (1890). The Principles of Psychology. Dover, New York, 1950.

Jones, R. T. (1978). Marihuana: Human effects. In Iverson, L., Iverson, S., Snyder, S. (Eds.), Handbook of Psychopharmacology Vol. 12. Plenum Press, New York, pp. 373-412.

Kafka, J. S. (1989). Multiple Realities in Clinical Practice. Yale University Press, New Haven.

Kant, I. (1965). Critique of Pure Reason, Trans. N. K. Smith. St. Martin's Press, London. References given to the second ("B") edition (1787) pagination.

Lehmann, A. and González, J. C. (2009). L'hallucination: entre rêve et perception. Cerveau & Psychologie, 31, 44-48.

Lieving, L. M., Lane, S. D., Cherek, D. R., & Tcheremissine, O. V. (2006).

Effects of marijuana on temporal discriminations in humans. Behavioural pharmacology, 17 (2), 173-183.

Mansky, P. A. (1978). Opiates: Human psychopharmacology. In Iverson, L., Iverson, S., Snyder, S. (Eds.), Handbook of Psychopharmacology Vol. 12. Plenum Press, New York, pp. 95-185.

Mathew, R. J., Wilson, W. H., G Turkington, T., & Coleman, R. E. (1998). Cerebellar activity and disturbed time sense after THC. Brain research, 797 (2), 183-189.

Meck, W. (1996). Neuropharmacology of timing and time perception. Cog Brain Res, 3, 227–242.

Meck, W. H. and Benson, A. M. (2002). Dissecting the brain's internal clock: how frontal–striatal circuitry keeps time and shifts attention. Brain and cognition, 48 (1), 195-211.

Merleau-Ponty, M. (1962). Phenomenology of Perception. Trans. C. Smith. Routledge and Kegan Paul, London.

Michaux, H. (1972). Misérable miracle: La mescaline. NRF, Gallimard. Paris.

Minkowski, E. (1970). Lived time: Phenomenological and Psychopathological Studies. Northwestern University Press, Evanston.

O'Boyle, D. J., Freeman, J. S. and Cody, F. W. J. (1996). The accuracy and precision of timing of self-paced, repetitive movements in subjects with Parkinson's disease. Brain, 119, 51–70.

Ornstein, R. E. (1969). On the Experience of Time. Penguin, Baltimore.

Rammsayer, T. (1989). Dopaminergic and serotoninergic influence on duration discrimination and vigilance. Pharmacopsychiatry, 22 (Suppl.1), 39–43.

Rammsayer, T. (1999). Neuropharmacological evidence for different timing mechanisms in humans. Q J Exp Psych, 52B, 273–286.

Reuter, M., Peters, K., Schroeter, K., Koebke, W., Lenardon,. D, Bloch, B. and Hennig, J. (2005). The influence of the dopaminergic system on cognitive functioning: a molecular genetic approach. Behav Brain Res, 164, 93–99.

Rorty, R. (2003). Out of the Matrix: How the late philosopher Donald Davidson showed that reality can't be an illusion. Boston Globe (10/5/2003). http://www.boston.com/news/globe/ideas/articles/2003/10/05/out_of_the_matrix/

Sass, L. (1992). Heidegger, schizophrenia and the ontological difference. Philosophical Psychology, 5, 109-132.

Schenk, T. and Zihl, J. (1997). Visual motion perception after brain damage: I. Deficits in global motion perception. Neuropsychologia, 35, 1289-1297.

Schutz, A. (1974). Collected Papers Vol. 1. The Problem of Social Reality. Springer, Dordrect.

Shanon, B. (2002). The Antipodes of the Mind: Charting the Phenomenology of the Ayahuasca Experience. Oxford University Press, Oxford.

Tenckhoff A, Tost H, Braus D F (2002) Altered perception of temporal relations in schizophrenic psychoses. Nervenarzt, 73, 428–433.

Tinklenberg, W. J., Kopell, B. S., Melges, F. T., & Hollister, L. E. (1972). Marihuana and alcohol: Time production and memory functions. Archives of General Psychiatry, 27, 812-815.

Vollenweider, F. X. (1998). Advances and pathophysiological models of hallucinogen drug actions in humans: A preamble to schizophrenia research. Pharmacopsychiatry, 31, 92–103.

Vollenweider, F. X., & Geyer, M. A. (2001). A systems model of altered consciousness: integrating natural and drug-induced psychoses. Brain research bulletin, 56(5), 495-507.

Volz H P, Nenadic I, Gaser C, Rammsayer T, Hager F, Sauer H. (2001). Time estimation in schizophrenia: an fMRI study at adjusted levels of difficulty. Neuroreport, 12, 313–316.

Wittmann, M., Carter, O., Hasler, F., Cahn, B. R., Grimberg, U., Spring, P. & Vollenweider, F. X., et al. (2007). Effects of psilocybin on time perception and temporal control of behaviour in humans. Journal of Psychopharmacology, 21(1), 50-64.

Zihl, J., von Cramon, D., and Mai, N. (1983). Selective disturbance of movement vision after bilateral brain damage. Brain, 106, 313-340.

12. Consciousness of Continuity of Now
Steven Bodovitz

Principal, BioPerspectives,
1624 Fell Street, San Francisco, CA 94117 USA

Abstract

One of the defining characteristics, if not the defining characteristic of consciousness is the experience of continuity. One thought or sensation appears to transition immediately into the next, but this is likely an illusion. I propose that consciousness is broken up into discrete cycles of cognition and that the sense of continuity is the result of determining the magnitude and direction of changes between cycles. These putative consciousness vectors are analogous to motion vectors that enable us to perceive continuous motion even when watching a progression of static images. Detailed characterization of consciousness vectors, assuming they exist, would be a significant advance in the characterization of consciousness.

Key Words: brainstorming, consciousness, conscious vector, consciousness vector, continuity, creativity, delay, DLPFC, dorsolateral prefrontal cortex, motion vector, philosophical zombie, sports psychology

Time is the substance I am made of. Time is a river which sweeps me along, but I am the river; it is a tiger which destroys me, but I am the tiger; it is a fire which consumes me, but I am the fire. - Jorge Luis Borges (1946).

1. Introduction

To paraphrase the eloquence of Borges, we are made of time and consumed by time. The continuity of experience is one of the defining features, if not the defining feature of consciousness. To be more specific, as first explained by Karl Lashley, each thought or sensation is stable, but each is immediately present after the other, fully formed, with no experience of the underlying processing that led each to become conscious (Lashley, 1956).

Another way to think about the continuity is through a thought experiment of the inverse condition. Start by imagining a lower state of continuity, in which each individual thought is stable, but gaps are apparent. Each. Word. For. Example. In. This. Sentence. Is. Separated. The extra breaks affect your experience of reading the sentence, because you have to think about the flow of the words to get the meaning. Now jump to a complete loss of continuity. Each. Word. Is. Completely. Frozen. In. Time. For. A. Moment. Each. Pops. Into. Cognition. And. Is. Replaced. By. Another. Without temporal integrity, we are repeatedly frozen in time. Frozen memories can inform us where we've been, but not where we are going. We become biological computers without sentience.

The concept of separating information processing from sentience has been proposed in a much more colorful manner by David Chalmers, who describes a philosophical zombie that roughly appears to be human, but otherwise has no awareness (Chalmers, 1996). This is not as abstract as it sounds, because one aspect of this concept has been demonstrated by Hakwan Lau and Richard Passingham (Lau & Passingham, 2006). These researchers used a variant of the well-known paradigm of masking. In a simple version, subjects are briefly shown an image, known as the target, followed quickly by a second brief image, known as the mask, and if the timing falls into well-defined parameters, the target is eliminated from conscious awareness (Koch, 2004a). Lau and Passingham used a more complex version known as a type II metacontrast masking, but the underlying principle is the same, and they tested the accuracy of identifying the target, in this case a square or a diamond, followed by asking the subjects to press keys to indicate whether they actually saw the identity of the target or simply guessed what it was. By using different lengths of time between the presentation of the target and the mask, the researchers were able to identify two conditions in which the accuracy of identifying the target was statistically the same, but the subjective assessments of awareness were significantly different (Lau & Passingham, 2006). Cognition (defined as high-level biological computation) and awareness can be separated, although presumably only under certain circumstances and for brief periods of time.

Taken together, the thought experiment and the actual experiment suggest that the output of cognitive processing is transferred into consciousness in a continuous or seemingly continuous process. The normal limit for information transfer is the speed of light, which is clearly faster than human perception, but not instantaneous. True continuity would presumably require a mechanism based on quantum mechanics. The possibility that microtubules mediate coherent quantum states across large populations of neurons was proposed by Stuart Hameroff and Roger Penrose (1996). This hypothesis has received indirect support in recent years. Physicists at the University of Geneva, for example, demonstrated quantum entanglement by observing two-photon interferences well above the Bell inequality threshold (Salart et al. 2008), but this was with isolated pairs of photons. Moreover, physicists at the University of California at Santa Barbara were able to coax a mechanical resonator into two states at once, which showed for the first time that quantum events could be observed in complex objects, but this required cooling to near absolute zero (Cho, 2010; O'Connell et al. 2010). Notwithstanding this recent progress, whether microtubules, which undergo constant remodeling, can be islands of quantum events in the biochemical and electrical cauldron of the human brain at 37 degrees Celsius remains to be observed. I propose an alternative hypothesis that, rather than true continuity, consciousness is broken up into discrete cycles of cognition and that the sense of continuity is the result of determining the magnitude and direction of changes between cycles.

2. Consciousness Is Likely Discontinuous

Even though the experience of continuity is a defining characteristic of consciousness, it is likely an illusion. The experimental evidence for the discontinuity is largely based on the delay between sensory perception and conscious awareness. Briefly, the pioneering work on the delay was performed by Benjamin Libet, in which he and his colleagues showed an undetectable stimulus could become conscious after approximately 500 msec (Libet et al. 1964; Libet et al. 1967; Libet et al. 1991), but it is not clear whether the delay was due to the processing time to reach consciousness or the time for summation of the stimulus to reach threshold, and others have criticized Libet's conclusions (for example, see Gomes, 1998; Pockett, 2002). A better study was designed by Marc Jeannerod and colleagues, in which subjects were trained to grasp one of three dowels following the appropriate signal and performed the task with a reaction time of 120 msec. When the subjects were asked to verbalize when they first became aware of the signal, the response time was 420 msec, or 300 msec longer (Castiello et al. 1991). Even allowing 50 msec for the required muscle contraction for verbalization, the delay is still a quarter of a second (Koch, 2004b). Thus, according to this experiment, the frequency of cycles of cognition is roughly 4 per second. A larger and arguably more compelling body of evidence for the delay comes from the well-established phenomena of masking, as described above, in which a mask eliminates and replaces the awareness of a target. The elimination is not the result of interfering with the sensory input because if a second mask is presented, the first mask can be eliminated and the awareness of the target can be restored (Dember & Purcell, 1967). The elimination is only possible with a delay between sensory perception and consciousness. The delay, in turn, indicates that consciousness is discontinuous (Koch, 2004c; Libet, 1999).

3. Continuity and Consciousness Vectors

If consciousness is discontinuous, but appears to be continuous, then the problem of understanding consciousness becomes better posed: what creates the continuity? I propose that the sense of continuity is the result of determining the magnitude and direction of changes between cycles (Bodovitz, 2008). These putative consciousness vectors, however, are largely undefined. They presumably track the magnitudes and directions of multiple changes in parallel and/or in aggregate. Moreover, they presumably track changes in inherently qualitative information, such as words and concepts. While these open questions leave the key tenet of this hypothesis unsubstantiated, they create opportunities for breakthroughs by experts in advanced mathematics, physics and/or computer science. At the very least, efforts to model consciousness vectors may provide insights into the value of using the flow of information as feedback for better organizing complex and dynamic data.

The most significant substantiation of consciousness vectors is through analogy to motion vectors, which add motion to a series of otherwise discrete images. The standard speed for movies based on film is 24 frames per second, but rather than strobing, we perceive smooth motion. This is because motion vectors are calculated by the visual system, most likely in visual area V5, also known as visual area MT (middle temporal). Without motion vectors, vision becomes a series of still images, a condition known as akinetopsia or visual motion blindness (Shipp et al. 1994; Zihl et al. 1983; Zihl et al. 1991). The strobe effect makes otherwise simple tasks, such as crossing a street, extremely difficult. The cars are a safe distance away, then bearing down, without any sense of the transition. Even though there is memory of where the cars were, there is no sense of where they are going. Likewise, without consciousness vectors, simple cognitive tasks involving even a limited series of steps would be extremely difficult.

Motion vectors, like consciousness vectors, appear relatively simple at first approximation. In fact, the retina is arguably even a two-dimensional, Euclidian array of detectors, and most objects in motion follow standard trajectories. Yet, the exact neural computations to generate motion vectors have been difficult to determine and are the subject of decades of debate (for review, see Born & Bradley, 2005), although there is consensus that they involve the mapping of retinal activity onto higher-order visual processors such as those in V1 and V5. Thus identifying the calculations for motion vectors may provide the ideal model system for identifying the calculations for the more complex consciousness vectors.

4. Continuity and the Awareness of Change

If consciousness vectors and a sense of continuity are necessary for consciousness, then the corollary is that changes in cognition are necessary for awareness. This corollary is supported by the fact that we only see changes in our visual field. Even though an image may be static, our eyes never are, even during fixation. Our eyes are in constant motion with tremors, drifts and microsaccades. If these fixational eye movements are eliminated, then visual perception fades to a homogenous field (Ditchburn & Ginsborg, 1952; Riggs & Ratliff, 1952; Yarbus, 1967). The significance of these fixational movements for visual processing has long been debated, and no clear consensus has emerged (for review, see Martinez-Conde et al. 2004). The best correlation of neuronal responses to fixational eye movements are specific clusters of long, tight bursts, which might enhance spatial and temporal summation (Martinez-Conde et al. 2004); in addition, a more recent study showed that fixational eye movements improve discrimination of high spatial frequency stimuli (Rucci et al. 2007). But a lack of enhancement or improved discrimination does not explain the complete loss of visual perception. A better explanation is that changes in cognition are necessary for awareness.

5. Discussion

If cognition is broken up into discrete cycles and consciousness vectors create the illusion of continuity, then conscious feedback has pitfalls. It is slow, such that any action that occurs in less than approximately 250 msec will be over before reaching consciousness. Moreover, if events are happening too quickly and/or you are thinking too fast, your conscious feedback will miss changes in cognition, and, to make matters worse, you will have no immediate awareness of any deficiencies and will only be able to deduce the errors afterwards (see figure 1).

In addition, according to this theory, conscious feedback is inherently subtractive. The feedback tracks the magnitude and direction of what has already happened, thereby constraining the introduction of new ideas. All things otherwise being equal, turning down the feedback should inspire more creativity by allocating more energy to new information. Of course, all things otherwise being equal, without feedback, thoughts will be much more disorganized.

Ideally, these practical benefits are only the beginning. If consciousness vectors are real, and if we can begin to understand how they are calculated, we will have a much deeper knowledge of the highest functions of the human brain and possibly be able to apply our insights to artificial intelligence. Unlocking consciousness vectors may unlock human consciousness.

Acknowledgment: I would like to thank Aubrey Gilbert for her insights and careful review. Some of the material in this review was previously published in Bodovitz, 2008. Whereas the previous review included an argument about the possible localization of the brain region responsible for calculating consciousness vectors, this review is focused more on the significance of continuity and the role of putative consciousness vectors.

References

Bodovitz, S. (2008). The neural correlate of consciousness. Journal of Theoretical Biology, 254(3), 594-598.

Borges, J.L. (1946). A New Refutation of Time.

Born, R.T., Bradley, D.C. (2005). Structure and Function of Visual Area MT. Annual Review of Neuroscience, 28, 157-189.

Castiello, U., Paulignan, Y., Jeannerod, M. (1991). Temporal dissociation of motor responses and subjective awareness. A study in normal subjects. Brain, 6, 2639-2655.

Chalmers, D.J. (1996). The conscious mind: in search of a fundamental theory. Oxford University Press, New York, US.

Cho, A. (2010). The first quantum machine. Science, 330(6011), 1604

Dember, W.N., Purcell, D.G. (1967). Recovery of masked visual targets by inhibition of the masking stimulus. Science, 157, 1335-1336.

Ditchburn, R.W., Ginsborg, B.L. (1952). Vision with a stabilized retinal image. Nature, 170, 36-37.

Gomes, G. (1998). The timing of conscious experience: a critical review and reinterpretation of Libet's research. Conscious and Cognition, 7(4), 559-595.

Hameroff, S.R., Penrose, R. (1996). Orchestrated reduction of quantum coherence in brain microtubules: a model for consciusness. In: Hameroff, S.R., Kaszniak, A.W., Scott, A. C. (Eds.) Toward a Science of Consciousness. MIT Press, Cambridge, MA, pp. 507- 540.

Koch, C. (2003). Lecture 15: On Time and Consciousness. In: California Institute of Technology course CNS/Bi 120. http://www.klab.caltech.edu/cns120/videos.php

Koch, C. (2004a). The quest for consciousness: a neurobiological approach. Roberts & Company, Englewood, Colorado, US, pp. 257-259.

Koch, C. (2004b). The quest for consciousness: a neurobiological approach. Roberts & Company, Englewood, Colorado, US, p. 214.

Koch, C. (2004c). The quest for consciousness: a neurobiological approach. Roberts & Company, Englewood, Colorado, pp. 249-268.

Lashley, K. (1956) Cerebral Organization and Behavior. In: Cobb, S. & Penfield, W. (Eds.) The Brain and Human Behavior. Williams and Wilkins Press, Baltimore, Maryland.

Lau, H.C., Passingham, R.E. (2006). Relative blindsight in normal observers and the neural correlate of visual consciousness. Proceedings of the National Academy of Sciences, 103(49), 18763-18768.

Libet, B. (1999). How does conscious experience arise? The neural time factor. Brain Research Bulletin, 50(5/6), 339-340.

Libet, B., Alberts, W.W., Wright, E.W., Jr., Feinstein, B. (1967). Responses of human somatosensory cortex to stimuli below threshold for conscious sensation. Science, 158, 2597-1600.

Libet, B., Alberts, W.W., Wright, E.W., Jr., Delattre, L.D., Levin, G., Feinstein, B. (1964). Production of threshold levels of conscious sensation by electrical stimulation of human somatosensory cortex. Journal of Neurophysiology, 27, 546-578.

Libet, B., Pearl, D.K., Morledge, D.A., Gleason, C.A., Hosobuchi, Y., Barbaro, N.M. (1991). Control of the transition from sensory detection to sensory awareness in man by the duration of a thalamic stimulus: the cerebral 'time-on' factor. Brain, 114, 1731-1757.

Martinez-Conde, S., Macknik, S.L. Hubel, D.H. (2004). The role of fixational eye movements in visual perception. Nat Rev Neurosci. 5(3), 229-240.

O'Connell, A.D., Hofheinz, M., Ansmann, M., Bialczak, R.C., Lenander, M., Lucero, E., Neeley, M., Sank, D., Wang, H., Weides, M., Wenner, J., Martinis, J.M., Cleland, A.N. (2010). Quantum ground state and single-phonon control of a mechanical resonator. Nature, 464, 697-703.

Pockett, S. (2002). On subjective back-referral and how long it takes to become conscious of a stimulus: a reinterpretation of Libet's data. Conscious and Cognition, 11(2), 144-161.

Riggs, L.A., Ratliff, F. (1952). The effects of counteracting the normal movements of the eye. J. Opt. Soc. Am. 42, 872-873.

Rucci, M., Iovin, R., Poletti, M., Santini, F. (2007). Miniature eye movements enhance fine spatial detail. Nature, 447, 851-854.

Salart, D., Baas, A., Branciard, C., Gisin, N., Zbinden, H. (2008). Testing the speed of 'spooky action at a distance'. Nature, 454, 861–864.

Shipp, S., de Jong, B.M., Zihl, J., Frackowiak, R.S., Zeki, S. (1994). The brain activity related to residual motion vision in a patient with bilateral lesions of V5. Brain, 117(5), 1023-38.

Yarbus, A.L. (1967). Eye Movements and Vision. New York: Plenum.

Zihl, J., von Cramon D., Mai, N., Schmid, C. (1991). Disturbance of movement vision after bilateral posterior brain damage. Further evidence and follow up observations. Brain, 114(5), 2235-2252.

Zihl, J., von Cramon, D., Mai, N. (1983). Selective disturbance of movement vision after bilateral brain damage. Brain, 106(2), 313-340.

13. The Observer's Now, Past and Future in Physics from a Psycho-Biological Perspective
Franz Klaus Jansen

126 chemin Fesquets, 34820 ASSAS, France

Abstract

The observer in physics makes observations and transforms them into fact and physical laws. Observations are based on perceptions and their transformations, which are influenced by biological and psychological functions. As argued by the philosopher Peirce, one might distinguish between extra-mental reality and its mental representation. An observer creates with his mental functions a mental representation of extra-mental reality due to perception based on specific sense organs. Extra-mental reality and its mental representation exist simultaneously, but are not always in direct contact with each other and can therefore diverge. Only during the NOW, the observer is through his sense organs in direct physico-neural contact with extra-mental reality. After interruption of this contact, observations belong to the past and the observer transforms with mental functions regularities of past observations into physical laws, which can be extrapolated into the far past and future. During the NOW, observations have precise time coordinates, but after interruption of the direct contact, memorized observations undergo transformations into abstract and often timeless concepts in classical and in quantum physics. In normal life, time is the perception of duration and its boundaries. In physics, time is reduced to the relation of its boundaries between different systems or can be completely discarded in timelessness. Whereas the NOW is a direct connection between extra-mental reality and its mental representation, past and future represent pure mental representations based on memorized NOWs. After their transformation, mental representation can predict future potentiality, which does not always correspond to extra-mental reality. Due to this reality-potentiality gap, physical laws created in mental representation need verification in a new experimental NOW, which alone assures direct contact to reality.

Keywords: physical observer; physical now; past; present; future; mental representation; extra- mental reality; time unreality; timelessness; Mc Taggert.

1. Introduction

Time is generally experienced or conceptualized as consisting of a "now" "past" and "future." The philosopher McTaggart (1908) distinguished between three series of time, the A-series as past, present and future, the B-series as earlier or later and the C-series as the order of events. However, he considered time as unreal and wrote "It may be the case that... the distinction of past, present and future -- is simply a constant illusion of our minds" (McTaggart, 1908, p.457). Einstein, came to similar conclusions:

"The distinction between past, present and future is only an illusion" (Einstein 1955).

Some theorists have argued, based on Einstein's theorems of relativity (Einstein 1905a,b,c, 1907, 1910, 1961), that the past, present and future overlap and exist simultaneously but in different distant locations in the dimension known as space-time (Joseph 2014). Experiments on quantum entanglement and what Einstein (1930) called "spooky action at a distance" all call into question the causal distinctions between past, present and future. Time may not even exist, except as a function of perception and the nature of consciousness which imposes temporal sequences on experience (Joseph 2011).

Contemporary physicists like Zeh (1998) also ague that time does not exist: "The quantum theory requires that paths can fundamentally no longer exist. Then any parameter of paths, such as the role of time ... cannot exist either." (Zeh, 1998, p. 15). Then what is time? Do clocks measure time, or is time a manifestation of clocks which are built according to the rules imposed by the conscious mind to account for the experience of change?

As demonstrated by Einstein and others, the experience of time and the dimension described as time-space, are effected by velocity and gravity. However, is what is experienced, time, or velocity and gravity all of which can effect even atomic clocks? The astronomer Barbour (2009, p. 1). writes "Duration and the behavior of clocks emerge from a timeless law that governs change" and Rovelli (2009, p. 1) that "... the best strategy for understanding quantum gravity is to build a picture of the physical world where the notion of time plays no role at all."

Thus, there is a general claim from philosophers, astronomers and physicists, that time does not exist and that the experience of time may be only an illusion of our minds imposed on the quantum continuum by consciousness.

"I regard consciousness as fundamental. I regard matter as derivative from consciousness" (Max Planck, 1931).

The "future" and the "past" are shaped and affected by consciousness which can effect events just by observing them; as illustrated by "entanglement" and Heisenberg's well established Uncertainty Principle (Heisenberg 1927). The Uncertainty Principle holds that we cannot know velocity, position and mass of any particle simultaneously. Moreover, uncertainties propagate into the future; that is small uncertainties becomes larger as distance from the present increases. Consciousness, therefore, is entangled with the quantum continuum (Joseph 2014).

The role of an observer is an essential part in quantum physics. Von Neumann included the observer as a quantum system (Zeh, 2013, p. 98). Wigner (1962) proposed an active influence of consciousness on the physical world. Zeh (2013, p. 99) himself insisted that the observer always had an essential role, since he performs all observations with physical instruments. In these theories, the observer is only a passive bystander for collecting observations. The recent form of quantum Bayesianism attributes to the observer a more active role, when he makes probability assignments to express his expectations (Fuchs et al. 2013).

However, if consciousness plays an active or passive role is debatable. Quantum physics, as exemplified by the Copenhagen school (Bohr, 1934, 1958, 1963; Heisenberg, 1925, 1927, 1930), like Einsteinian physics, makes assumptions about the nature of reality as related to an observer, the "knower" who is conceptualized as a singularity (Joseph 2014). As summed up by Heisenberg (1958), "the concepts of Newtonian or Einsteinian physics can be used to describe events in nature." However, because the physical world is relative to being known by a "knower" (the observing consciousness), then the "knower" can influence the nature of the reality which is being observed through the act of measurement and registration at a particular moment in time. The same principles must be applied to time. Time is effected by observation.

What is observed or measured at one moment can never include all the properties of the object under observation, and this includes the perception of time. In consequence, what is known vs what is not known becomes relatively imprecise (Bohr, 1934, 1958, 1963; Heisenberg, 1925, 1927). Moreover, as dictated by the "uncertainty principle" energy and mass can be time-independent (Heisenberg 1927, 1958). This is illustrated by evidence of entanglement where effects may occur simultaneously with causes, and take place at faster than light speeds (Francis 2012; Juan et al. 2013; Lee et al. 2011; Matson 2012; Plenio 2007). One might conclude, therefore, that if consciousness plays a passive or active role, time remains an illusion and there is no past, present, or future, as all are merely aspects of the quantum continuum which become subject to the passive or active observation of consciousness which perceives all experience in terms of temporal sequences (Joseph 2011).

From a psychological viewpoint, the observer has a much greater influence, since besides the observation of information, he also affects a transformation of information. Both functions are essential for the establishment of physical laws and are differently influenced by present, past and future. The present is essentially required for observation, the past for memorized observations and their comparison for detecting regularities and the future consists in the extrapolation of past regularities, which are then considered as general physical laws with validity from the past to the future. Thus, the observer's present, past and future exert a special role during the constitution of physical laws, which will be analyzed by distinguishing between the extra-mental reality and its mental representation, which is responsible for the transformation of regularities from past observations into general physical laws. Extra-mental reality concerns all physical factors outside the body and all biological factors within the body, like respiration or heart beats, whereas mental representation has the function to represent the extra-mental reality in the human mind. The NOW is only present when both are in direct contact, whereas the past and the future are pure mental representations no longer linked to the extra-mental reality and require verification in a new NOW.

2. Physical Reality and its Mental Representation

As summed up by the philosopher CS Peirce reality exists independent of the time "... it is out of the mind, is independent of how we think it, and is, in short, the real." (Hookway, 2013, paragraph 3.1). In the same way mental representation could be distinguished from extra-mental reality as different independent entities. Extra-mental reality would comprise all physical events outside the body and all biological events inside the body, whereas mental representation concerns sensation, perception and cognition. The only connection between both realities are sense organs permitting direct physico-neural contact for representing the extra-mental reality within mental representation.

Humans lack the sensory perceptual capability to perceive X-rays, radioactivity, radio waves and others. In the same sense, blind and deaf people have a different representation of the extra-mental reality which is shaped by their sensory experiences. Certain animal species possess additional sense organs permitting a richer representation of extra-mental reality. Pigeons have sense organs for the detection of magnetic fields (Wu & Dickman, 2012), bees perceive ultra-violet light (Frisch, 1963) and some fish species can feel electric fields (Pusch et al. 2008). For these reasons, the representation of extra-mental reality is necessarily incomplete and can therefore only be considered as a model representation.

In addition sense organs only detect a small part of the universe and the finite memory only retrieves part of all encoded observations. The view of the human eye is limited by a perspective, not allowing to see objects at long distances with precision. Hearing and smelling are also limited to certain distances. Nevertheless, science allowed indirect knowledge on imperceptible physical events by transforming them to perceptible manifestations, for instance a radio can transform inaudible radio waves into audible mechanical waves. Thus, the extra-mental reality is necessarily different from its mental representation and can only be modeled with respect to the information provided by sense organs.

Since the mental representation can be different from the extra-mental reality, it has to be verified. David (2009, p. 1) described the correspondence theory of truth and defined that "... truth is a relational property involving a characteristic relation (to be specified) to some portion of

reality ...". The correspondence may not always be evident for events lying beyond the perspective of sense organ at physical scales not directly accessible. However, if the correspondence is not verified, one could believe in physical laws established in mental representation, which after extreme extrapolation into the far past or the final future may no longer correspond to their extra-mental reality.

3. Present, Past and Future in Natural Science

Corballis (2014) and Tulving's theory (1983) argue that memory serves not only to remember and reconstruct the past but also creates the future by planning for or anticipating the future episodes all of which contributes to the subjective sense of time. Memory, therefore, serves consciousness by providing a foundation for the perception of time as consisting of a future, past, or present. Memory, however, is therefore subject to observation and observation may effect memory; and which may explain why so many people remember things differently, even eye-witnesses to a crime (Joseph 2014).

The philosopher McTaggart (1908) argued that "We perceive events in time as being present, and those are the only events which we perceive directly. And all other events in time which, by memory or inference, we believe to be real, are regarded as past or future".

Zeh (2013,) argues that "... The observer has always played an essential role in the empirical sciences, simply because they are based precisely on observations performed by humans by physical means."

The acquisition of knowledge through observation by an observer is subjected to its position in time: the present, past and future. The present allows acquiring new knowledge, the past is limited to knowledge already acquired in a former present, whereas the not yet existing future can only be imagined with uncertainty by projection of past regularities into the future. Regularities found in observations allow the conception of physical laws under three basic conditions dependent on present, past and future:

a) observations in the present are the bases for all physical laws (section 3.1),

b) memorized observations of the past are transformed into abstract regularities (section 3.2),

c) abstract regularities are extrapolated into the far past and future (section 3.3).

Although physical laws seem to possess universal validity from the past to the future, the initial acquisition of information by an observer was only possible in a limited present, but the past and future have still important consequences, which will be discussed.

3.1 The "Now" or the Present. Observation consist in a mental perception of information provided by all sense organs communicating by physico-neural contact with the extra-mental reality. Information from extra-mental reality comprises physical events outside the body as well as physiological events inside the body.

Sense organs are an intermediary to extra-mental reality, in which physical stimulation of the sensory surface followed by sensory transduction lead to perception in mental representation (O'Callaghan 2012). For vision, physical factors such as photons are directly transmitted to the retina of the eye, where they activate light sensible neurons by depolarization in a bottom-up direction to specialized brain regions. The ear receives mechanical waves, which induce neurological activations transmitted to other specialized brain regions. Heat activated skin receptors also transmit their information by neurons to the corresponding brain regions. Thus in the present, there is a direct physico-neural information influx between the extra-mental reality and its mental representation in the brain. However, as soon as sense organs interrupt the direct physical contact, for instance by closing the eyes or the ears, any communication with the extra-mental world is broken, whereas the already obtained observations are partially encoded in the episodic long-term memory and after retrieval represent the past.

The direct contact of extra-mental physical factors with biological sense organs lead to sensory transduction with transmission to specific brain regions, where mental representation takes place and explains why this physico-neural contact can only be found in the present. The notion of "NOW" could be psychologically defined and describe time as duration limited by boundaries and unified by a common aspect, which can be extremely short, when hearing the thunder of an unexpected lightning, but can also take minutes to hours when seeing a film or reading a book. It corresponds always to a physico-neuronal information influx, in which extra-mental reality and mental representation are in direct contact. However, when thinking on physical laws, the physico- neural contact can be interrupted. Time is then reduced to its quantitative aspect as relation between boundaries, whereas its qualitative aspect as psychological duration disappears. In the general relativity theory, time is conserved as relation with respect to a reference frame.

3.2 Perception during the NOW. In contrast to pure elementary sensation, during the construction of mental representation from extra-mental reality by perception (O'Callaghan 2012), top down activities are simultaneously activated. They allow binding of the information from different sense organs into a unified multi-modal experience (Lycan 2012), such as simultaneously viewing a fire in the chimney, hearing its sound and feeling its heat. Other cognitive activities are categorization such as grouping animals into classes (Pothos, 2011) or reactivation of visual entities from long-term memory when perceiving faces. In neuroscience visual mental imagery with pure top-down activation mechanisms has to be distinguished from visual percepts including bottom-up mechanisms, which explains why people in general do not confuse visual percepts with visual mental images (Ganis & Schendan, 2008). Mental imagery with pure top down activation no longer necessitates direct elementary sensation from extra-mental reality.

Corballis (2014) distinguishes the psychological present and the past with the concepts of anoetic consciousness, as awareness of the present and noetic consciousness as semantic memory, representing events not tied to the immediate environment and allowing mental time travel between the past and the future. Finally autonoetic consciousness allows episodic memory containing events already experienced by the self and later reminded as the past.

3.3 The Past as Memorized Observation. The NOW allows direct contact by observation between extra-mental reality and its mental representation, which is then encoded in an episodic long-term memory and can later be retrieved as the past. Encoding is an automatic by-product of attending to or processing a stimulus and therefore differs according to the level of attention and processing (Craik & Lockhart, 1972). As soon as observations are memorized, they already represent the past. Thus, immediately after the break down of direct physico-neural contact between extra-mental reality and mental representation, characterized as elementary sensation with sense organs, a partial copy of the prior observations remains. Thereby observations in the present are not lost and can be used for further analyses with cognitive functions. Past observations have the great advantage of preserving copies from multiple observations, which allow their comparison for exploring regularities.

Thus, only the past containing multiple memorized observations allows comparisons, although it is disconnected from any direct physical contact with the extra-mental reality. This complete detachment of memorized observations also allows the transformation of information. In contrast to unchangeable "elementary sensations" in the present, mental cognition is able to modify information of past observations through: 1. association, 2. dissociation or 3. modeling. Besides the representation of extra-mental reality as past observations, their cognitive transformation allows a total independence of mental representation from extra-mental reality. This kind of representation could be called "potentiality" or "possible worlds" as claimed by philosophers like Dennett (2004). Thus, there is in mental representation a gap between its representation of observed reality and its representation of imagined potentiality (i.e. Reality Potentiality Gap or "RePoGap"). This gap is permanently experienced by everybody, when one's imagination does not correspond to one's observation of extra-mental reality.

3.4 Transformation of Past Observations. There are three main transformations of past observations : association, dissociation and modeling. A first transformation by the observer with his cognitive functions is the association of interpretations to perceptions, similar to the model of perception from Bruner (2011). The sun is daily perceived as rising in the east and setting in the west. Although the corresponding interpretation of geocentrism was believed for many centuries, more precise information imposed a different association with the interpretation of heliocentrism for exactly the same perceptions. Thus, insufficient information may induce incorrect interpretations, which have to be changed after more detailed information becomes available.

A second transformation by cognitive functions is the dissociation of certain properties from the complexity of memorized observations. In Galilee's (1638) experiment on the inclined plane, repetitive time measurements had to be compared, but only the relations of the boundaries of time could be used in mathematical calculations and time as duration was dissociated from the relation of its boundaries. Physical laws thereby become invariant with respect to the observed individual time measurements. Thus, our knowledge on the whole universe can be considered as relational and only allows correlations between different systems (Baird, 2013). When only one aspect of highly complex past observations is under study, the other aspects are discarded. Thereby, complexity becomes reduced and looses all other properties, which are not under study.

A third transformation of memorized observations is modeling of past observations with visual models from the macrocosm, such as atoms as billiard balls or with mathematical models by following Galilee. Mathematical models can be of a different nature, either describing, or simplifying or approximating past observations. Galilee's mathematical models directly described the behavior of bronze balls on the inclined plane. Statistical methods simplify the behavior of multiple individual events by calculating one virtual mean value and its standard deviation (sigma) with arbitrary confidence levels such as 95%. In quantum physics probability estimations approximate the characteristics of the behavior of elementary particles. According to the mathematical model applied, past observations are more or less precisely represented by mathematical models.

As expressed by the Heisenberg uncertainty principle (Heisenberg, 1927), the more precisely one physical property is known the more unknowable become other properties. The more precisely one property is known, the less precisely the other can be known and this is true at the molecular and atomic levels of reality (Bohr, 1934, 1958, 1963). Heisenberg's principle of indeterminacy focuses on the relationship of the experimenter to the objects of his scientific scrutiny, and the probability and potentiality, in quantum mechanics, for something to be other than it is. Einstein objected to quantum mechanics and Heisenberg's formulations of potentiality and indeterminacy by proclaiming "god does not play dice."

According to Heisenberg (1925, 1927, 1930), chance and probability enters into the state and the definition of a physical system because the very act of measurement can effect the system. No system is truly in isolation. No system can be viewed from all perspectives in totality simultaneously which would require a god's eye view (Joseph 2014). As summed up by Joseph (2014) "Only if the entire universe is included can one apply the qualifying condition of "an isolated system." Simply including the observer, his eye, the measuring apparatus and the object, are not enough to escape uncertainty. Results are always imprecise. Heinsenberg puts it this way:

"What one deduces from an observation is a probability function; which is a mathematical expression that combines statements about possibilities or tendencies with statements about our knowledge of facts....The probability function obeys an equation of motion as the co-ordinates did in Newtonian mechanics; its change in the course of time is completely determined by the quantum mechanical equation but does not allow a description in both space and time....The probability function does not describe a certain event but a whole ensemble of possible events" whereas "the transition from the possible to the actual takes place during the act of observation... and the interaction of the object with the measuring device, and thereby with the rest of the world... The discontinuous change in the probability function... takes place with the act of registration, because it is the discontinuous change of our knowledge in the instant of registration that changes the probability function." "Since through the observation our knowledge of the system has changed discontinuously, its mathematical representation has also undergone the discontinuous change and we speak of a quantum jump" (Heisenberg, 1958).

The same principles can be applied to the experience of time by the conscious mind (Joseph 2014). The mental transformation of past observations through association, dissociation and modulation, shows that memorized past observations can be completely transformed by cognitive mental functions into virtual concepts, corresponding only partially to the initial observations. Then, the mental representations of physical laws could only be considered as models for extra-mental reality.

3.5 Timelessness of Past Observations. As detailed by Joseph (2014): The wave function describes all the various possible states of the particle. Rocks, trees, cats, dogs, humans, planets, stars, galaxies, the universe, the cosmos, past, present, future, as a collective, all have wave functions. Waves can also be particles, thereby giving rise to a particle-wave duality and the Uncertainty Principle. Particle-waves interact with other particle-waves. The wave function of a person sitting on their rocking chair would, within the immediate vicinity of the person and the chair, resemble a seething quantum cloud of frenzied quantum activity in the general shape of the body and rocking chair. This quantum cloud of activity gives shape and form to the man in his chair, and is part of the quantum continuum, a blemish in the continuum which is still part of the continuum and interacts with other knots of activity thus giving rise to cause and effect as well as violations of causality." As summed up by Joseph (2014) "in a quantum universe all of existence consists of a frenzy of subatomic activity which can be characterized as possessing pure potentiality and all of which are linked and entangled as a basic oneness which extends in all directions and encompasses all dimensions including time (Bohr, 1958, 1963; Dirac, 1966a,b; Planck 1931, 1932, Heisenberg 1955, 1958; von Neumann 1937, 1955). The act of observation be it visual, auditory, tactile, mechanical, digital, is entangled with the quantum continuum and creates a static impression of just a fragment of that quantum frenzy that is registered in the mind of the observer as length, width, height, first, second, and so on; like taking a single picture of something in continual motion, metamorphosis, and transformation. That is, the act of sensory registration, be it a function of a single cell, or the conscious mind of a woman or man, selects a fragment of the infinite quantum possibilities and experiences it as real, but only to that mind or that cell at the moment of registration."

Kubs et al. (1998) demonstrated the direct interaction of instrumental observation on the behavior of elementary particles, when an increasing "which-way detector " changed the interference pattern of electrons. Stapp (2004) conceives an active observer, who has the freedom of choice and intentionally prepares experimental actions, some of which lead to a wave like behavior and others to the particle structure of elementary particles. His theory is based on the interpretation of von Neumann, who included the observer in the quantum system and the collapse of the quantum system was considered as a psycho-physical parallelism (Zeh, 2013 p 98). The observer collects information by observation, thereby inducing the wave function to collapse. Wigner also suggested a participation of consciousness in the physical world (Wigner, 1962). According to Everett, the subjective observer may simultaneously exist in various versions in his multi-world theory. Zeh extended this theory by including the mind of the observer as an essential factor in his multi-mind theory (Zeh, 1970). Besides Stapp, these theories considered the observer as a passive collector of information. Fuchs et al. (2013) gave the observer a more active role by creating with probability assignments a belief in future experimental outcomes and Stapp proposed that the observer by his freedom to choose experiments actively interfered with physical reality (Schwarts et al 2005).

Reducing individual observations only to one property, such as time, space, weight, height or others, allows the elimination of other properties not under study. Besides others, time coordinates can be eliminated, which leads to partial or total timelessness. Partial timelessness can be represented by astronomical constellations only requiring time as relations. The Antikythera Mechanism, a mechanical computer for the calculation of astronomical positions, was used by the ancient Greeks already in the first century BC (Price, 1975) and indicated the positions of the sun, the moon and the planets Venus and Mars. A modern planetarium fulfills the same function, but with a digital computer. Such mechanisms can be turned forwards but also backwards in time for other astronomical positions relative to another time point. Thereby the time arrow is no longer represented and only time relations indicate the corresponding positions of planets. Thus, already classical physical formalism concerning the relation between planets could reduce time only to relations corresponding to partial timelessness.

Time is entangled and is affected by consciousness and relative to and effected by the act of observation and measurement--as predicted by quantum mechanics

(Bohr, 1958, 1963; Dirac, 1966a,b; Planck 1931, 1932, Heisenberg 1927, 1958; Neumann 1937, 1955). Complete timelessness is found in quantum physics, when time measurements become uncertain, due to variable outcomes under identical experimental conditions. The uncertainty of time relations only allows probability estimations and corresponds to complete timelessness as claimed by physicists (Zeh, 1998). Nevertheless, complete timelessness can also be found in classical physics, when the time relations are without interest, for instance when only variations are calculated. Already in Galilee's experiments the grouping of individual measurements showed a certain variability of outcomes, which corresponds to a timeless concept even in classical physics.

When timeless values are represented in a graph, they could be interpreted, as if they were in superposition or obtained at an identical virtual time point. Although the physical formalism is completely different, the general idea of superposition of values in classical physics for an identical virtual time point might be seen in analogy to superposition of wave structures in quantum physics. The principle of superposition was applied in the mathematical formalism of the wave function with complex amplitudes by Schrödinger, which differs from weighted sums in classical theory. The wave function also proved to be a better model for human decision making (Pothos & Busemeyer, 2009) and could be interpreted as a superposition of mental functions. Thereby superposition in quantum mechanics seems to be isomorphic to superposition of cognitive mental functions (Jansen, 2008, 2011b).

Highly varying observations can only be extrapolated into the future as timeless group values with their means and variations, but not predict individual time coordinates. Nevertheless, in new measurements, extrapolated mean values will necessarily acquire new individual outcomes with their time coordinates. Since for irregular outcomes only their group behavior is predictable, individual behavior appears to be indeterminist. Thus, the extrapolation of timeless values into the future will acquire new precise time coordinates in new observations.

3.6 The Future as Extrapolation from the Past. Comparison between multiple observations can only be obtained with memorized past observations, since present observations are not finished and future observations are not yet existing. The only way to have a guess of the future is the projection of past events into the future. Patients with a bilateral medial temporal ablation of the brain for the reduction of epilepsy are unable to look neither into the past nor into the future (Berlucchi, 2014). However, the future is of great importance for normal life, and imagining the past is probably not so essential for life as projecting the future (Corballis, 2014). The scientific aim for prediction of the future could be realized by a "physical theory of everything", which covers the past, the present and the future and can be established with past regularities under the condition that they are also reliable in the future. In physics, this could be achieved by the extrapolation of mathematical formalism obtained from the past and projected into the future.

The reliability of any projection from the past into the future essentially depends on the regularity of past events. The revolution of the earth around its axes providing the time unite of one day with 24 hours is highly regular. In contrast, irregular weather conditions in the past projected in the future remain extremely hazardous. Nevertheless, even the extrapolation of apparently regular events, like the revolution of the earth, can also lead to errors at the long term, due to tidal forces of the moon rendering the revolution of the earth more slowly, although only significant in billions of years. Thus, any extrapolation from past events after a relatively short time of observation into the far future of billions of years may become doubtful. The extrapolation to the far past is also uncertain. Under the condition that physical laws are constant, one can calculate that the Big Bang occurred about 14 billion years ago. However, how can the constancy of physical laws and constants be verified over such long time periods of billions of years ? The opposite hypotheses that physical laws could have changed, especially during the first periods after the Bib Bang, would allow another interpretation of the anthropic principle, which indicates that the actual fine tuned universe seems to be required for allowing life to emerge. Magueijo (2004) proposed a change of the speed of light during the early universe, which is in opposition to Einstein's constancy of the speed of light. If physical laws had changed in the past, the actual fine tuned universe could have been obtained through evolution, whereas with constant laws, a fine tuned universe seems to need a prior design.

3.7 Projection of Irregular Outcomes into the Future. Only for extremely regular past observations in classical physics, like ocean tides, individual outcomes will be precisely predictable for the future. The situation is different for the extrapolation of highly varying observations grouped around means with timeless values, which can only predict mean future values. However, in new experiments timeless values necessarily acquire new time coordinates, which could not be predicted with timeless extrapolations. In classical physics mathematical equations do not take into account the variability of experimental manipulation errors and only calculate theoretically modeled outcomes. Thereby the precise practical outcomes induced by manipulation errors are unpredictable. Thus, already in classical physics, there is a limited uncertainty for practical outcomes, although it remains relatively small, not producing practical problems. However, in biology variations can become considerable and sometimes threaten an unambiguous interpretation. The chaos theory based on small initial variations might explain some of such variations.

In quantum mechanics, variations are also considerable, but for different reasons due to Heisenberg's uncertainty principle as well as to instrumental variations. The measurement of location of elementary particles can no longer provide a precise measurement of their velocity and leads to variations. The physical wave function allows the extrapolation of timeless values only as probabilities into the future and the variability of outcomes is mathematically calculated by superposition of timeless values. Even in classical physics, precise time coordinates including manipulation errors cannot be predicted, so that the expected values and their time coordinates remain uncertain. In quantum mechanical experiments the superposition of outcomes cannot be observed, only individual outcomes are found, which correspond to their probable timeless values and seem to appear randomly. Therefore, multiple repetitions are needed for statistically significant outcomes.

Complete knowledge on all past and future events from the initial Bib Bang to the final Big Freeze seems to be achieved by the extrapolation of physical formalism into the far past and the final future. A physical "theory of everything", unifying all physical forces in one unique theory, could be considered as complete knowledge on the whole universe including the whole past and future. Such a view on the universe resembles the philosophical theory of eternalims, already evoked by the Greek philosopher Parmenides and is also called Block-universe (Savitt, 2013). However, the block-universe can only be a model in mental representation and its exact correspondence to extra- mental reality can never be verified for the far past and the final future and necessarily remains a mathematical model with uncertainty.

4. McTaggert's Unreality of Time

The philosopher McTaggart (1908) distinguished between the positions in time, the A-series as past, present and future, the B-series as earlier and later and the C-series as ordering of events. The A- series was explained with the example of the death of Queen Anne, which is for an observer first a future event, then becomes a present and finally a past event. Thus, the same event acquires a different character, which seems to be in contradiction with the unicity of the physical event of the death of Queen Anne happening only once. However, an observer can in his mental representation position the death of the Queen with respect to his own NOW. Then the death could be expected in the observer's future, or happening in his present or was already finished in his past. Thereby the A- series of past, present and future comprises in addition to the unique biological fact a relation to the observer's NOW, whereas the B series concerning earlier and later and the C-series considering an ordering are no longer positioned in relation to the observer, but with respect to each other. Therefore the A, B and C series as relations belong to mental representations.

Mc Taggert argues that the A-series is unreal. The notion of "real" is generally considered to distinguish between the extra-mental and the mental representation world. In this sense, Mc Taggert's seems to interpret the A-series as a mental manifestation, which he considers as unreal and thereby an illusion. There is agreement with Mc Taggert, that the A-series corresponds to mental representation and relations to the personal NOW. However, why should mental representation be an "illusion"? Mental representation reality in the mind can be considered as real and existent in the same sense as the extra-mental world. Thus, there is no illusion, there are two realities with different properties, but not always corresponding to each other.

In mental representation, the notion of time has two aspects, a more qualitative biological and a pure quantitative physical aspect. In biology, time can be considered as a perception of duration with boundaries (Bergson 1992), which is more qualitative and corresponds to the inner time clock, also necessary in biology for the circadian rhythm (Wearden, 2005). Time evaluated with the inner time clock can be considered as perception, similar to visual, auditive or temperature perception depending on specialized sense organs. All sense organs are activated by a physical factor, such as seeing by electromagnetic waves, hearing by mechanical waves and heat feeling by molecular agitation. Eagleman (2009, p 1841) proposed a physical factor for time perception: "the experience of duration is a signature of the amount of energy expended in representing a stimulus, i.e. the coding efficiency." Psychological time perception is opposed to the fixity of physical time and therefore allows a mental time travel between the past, the present and the future (Corbalis 2014).

The physical aspect of time is necessarily only quantitative and therefore restricted to the boundaries of durations, since they can be set in relation to intervals of other systems in extra-mental reality. Thus, the qualitative aspect of time in biology as perception of duration becomes in physics reduced to its boundaries, which can be compared to other time intervals, such as the revolution of the earth, counted as one day with its fractions. The more qualitative biological perception of the inner time clock is not identical to physical relations. For short durations, less than a second, time perception can resemble physical relations, although duration estimations in psychological tests can be shorter or longer than the corresponding physical relations (Eagleman, 2009). This proves that the inner clock is a biological perception of time, which is different from physical relations, but it has biological functions, like the circadian rhythm adapting biological activity to the day and night rhythm. For short durations the inner time clock and physical clocks seem to have good correspondence, but for durations shorter than the flicker frequency or very long in the range of years or centuries, the inner time perception is no longer adapted and only relations to extra-mental systems like clocks can replace time perception in mental representation. The high concentration of information in physical formalism is also unadapted for time representation, however, time reappears, when the abstract formalism of physical laws is applied to new concrete experiments (Jansen 2011, a).

6. Conclusions

The observer creates within his mental functions a representation of extra-mental physical reality. Both exist simultaneously, but are not always in contact with each other and can diverge. Only during the present, the observer is in direct physical contact with extra-mental reality due to "elementary sensation" with physico-neural contact through his sense organs. After interrupting the contact, the observer can with his mental functions transform regularity of past observations to physical theories and extrapolate them to the far past and final future. In the present, during the direct contact with physical reality, all observations have precise time coordinates. Once the direct contact is interrupted, memorized observations undergo important transformations into abstract, often timeless concepts for classical as well as for quantum physics. Thereby, timeless mathematical formalism becomes the bases of physical theories for the prediction of an unobservable past or future. Between the reality of the NOW and the potentiality of the future there is an important gap, the Reality-Potentiality Gap (RePoGap), which can be permanently verified by everybody, when potential projects imagined in mental representation cannot always be realized in extra-mental reality.

Time defined as duration with boundaries exists in both, mental representation and extra-mental reality during direct contact in the NOW. Partial physical timelessness corresponds to the reduction of duration to its boundaries, only considering the relations between systems in physical equations. However, time reappears, when physical laws are again verified in new experiments in a new present. The NOW is the only access of mental representation to extra-mental reality, whereas past and future are only mental representation models for extra-mental reality. Thus, physical formalism for the whole universe can only be considered as a "mathematical model universe".

All physical theories are necessarily based on an observer, who is the only one to collect observations and thereafter to transform them into physical theories. Only when the observer is not considered, a "theory of everything" could give the impression of complete knowledge similar to the philosophical concept of eternalism of Parmenides. However, if the observer is still considered as an important factor in the conception of physical theories, his uncertainty about the unverifiable extrapolation to the far past and the final future has also to be respected. There seems to be a semantic problem with the concept of time. Time perception could only be an illusion, if visual perceptions reduced to physical wave lengths is also considered as illusion. But then every reduction to more basic physical entities should be an illusion and the word losses its original sense. Misinterpretation after unclear sensory experience remains the essential meaning of illusion, whereas any reduction to lower level physical factors should not be considered as illusion. If a future "theory of everything "only considers relations between systems, the notions of present, past and future will disappear in the physical formalism. In this sense Barbour (2009, p.1) might be interpreted when claiming "Duration and the behavior of clocks emerge from a timeless law that governs change". In agreement with Barbour, physical equations essentially consider timeless relations conceived in mental representation, but during the verification of timeless laws in new experiments with a new NOW, time emerges again with duration and boundaries and the behavior of clocks indicates time again.

REFERENCES

Bohr, N. (1958/1987), Essays 1932–1957 on Atomic Physics and Human Knowledge, reprinted as The Philosophical Writings of Niels Bohr, Vol. II, Woodbridge: Ox Bow Press.

Baird, P. (2013). Information, universality and consciousness: a relational perspective. Mind and Matter, 11.1, 21-43.

Barbour, J. (2009). The Nature of Time. arXiv:0903.3489 (gr-qc)

Barbour, J (2008). Youtube, a video on "Killing time " http://www.youtube.com/watch?v=WKsNraFxPwk

Bergson, H. (1992). The Creative Mind. tr., Mabelle L. Andison, The Citadel Press, New York. translation of "La Pensée et le mouvant" [1946].

Berlucchi, G. (2014). Mental Time Travel: How the Mind Escapes from the Present. Commentary on Michael C. Corballis's Essay. Cosmology, 18. 146-150.

Bitbol, M. (2008). Is Consciousness Primary? NeuroQuantology, 6.1, 53-71.

Bruner, J. (2011) in Alan S. & Gary J. Perception, Attribution, and Judgment of Others. Organizational Behaviour: Understanding and Managing Life at Work Vol. 7

Buks, E., Schuster, R., Heiblum, M., Mahalu, D., Umansky, V,. (1998) Dephasing in electron interference by a 'which-path' detector. Nature 391, 871-874.

Chalmers, D.J. (2003) Consciousness and its Place in Nature. in (S. Stich and F. Warfield, eds) Blackwell Guide to the Philosophy of Mind , Blackwell, Oxford, UK. 102 – 142.

Corbalis, M.C., (2014). Mental Time Travel: How The Mind Escapes From The Present. Cosmology, 18. 139-145.

Craik & Lockhart (1972) Levels of Processing: A Framework for Memory Research. J. Verbal Learning and Verbal Behavior, 11, 671-684.

David, M. (2009). The Correspondence Theory of Truth. (Stanford Encyclopedia of Philosophy)

Dennett, D. C. (2004). Freedom evolves. London: Allen Lane.

Eagleman, D. M. and Pariyadath V. (2009). Is subjective duration a signature of coding efficiency? Phil. Trans. R. Soc. B, 364, 1841–1851.

Frisch, C. von (1963). Bees: Their Vision, Chemical Senses, and Language. Vail-Ballou Press, Inc. Cornell University Press, Ithaca, New York.

Fuchs, C.A., Mermin N.D. and Schack R. (2013). An introduction to QBism with an application to the locality of quantum mechanics. arXiv: 1311.5253v1 (quant-ph).

Ganis, G. and Schendan, H. E. (2008). Visual mental imagery and perception produce opposite adaptation effects on early brain potentials. NeuroImage, 42, 1714–1727.

Hamilton, D. L. (1996). The Geophysical Effects of the Earth's Slowing Rotation. The Mind of Mankind, Suna Press, New York.

Heisenberg, W. (1927), "Über den anschaulichen Inhalt der quantentheoretischen Kinematik und Mechanik", Zeitschrift für Physik 43 (3–4): 172–198.

Hookway, C. (2013). Pragmatism. Stanford Encyclopedia of Philosophy.

James, W. (1907). Pragmatism. In: Pragmatism and The Meaning of Truth. Harvard University Press, 1975. Cambridge, Mass.

Jansen, F. K. (2008). Partial isomorphism of superposition in potentiality systems of consciousness and quantum mechanics. NeuroQuantology, 6.3, 278-288.

Jansen, F.K. (2011a). Isomorphism of hidden but existing time in quantum mechanical formalism and human consciousness. NeuroQuantology, 9.2, 288-298.

Jansen, F.K. (2011b). Isomorphic concepts for uncertainty between consciousness and some interpretations of quantum mechanics. NeuroQuantology, 9.4, 660-668.

Jansen, F.K. (2014). Elementary and memory perception versus cognition with implications for the brain-consciousness problem. Congress : Towards a Science of Consciousness, Tucson AZ.

Joseph (2011). Quantum Physics and the Multiplicity of Mind: Split-Brains, Fragmented Minds, Dissociation, Quantum Consciousness. In Consciousness and the Universed, Penrose, R. (ed). Cosmology Science Publishers.

Joseph (2014) The Time Machine of Consciousness, Cosmology 14, 400-450

Lykan W.G. (2012). Consciousness. in The Cambridge Handbook of Cognitive Science (Frankish K. and Ramsey W M eds) p 213, Cambridge University Press, Cambridge.

Magueijo, J. (2004). Faster than the speed of light: the story of a scientific speculation. Arrow Books, London. ISBN:9780099428084.

McTaggart, J. (1908). The unreality of time. Mind, 17, 457-473.

O'Callaghan, C. Perception (2012) in The Cambridge Handbook of Cognitive Science (Frankish K. and Ramsey W.M. eds) p 76, Cambridge University Press, Cambridge.

Pothos, E. M. and Wills, A. J. (Eds. 2011). Formal approaches in categorization. University Press, Cambridge.

Pothos, E.M. and Busemeyer, J.R. (2009). A quantum probability explanation for violations of 'rational ' decision theory. Proc. R. Soc. B rspb. 2009.0121.

Price, D. J. deSolla (1975) Gears from the Greeks: The Antikythera Mechanism – A calendar computer from ca. 80 BC. Science History Publications, New York

Pusch, R., von der Emde, G., Hollmann, M., Bacelo, J., Nöbel, S., Grant, K., Engelmann, J. (2008). Active sensing in a mormyrid fish: electric images and peripheral modifications of the signal carrier give evidence of dual foveation. J Exp Biol., 211 921-34.

Rovelli, C. (2009). Forget time. arXiv:0903.3832v3 [gr-qc] 27 Mar 2009 (Essay for FQXi). Savitt, S. (2013). Being and Becoming in Modern Physics (Stanford Encyclopedia of Philosophy).

Schwartz, J.M., Stapp,H.P, and Beauregard, M. (2004). Quantum theory in neuroscience and psychology: a neurophysical model of mind/brain interaction. Phil. Trans. Royal Society, B 360(1458) 1309-27 (2005)]

Straulino, S. (2008). Reconstruction of Galileo Galilei's experiment: the inclined plane. Physics Education, 43.3, 316-321.

Trueblood, J.S. and Busemeyer, J.R. (2012). Quantum Information Processing Theory. In: D. Quinones (Vol. Ed.) and N. M. Seel (Ed. in Chief) Encyclopedia of the Sciences of Learning (pp. 2748-2751). Springer, New York.

Wearden, J. H. (2005). Origines et développement des théories d'horloge interne du temps psychologique [Origin and development of internal clock theories of psychological time]. Psychologie Francaise, 50, 7 – 25.

Wigner, E.P. (1962). Remarks on the mind-body question. In: Good LG. The scientist speculates. Heinemann, London , 284-302.

Wu, L.-Q. and Dickman, J.D. (2012). Neural Correlates of a Magnetic Sense, Science, 336. 6084, 1054-1057, http://dx.doi.org/10.1126/science.1216567

Zeh, H.D. (1970). On the interpretation of measurements in quantum theory. Found. Phys.1, 69-76 Zeh, H.D. (1998). Über die Zeit in der Nature. In Evolution und Irreversibilität H.J. Krug & L.Pohlmann, Dunker und Humblot. Berlin.

Zeh, H.D. (2013). The role of the observer in the Everett interpretation. NeuroQantology, 11, 97-105.

14. Mental Time Travel: How The Mind Escapes From The Present
Michael C. Corballis

School of Psychology University of Auckland

Abstract:
Although physical time is unidirectional and physical events are pinned to the present, we can mentally go back and forth in time to "replay" past episodes or imagine future ones. Mental time travel is imprecise, since we can never recover past events with complete fidelity or know exactly what will happen, but it is adaptive in that it enhances future planning and creates a sense of personal continuity through time. Recording of brain activity suggests that mental time travel is not unique to humans, as often supposed, but is an adaptation that goes far back in evolution—although in humans it probably allows greater detail and narrative structure than even in our great-ape cousins. Language may have evolved to allow us to share our mental time travels, and this sharing does appear to be uniquely human, an adaptation to the intense social structure of our lives.

Keywords: evolution, language memory, mental time travel, time

Introduction

Einstein is supposed to have said that the only reason for time is to prevent everything from happening at once. Time also proceeds in one direction only, which may be nature's way of ensuring that things don't happen more than once. Of course there are repeated events—the rising of the sun, the phases of the moon, the tax return—but these are never exact repetitions; each event is a singular happening. As the Greek philosopher Heraclitus put it, you can never step twice into the same river. The impossibility of time reversal is sometimes demonstrated by a thought experiment, known as the grandfather paradox, and attributed to the science fiction writer René Barjavel (1943). Suppose reversal were indeed possible, so that a time traveller could then go back in time to when his grandfather had not yet produced offspring, and kill him. This would mean that he, the time traveller, would never have been born, and so could not have travelled back in time. Nature, it might be said, abhors a paradox.

In our minds, though, we can escape the constraints of physical time by imagining going back in time to replay past events. The capacity to do this depends on what is known as episodic memory, the memory for specific episodes in our lives, allowing us to go back in time and mentally relive those episodes. Episodic memory is distinguished from semantic memory, which comprises our knowledge about the world and is not associated with specific points in the past; it is also differentiated from memories that are largely unconscious, such as the memory for skills like riding a bike or tying shoe laces.

It is through our episodic memories that we indeed seem able to step, mentally if not physically, into the same river twice—or indeed as often as we like.

Our ability to mentally go back into the past nevertheless does not result from a simple reversal of time. We may go back to a memory of some earlier event, but we then mentally play it forwards, not backwards. We seem to reverse time in discrete leaps rather than in continuous fashion. For instance, people can count backwards, perhaps hesitantly, but each number is repeated forwards, and we can't even begin to imagine how they sound backwards. Even when played backwards, speech is almost entirely unintelligible, with perhaps two dubious exceptions: The Led Zeppelin song Stairway to Heaven, when played backwards, is said to contain intelligible satanic messages, and Johnny Weissmuller's patented Tarzan yell is identical when played forwards or backwards (this was evidently contrived artificially by recording the second half as the reverse of the first half). In our imagined movements about the world we may sometimes retrace our steps, perhaps to figure out where we might have dropped a pair of spectacles, but again we don't imagine actually walking backwards, or the spectacles leaping from the ground back into a pocket.

Moving Forward

We can also go forwards in time in our minds, imagining future events. Of course, we cannot know for sure that these events will actually occur, and indeed we might imagine several different possibilities so that we can choose which one to plan. Constructing future events in our minds is therefore important in designing our futures. We may for example conjure up different scenarios for a future event such as an interview or a wedding, examining the different possibilities as to who to invite or what drinks to serve. And there is a natural continuity between forward and backward time travels, as anticipated events proceed from future to past, via the present.

The ability to imagine events at other points in time, whether past or future, has been dubbed mental time travel, a term first used by Tulving (1985) and elaborated by Suddendorf and Corballis (1997). Brain-imaging shows that imagining past and future episodes activate a common network in the brain (Addis et al., 2007). This is perhaps not surprising, since our ability to mentally conjure a future event depends heavily on our memory for events in the past. Indeed, the primary function of memory is probably not to serve as a repository of the past but rather to support projections into the future; as Klein et al. (2010) put it, memory is "an evolved system for planning future acts" (p. 13). The imagining of future episodes typically involves people we already know in places we know, and doing things we have witnessed them doing in the past. In most cases, an imagined future event is a recombination of elements experienced in the past, although we may of course introduce new elements, perhaps imagining (often wrongly) what the bride's new husband looks like.

The brain network activated during mental time travel covers broad regions of the frontal, temporal, and parietal lobes, but the critical area is the hippocampus, lying on the underside of the temporal lobe. Destruction of the hippocampus in humans not only destroys the ability to remember past events, but also renders the victim unable to envisage what might happen in the future. One famous example is the patient H.M., whose hippocampus was surgically removed in an attempt to control his intractable epilepsy. Following the operation, H.M. proved unable to recall episodes that occurred following his operation, and also unable to recall all but a few events that had occurred in his earlier life. It has also transpired that he could not imagine future events. His plight is captured by the title of Suzanne Corkin's (2013) book, Permanent Present Tense. A very similar case is that of the English musician Clive Wearing, whose hippocampus was destroyed as a result of a viral infection. The title of his wife Deborah Wearing's (2005) book, Forever Today, encapsulates his predicament.

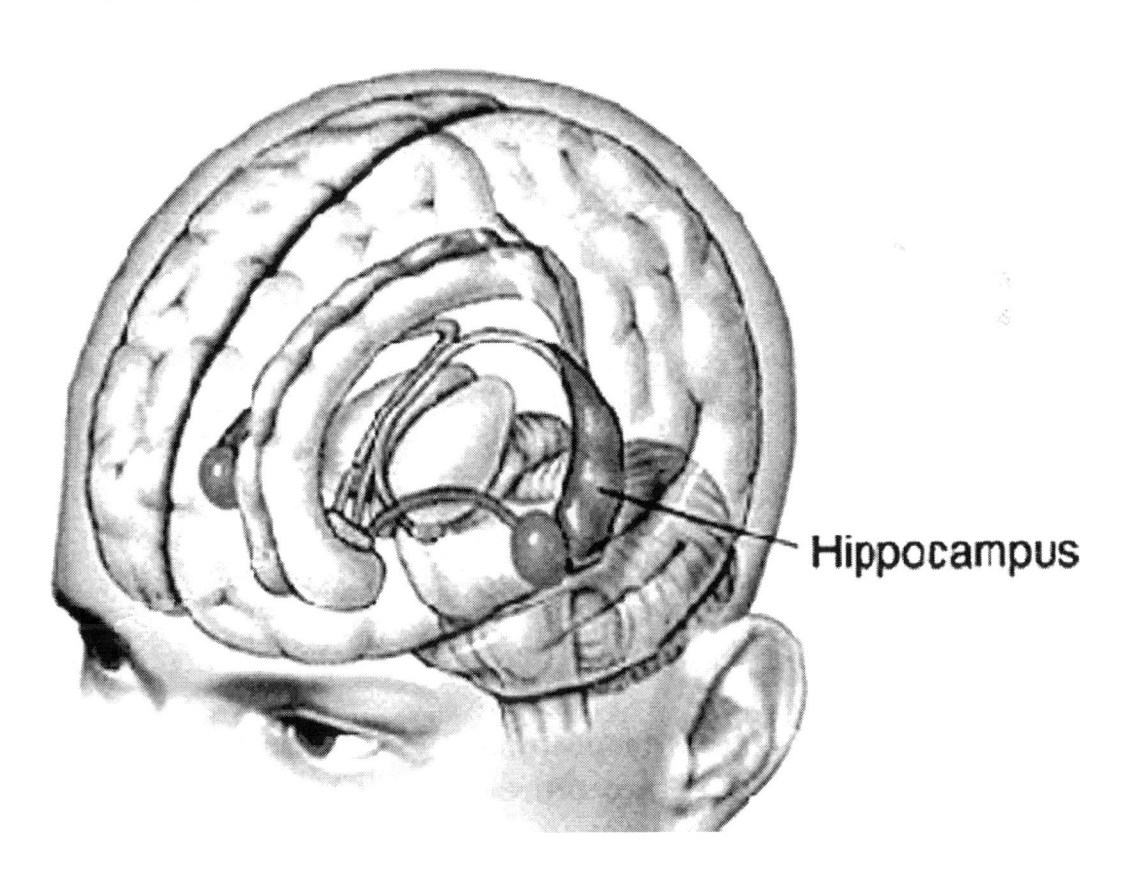

Mental time travel does not represent events with great accuracy. Of course, imagined future events often turn out to be wrong, but imagined past events are also very often distorted or incomplete. Indeed we probably don't remember most of the events that have occurred in our lives, as a school reunion or browsing through an old photograph album may reveal; conversely, we may sometimes remember events that did not actually happen. Mark Twain is reported to have said, "I have lived through some terrible things in my life, some of which actually happened"—although this itself may be false, since the remark, or ones similar to it, apparently has a long history, and was probably never uttered by Twain. The pioneering work of Elizabeth Loftus shows that false memories are easily implanted by giving cues as to what might have occurred on a past occasion, but in fact did not (Loftus & Davis, 2006). The gaps, distortions and intrusions in our memories are the bane of courts of law, where accounts of events often depend critically on eye-witness testimonies (Loftus et al., 2008).

The adaptiveness of episodic memory may lie, not in its accuracy, but rather in its potential for future planning and personal fulfillment. We may remember our past actions as more heroic or more generous that they actually were. Brain imaging shows that imagined events may be later recorded as though they actually happened (Martin et al. 2011), suggesting that our plans may become entangled with our memories. It is as though the brain itself has difficulty distinguishing the real from the imagined. Dreams, too, are sometimes confused with reality.

Our mental travels occur not only in time, but also in space, since the events we remember take place in four-dimensional space-time. We might remember what we did last summer in the South of France, or what we might do next winter in the Swiss Alps. Indeed, we can also mentally inhabit the minds of others, imagining how one's children are enjoying their time on holiday. The novelist Ian McEwan (2001), in an essay published shortly after the destruction of the twin towers in New York, wrote that "Imagining what it's like to be someone else is at the core of our humanity."

The human capacity for imagining events is exploited in storytelling. Much of our lives is taken up with fiction—novels, plays, movies, television soaps. Even in preliterate times, stories were passed on through word of mouth, and carried through the generations. These are typically stories of heroes and often miraculous events, serving to both bond and boost the cultures we inhabit. From Hercules to Harry Potter, stories often stretch the bounds of the possible.

Is Mental Time Travel Uniquely Human?

It is often suggested that mental time travel is unique to our species. In his poem A Grammarian's Funeral, published in 1855, the English poet Robert Browning wrote: "'What's time? Leave Now for dogs and apes! Man has Forever!'" Wolfgang Köhler, the famous German gestalt psychologist, was stationed at a primate research institute in the Canary Islands during World War I, and took the opportunity to test the problem-solving abilities of nine chimpanzees. He concluded that although these animals, our closest nonhuman relatives, were often capable of insight, they had little conception of past and future (Köhler, 1925). Endel Tulving, the Canadian cognitive scientist who first distinguished episodic from semantic memory, has long maintained that episodic memory is unique to humans (e.g., Tulving, 1972, 1984). Thomas Suddendorf and I expanded this claim by arguing that mental time travel, both back and forward in time, is a uniquely human capacity (Suddendorf & Corballis, 1997, 2007).

One reason to suppose that mental time travel might be restricted to humans is that language itself widely regarded as unique to our species (e.g., Chomsky, 2007; Hauser et al., 2002), and language may well have evolved precisely to enable the communication of the non-present (Corballis, 2009, 2013). Indeed episodic memory, along with semantic memory, is often dubbed "declarative memory" (e.g., Squire, 2004)—or memory that can be declared, implying a dependence on language. Language equips us with symbols to represent aspects of the world that need not be accessible to the senses, along with rules to enable us to combine these components, and so tell of past events and future plans, gossip about people in their absence, and of course make up stories about imaginary people in imaginary places at imaginary times. Animal vocalizations, in contrast, seem limited to fixed calls, with little evidence for learning or modification, dealing with events in the present, and lacking the grammatical rules to communicate about ordered episodes. Captive chimpanzees, bonobos and gorillas have been taught rudimentary forms of visual language that show limited evidence of rules, but these are for the most part limited to requests, with little evidence of reference to future episodes or to events from the past.

But mental time travel itself need not depend on the ability to communicate our thoughts to others. Some animal researchers have been quick to challenge the idea that nonhuman animals are incapable of episodic memory, or of mental time travel. Scrub jays, for example, have been shown to recover food they have cached in relation to when it was cached, preferring to recover worms if they were recently cached, but the less palatable peanuts if they were cached earlier, presumably on the grounds that the worms will have decayed and become unpalatable if stored for too long. This has been taken to imply that the birds remember the caching event itself. Again, if another jay watches birds caching food, the watched bird will sometimes later re-cache the food, seemingly anticipating that the watcher may later steal the food, suggesting mental time travel into the future. These studies, and others involving different species, are summarized by Clayton et al. (2003). In another recent study, a male chimpanzee called Santino in Furuvik Zoo in Sweden has been observed to collect stones and throw them at visitors. He gathers them well in advance of visitors arriving, and hides them so the visitors won't see them (Osvath & Karvonen, 2012). It is difficult to avoid the impression that Santino is planning a mischievous future event.

These and other studies can generally be explained without supposing that the animals in question mentally actually "relive" the past episode, or actively imagine the future one. The results can often be interpreted in terms of associative learning, perhaps based on learning that results from a single experience, or perhaps based on a "time tag" in the brain that records how long an experience has been held in memory. Thus the scrub jays may know that worms have passed their use-by date without necessarily recalling the act of caching them. Some claims may be based on questionable methodology (for detailed critiques see Suddendorf & Corballis, 2008, 2010; Suddendorf et al., 2009). But perhaps the main issue is whether we can really ever know whether nonhuman animals truly experience past or future episodes. What can we know about animal consciousness?

Tulving (1984) distinguished three levels of consciousness. First is anoetic consciousness, which is awareness bound to the present, in both space and time. It is what allows animals to respond to the immediate environment. Second is noetic consciousness, which applies to knowledge, as contained in semantic memory, and includes representation of objects or events that are not tied to the immediate environment. This does allow animals to think about events in their absence, and may be sufficient to explain some of the results of experiments claimed to show mental time travel. The third level is autonoetic consciousness, which is self-knowing, and is the level required for episodic memory. It allows one to remember episodes experienced by the self, so that these episodes can be re-experienced. It is this level of consciousness, according to Tulving, than nonhuman species lack; they do not know themselves.

Autonoetic consciousness is readily accessed through language—one can easily ask people what they did yesterday or what they plan to do tomorrow, and be reasonably sure that they are indeed revealing personal events, even if inaccurately. The absence of language in other species has suggested to some that animal consciousness may be unknowable, and some cautiously refer to memory as inferred from behaviour as "episodic-like" (e.g., Clayton et al., 2003) rather than episodic. Nevertheless there may well be a means of access to animal consciousness that does not depend on language, and that may indeed suggest a capacity for autonoetic consciousness, and mental time travel.

Hippocampal Recording

In humans the hippocampus is critical to mental time travel, as we saw from the cases of H.M. and Clive Wearing. The hippocampus is also widely documented across other species, including birds and mammals, as constituting a "cognitive map" (O'Keefe & Nadel, 1978). For instance, recordings from single neurons on the hippocampus of the rat show that different cells correspond to particular locations in a spatial environment, such as a maze. We now know, though, that these cells are not dependent on the rat being actually located in the maze. They sometimes fire in fast volleys, known as sharp-wave ripples (SWR), when the map is outside of the environment, sometimes when the animal is asleep (Wilson & McNaughton, 1994), and sometimes when it is awake but immobile (Karlsson & Frank, 2009). These SWRs often correspond to trajectories in a maze, or some other known environment, as though the animal is consciously "imagining" episodes of activity.

Sometimes, these imagined trajectories correspond to actual paths taken in the past, but sometimes they are the reverse of earlier trajectories (e.g., Foster & Wilson, 2006), and sometimes they correspond to trajectories that were never actually taken (Gupta et al., 2010). In some cases, too, hippocampal recording seems to anticipate trajectories taken in the future (Pfeiffer & Wilson, 2013), and both forward and backward trajectories as indicated by SWRs are greatly compressed in time relative to actual ones (Buzsáki & Lopez da Silva, 2012). In short, SWRs seem to exhibit mental time travel. One explanation is that SWRs are critical to the consolidation of memories, perhaps to include expansion of the mental map to regions not yet explored, but it is difficult to avoid the sense that they also correspond to mental time travels.

This interpretation is reinforced by brain-imaging studies, which show that the hippocampus operates as a cognitive map in humans much as it does in animals (e.g., Maguire et al., 1998), and that hippocampal activity occurs when participants actively remember past events or imagine future ones (e.g., Addis et al., 2007). Of course, the imagined episodes in these studies are more complex than wanderings in mazes; they include people, objects, specific actions, as well as locations—and indeed may have a narrative structure of the sort that underlies human storytelling. As far as we yet know, the mental trajectories indicated by hippocampal recordings in the rat do not include elements other than locations. But as Darwin (1871) famously wrote, "The difference in mind between man and the higher animals, great as it is, certainly is one of degree and not of kind" (p. 126), and it seems that the rat's mental wanderings may well be of the same kind as the more complex mental time travels enjoyed by humans. The basic capacity to imagine oneself in different places at different points in time, past or future or "once upon a time," may well go far back in evolution, perhaps even to the common ancestry of mammals and birds. It may have been an early adaptation associated with the capacity to move through space and time—to remember where one has been and plan where to go in the future.

Whether one accepts the evidence that hippocampal recordings correspond to autonoetic consciousness is something of a philosophical question, depending on what is taken to be the default position. On the one hand, animal researchers are urged to adhere to what is known as Lloyd Morgan's canon: "In no case may we interpret an action as the outcome of a higher mental faculty, if it can be interpreted as the exercise of one which stands lower in the psychological scale" (Morgan, 1903, p. 59). This perhaps underlies attempts to explain behavioural experiments on episodic memory or mental time travel in nonhuman animals in terms of semantic rather than episodic memory, or in terms that do not admit to autonoetic awareness in the animal mind. On the other hand, Darwin reminded himself in the margin of his notebook to "never say higher or lower"—albeit in contradiction to the quote in the previous paragraph. In the Darwinian view, then, the default position is that animal minds, including the human mind, differ only in complexity. It is in this spirit that Smith and Mizumori (2006), summarising some of the work on hippocampal recording in the rat, wrote that "the most conservative position is to assume that rats possess an episodic memory system that is qualitatively similar to that of humans" (p. 720).

In the case of episodic memory and mental time travel, there no longer seems a compelling reason to adopt the parsimony of Lloyd Morgan's canon over Darwin's perhaps equally parsimonious assertion of continuity between species. When combined with the behavioural evidence, hippocampal recordings provide a strong case for mental time travel, albeit at a less complex level than experienced by humans. Nevertheless there remains a difficulty in the evolutionary case for continuity. In humans, language is the device that enables us to share our mental travels in space and time, and most are agreed that language itself is uniquely human.

The Problem of Language

The most trenchant view is that of Chomsky, who argues that language evolved in a single step within the past 100,000 years, endowing us humans with a uniquely symbolic concepts, along with recursive mechanisms for combining symbols into grammatical structures (Hauser et al., 2002). In this view, the communicative function of language is secondary to its function as the mechanism of thought; moreover, it could not have evolved through natural selection, since the abstract symbols have no relation to objects in the external world, and must therefore be innate. Chomsky (2011) writes that "... one basic problem facing the study of evolution of the human capacity is to account for the emergence of these universal and innate properties of language and thought; not a trivial matter, as we quickly discover when we take words and ideas seriously" (p. 10).

This view is supported by some archaeologists, based on an apparent discontinuity in the fossil record within the past 100,000 years, suggesting the sudden emergence of symbolic thought. For instance, Tattersall (2012) writes as follows:

Our ancestors made an almost unimaginable transition from a non- symbolic, nonlinguistic way of processing information and communicating information about the world to the symbolic and linguistic condition we enjoy today. It is a qualitative leap in cognitive state unparalleled in history. Indeed, as I've said, the only reason we have for believing that such a leap could ever have been made, is that it was made. And it seems to have been made well after the acquisition by our species of its distinctive modern form (p. 199).

Could this "great leap forward" (Chomsky, 2011) be an exception to Darwin's theory of evolution? Not all are agreed. Pinker and Bloom (1990) argued for the evolution of language by natural selection, and others, including myself, have proposed that language evolved from manual gestures, perhaps rooted in the so-called "mirror system" in the primate brain. In monkeys, this system maps the perception of manual gestures by another individual onto the production of those same gestures by the monkey itself, and overlaps with the homologues of language areas in the human brain. The manner in which this system might have formed the basis for the development of the complex communication system we call language has been developed by a number of authors (e.g., Arbib, 2012; Corballis, 2002; Rizzolatti & Sinigaglia, 2007). The notion of a sudden discontinuity in the fossil record is also questioned, with suggestions that the complexity of hominin culture and artifacts has risen gradually since the middle Pleistocene (McBrearty, 2007; Shea, 2011) rather than in a great leap forward a mere 90,000 or so years ago.

There seems little reason also to believe that human concepts differ fundamentally from those of other animals, as Chomsky suggests. Chimpanzees appear able to understand simple spoken commands (Savage-Rumbaugh et al., 1998), and a border collie known as Chaser seems to have a receptive vocabulary of over 1000 words for common objects (Pilley & Reid, 2011). The limitation seems to lie in production, not in reception. With the exception of some birds, such as the parrot, nonhuman animals, including our closest nonhuman relatives the great apes, are extremely limited in their capacity for vocal learning (Petkov & Jarvis, 2012). But this is a limitation in the capacity to learn and produce word-like sounds—it is not a limitation in the capacity to understand such sounds and associate them with real-world objects. Of course this limitation in production is overcome to some extent if we consider communication through visible movements of the body, and especially the hands, and communication through gestures and visible signals in great apes is more language-like than their vocalizations.

The difference between humans and other mammals therefore seems to lie at least in part in the human ability to produce spoken words to represent concepts, and not in the presence of concepts themselves, nor even in the ability to attach words spoken by humans to concepts. Contrary to the Chomskyan view, language seems not to provide the basis for conceptual thought itself, but is better regarded as a means of transmitting thoughts from one individual to another.

Another likely precursor to language is mental time travel itself. Language may indeed have evolved to enable communication about the non-present, and, as suggested earlier, mental time travel may go far back in mammalian and perhaps avian evolution. Some nonhuman species, at least, appear able to represent objects that are not present; the border collie Chaser, for example, can fetch a named object from another room when verbal requested to do so (Pilley &

Reid, 2011), and many birds have prodigious memory for the locations of cached items of food. Curiously, though, there is little evidence that nonhuman species intentionally communicate about non-present items. For instance, Liszkowski et al. (2009) tested the abilities of chimpanzees and prelinguistic human infants to request desirable objects by pointing to them. Both could learn to point to desirable objects in front of them. The 12-month-old infants readily learned also to point to locations in which known objects were hidden, but the chimpanzees were unable to do this. Mental time travel itself may well be an ancient adaptation, but the imperative to communicate about the remembered non-present seems not to have emerged until after the split of hominins from great apes some 6 million years ago, and was perhaps an adaptation to enhanced pressure toward sociality and sharing of information.

Perhaps the problem lies in an unwillingness to share internal secrets, rather than the failure to harbor the secrets themselves. The scrub jays mentioned earlier were at pains to re-cache food if another jay watched them caching it, and we humans are also as often at pains to preserve our secrets as to share them, even in the age of Twitter.

Conclusions

Our ability to mentally escape the present, reverse time and project into the future, stands in contrast to the fixity of physical time. This does not mean that we have ability to perceive what actually happened or will happen at times other than the present. It is sometimes claimed, for example, that dreams provide us with precognition, the capacity to foresee precisely what is about to happen, but I am not claiming any such capacity. Mental time travels, and the ability to take the perspectives of others, need not imply any form of mind over matter, be it precognition, clairvoyance, or communication with the dead.

Although physical time is unidirectional and irreversible, it does operate in such a way as to leave traces of the past. The physical past can therefore be reconstructed, but not physically replayed; the future can be predicted but not actually visited. Recording devices can be used to replay past events with high fidelity, but this can never be the same as actual replay. So it is with mental time travel. It is based on physical traces in the brain, which can then be programmed to reproduce at least some internal sense of earlier experiences, as well as of possible future ones. There is no good reason to suppose that the mind is more than a computational device that exists in, and is indeed part of, a physical universe.

Although a possible basis for mental time travel can be detected from hippocampal recordings in non-human species, we humans seem especially obsessed with time, recording it in units ranging from nanoseconds to eons. We are relentlessly driven by clocks, timetables, appointments, dates, deadlines, anniversaries. The understanding of time leads to the awareness of death, underlying the grief we feel when a friend or family member dies, and of course leading to fear of the oblivion that follows when we ourselves perish. Understanding of death is a seed for religions that proclaim life after death, and the notion of a spirit that transcends the physical world. We are even gaining increasing understanding of what happened at the origin of the universe itself, as well as what will happened some 7.5 billion years from now when the sun will grow so big that it will subsume the earth. This extraordinary obsession with time, and awareness of the consequences of the passage of time, almost certainly exceed the proclivities of non-human species, but perhaps only in complexity but not in kind. (Perhaps the understanding of time has such dire implications that our great-ape cousins just don't want to go there!).

And of course the inventions of language and recording instruments have fed into our communal understanding of time, and enabled us to explore the passage of time as comprehensively as we explore space. But the extent to which language is responsible for mental time travel itself remains a matter of speculation. My own view is that recognizing that mental time travel has a long evolutionary history may be a further step toward the understanding that we humans are not unique in the animal kingdom. It may also be a step toward the understanding that language itself may be not as special as we like to think, but evolved incrementally in Darwinian fashion.

REFERENCES

Arbib, M.A. (2012). How the Brain Got Language: The Mirror System Hypothesis. Oxford University Press, Oxford.

Addis, D.R., Wong, A.T. & Schacter, D.L. (2007). Remembering the past and imagining the future: common and distinct neural substrates during event construction and elaboration. Neuropsychologia, 45, 1363-1377.

Barjavel, R. (1943). Le Voyageur Imprudent. Portulan, Paris.

Buzsáki, G. & Lopez da Silva, F. (2012). High frequency oscillations in the intact brain. Progress in Neurobiology, 98, 241-249.

Chomsky, N. (2007). Biolinguistic explorations: Design, development, evolution. International Journal of Philosophical Studies, 15, 1-21.

Clayton, N., Bussey, T.J. & Dickinson, A. (2003). Can animals recall the past and plan for the future? Nature Reviews Neuroscience, 4, 685-691.

Corballis, M.C. (2002). From Hand to Mouth: The Origins of Language. Princeton University Press, Princeton, NJ.

Corballis, M.C. (2009). The evolution of language. Annals of the New York Academy of Sciences, 1156, 19-43.

Corballis, M.C. (2013). Wandering tales: Evolutionary origins of mental time travel and language. Frontiers in Psychology, 4, Article 495.

Corkin, S. (2013). Permanent Present Tense: The Man With No Memory, And What He Taught The World. Allen Lane Press, London.

Darwin, C. (1871). The Expression of the Emotions in Man and Animals. London: John Murray.

Foster, D.J. & Wilson, M.A. (2006). Reverse replay of behavioural sequences in hippocampal place cells during the awake state. Nature, 440, 680-683.

Gupta, A.S., van der Meer, M.A.A., Touretzky, D.S. & Redish, A.D. (2010). Hippocampal replay is not a simple function of experience. Neuron, 65, 695-705.

Hauser, M.D., Chomsky, N. & Fitch, W.T. (2002). The faculty of language: What is it, who has it, and how did it evolve? Science, 298, 1569-1579.

Karlsson, M.P. & Frank, L.M. (2009). Awake replay of remote experiences in the hippocampus. Nature Neuroscience, 12, 913–918.

Klein, S.B., Robertson, T.E. & Delton, A.W. (2010). Facing the future: memory as an evolved system for planning future acts. Memory & Cognition, 38, 13-22.

Köhler, W. (1925). The Mentality of Apes. Routledge & Kegan Paul, New York (Originally published in German in 1917).

Liszkowski, U., Schäfer, M., Carpenter, M. & Tomasello, M. (2009). Prelinguistic infants, but not chimpanzees, communicate about absent entities. Psychological Science, 20, 654-660.

Loftus, E.F. & Davis, D. (2006). Recovered memories. Annual Review of Clinical Psychology, 2, 469-498.

Loftus, E.F., Doyle, J.M. & Dysert, J. (2008). Eyewitness Testimony: Civil and Criminal (4th Edition). Lexus Law Publishing, Charlottesville, VA.

McBrearty, S. (2007). Down with the revolution. In Mellars, P., Boyle, K., Bar-Yosef, O., & Stringer, C. (Eds.), Rethinking the Human Revolution (pp.133–152). McDonald Institute Monographs. Cambridge: McDonald Institute for Archaeological Research.

McEwan, I. (2001). Only love and then oblivion. Essay published in The Guardian on 15 September.

Maguire, E.A., Burgess, N., Donnett, J.G., Frackowiak, R.S., Frith, C.D. & O'Keefe, J. (1998). Knowing where and getting there: a human navigation network. Science, 280, 921–924.

Martin, V.C., Schacter, D.L., Corballis, M.C. & Addis, D.R. (2011). A role for the hippocampus in encoding simulations of future events. Proceedings of the National Academy of Sciences, 108, 13858-13863.

Morgan, C.L. (1903). An Introduction to Comparative Psychology (2nd Edition). W. Scott, London.

O'Keefe, J. & Nadel, L. (1978). The hippocampus as a Cognitive Map. Oxford: Clarendon Press.

Osvath, M. & Karvonen, E. (2012). Spontaneous innovation for future deception in a male chimpanzee. PLoS ONE, 7, e36782.

Petkov, C. I. & Jarvis, E.D. (2012). Birds, primates, and spoken language origins: Behavioral phenotypes and neurobiological substrates. Frontiers in Evolutionary Neuroscience, 4, article 12.

Pfeiffer, B.E. & Foster, D.J. (2013). Hippocampal place-cell sequences depict future paths to remembered goals. Nature, 497, 74-79.

Pilley, J. W. & Reid, A.K. (2011). Border collie comprehends object names as verbal referents. Behavioural Processes, 86, 184-195.

Pinker, S. & Bloom, P. (1990). Natural language and natural selection. Behavioral & Brain Sciences, 13, 707-784.

Rizzolatti, G. & Sinigaglia, C. (2008). Mirrors in the Brain. Oxford University Press, Oxford.

Savage-Rumbaugh, S., Shanker, S.G. & Taylor, T.J. (1998). Apes, Language, and the Human Mind. New York: Oxford University Press.

Shea, J.J. (2011). Homo sapiens is as Homo sapiens was. Current Anthropology, 52, 1-35.

Smith, D.M. & Mizumori, S.J.Y. (2006). Hippocampal place cells, context, and episodic memory. Hippocampus, 16, 716-729.

Squire, L. (2004). Memory systems of the brain: A brief history and current perspective. Neurobiology of Learning & Memory, 82, 171-177.

Suddendorf, T. & Corballis, M.C. (1997). Mental time travel and the evolution of the human mind. Genetic, Social, and General Psychology Monographs, 123, 133-167.

Suddendorf, T. & Corballis, M.C. (2007). The evolution of foresight: What is mental time travel, and is it unique to humans? Behavioral and Brain Sciences, 30, 299-351.

Suddendorf, T. & Corballis, M.C. (2008). New evidence for animal foresight? Animal Behaviour, 75, e1-e3

Suddendorf, T. & Corballis, M.C. (2010). Behavioural evidence for mental time travel in nonhuman animals. Behavioural Brain Research, 215, 292-298.

Suddendorf, T., Corballis, M.C. & Collier-Baker, E. (2009). How great is great ape foresight? Animal Cognition, 12, 751-754.

Tattersall, I. (2012). Masters of the Planet: The Search for Human Origins. Palgrave Macmillan, New York.

Tulving, E. (1984). Elements of episodic memory. Behavioral & Brain Sciences, 7, 223-268.

Tulving, E. (1985). Memory and consciousness. Canadian Psychology, 26, 1-12.

Tulving, E. (2002). Episodic memory: From mind to brain. Annual Review of Psychology, 53, 1-25.

Wearing, D. (2005). Forever Today. Doubleday, New York.

Wilson, M.A. & McNaughton, B.L. (1994). Reactivation of hippocampal ensemble memories during sleep. Science, 265, 676-679.

15. How the Mind Escapes from the Present
Giovanni Berlucchi

National Institute of Neuroscience – Italy.
Dipartimento di Scienze Neurologiche e del Movimento Sezione di Fisiologia e Psicologia Università di Verona

Abstract:
Temporal cognition and timing of behaviour are essential ingredients of animal life, and the ability to use past experiences for predicting the future most probably had an important survival value in human evolution and history. This commentary on the elegant and instructive essay by Michael Corballis (2014) on mental time travel takes off from the classical psychological and neurological notion that there can be no future without a past to remember. It then touches briefly on some of the points elaborated by Corballis, including the difference between semantic and episodic time cognition, time as an urge for executive behaviour, the role of the body in time metaphors, knowledge of one's own mortality in humans and animals, and brain activity in remembering and imagining the future.

Keywords: evolution, language memory, mental time travel, time

1. The Tight Link Between Time Past and Time Future.

This subject has some personal flavor for me. In October 1952 a young epileptic male patient from the city where I live, Verona, was submitted in Padua to a two-stage extensive bilateral removal of the temporal lobe of the brain. This was an extreme attempt to hold in check his repeated bouts of potentially murderous and suicidal behaviour which posed a constant threat to his family, the attending medical personnel and himself. The case was followed and later reported by the neurologist Hrayr Terzian and the neurosurgeon Giuseppe Dalle Ore, both of whom were then working at the University of Padua. Several years later they would become professors at the University of Verona, since 1983 my own University, where I was privileged to be their colleague and friend. In the 1950s era of rampant psychosurgery, the main justification for the operation was the finding by Klüver and Bucy (1939) that wild rhesus monkeys were "tamed" by a similar bilateral temporal ablation within the framework of a complex syndrome including peculiar alterations of visual perception and memory, and exploratory, alimentary and sexual behaviour. The Verona patient also exhibited a complex post-operatory pattern of behavioral and psychological symptoms, partly comparable with those of the experimental monkeys, and including the hoped-for tameness; of these, for the purposes of the present discussion, I will only mention a dense amnesia involving past and future episodes alike. Not only did the patient fail to remember what had recently happened, he did not remember anything of his past. He did not seem to understand questions about his family, his house, his city of birth and residence, as if their object was unknown to him. Although Terzian and Dalle Ore were not in the position to examine the patient's memory with formal tests, they were convinced that he felt completely isolated, in their words "without a past to remember and consequently without any future what(so)ever" (Terzian and Dalle Ore, 1955). Terzian was extremely well read in the classical neurological and psychological literature, and I suspect that he got from William James the idea that memory of the past is a route to foreknowledge of the future. In the chapter on the sense of time in the Principles of Psychology (James, 1890) James had written that "the practically cognized present is no knife-edge, but a saddle-back, with a certain breadth of its own on which we sit perched, and from which we look in two directions into time. The unit of composition of our perception of time is a duration, with a bow and a stern, as it were -- a rearward- and a forward-looking end." And in the chapter on memory James had also distinguished this "short-lasting intuited duration – the specious present – from the immense region of conceived time, past and future, into one direction or another of which we mentally project all the events we think of as real, and form a systematic order of them by giving to each a date."

The Terzian-Dalle Ore report (1955), and particularly its amnesia findings, did not attract much attention from the scientific community, even though the paper was published in English in an international neurology journal. A major permanent place in the history of amnesia and the cerebral bases of memory has instead been gained by a similar operation performed at the Hartford Hospital in Connecticut in 1953 on a young male patient with the hope to reduce the severity of his epileptic condition. As mentioned by Corballis (2014), this is the patient who is known in the international neuroscientific and bellettristic literature as H.M, the purest case of amnesia caused by a bilateral medial temporal ablation. H.M. died a few years ago, and as recounted by his biographer Suzanne Corkin, who studied him for decades, he appeared to live forever in the present: perched on James's saddleback of the specious present, he could not look into either time past or time future, and like the Padua patient, he was unable to think of his future because he did not remember his past (Corkin, 2013).

2. Chronesthesia and Mental Time Travel

With Suddendorf and Addis, Corballis (Suddendorf and Corballis, 1997, 2007; Suddendorf et al., 2009) has elaborated the concept of mental time travel based on the dissection of consciousness and memory into various types proposed by Tulving (2002a). According to Tulving (2002b), chronesthesia is an aspect of human self-consciousness that allows individuals to think about subjective time and to travel in it. Chronesthesia is strongly linked with but not identical to autonoetic episodic consciousness, the intimate sense of the self as the subject of all past, present and future experiences. In both autonoetic consciousness and chronesthesia, the present moment is felt as a continuation of the past and a prelude to the future, but in being aware of self in time the emphasis is on the self in autonoetic consciousness, while it is on subjective time in chronesthesia. In severe amnesics a deficient chronesthesia and an incapacity for mental time travel may blur, but do not destroy the sense of the self, which expresses itself in the appropriate use of personal pronouns and adjectives in verbal communication. Nor do amnesics lose the objective cognition of physical time as measured by clocks and calendars, since they can conceive about the past and the future, although for them both are empty of personal episodes. Patient K.C., suffering from a severe episodic amnesia from brain trauma, has been shown to possess a correct temporal orientation toward the past and the future, and to be able to make future oriented decisions, such as choosing between smaller immediate rewards and larger later rewards. Amnesic patients appear to be stuck in time as far as their episodic memory is concerned, because they cannot retrieve personal episodes from their past or imagine them in their future. Yet they possess a factual, semantic knowledge of a physical time, whereby present is preceded by past and followed by future, despite being unable to travel in it with their mind (Kwan et al., 2012, 2013; Craver et al., 2014).

3. Time As An Urge.

The so-called civilized life is, to use Corballis' (2014) words, relentlessly driven by clocks, timetables, appointments, dates, deadlines, anniversaries. Many of our actions are to be initiated and terminated at or before a particular future time, whether proximal or remote. Remembering to carry out a specific action in due time requires a memory which has been called memory for the future or time-based prospective memory. Time-based prospective memory can be habitual, as exemplified by remembering to go to the station to catch a train at 7 am on working days, or episodic, as exemplified by remembering to go to the station the day after tomorrow at 4 pm to pick up an arriving friend. In both cases time can be seen as an urge, that is a felt need for performing an action, comparable in some respect to typical urges constituted by bodily stimuli for simple reflexes, such as coughing, yawning, urinating, defecating and so on, which can be voluntarily allowed to occur or inhibited (Jackson et al., 2011). The mechanisms by which the lapsing of time leads to action execution may be multiple and are not known precisely. In a study devoted to the subject (Kvavilashvili and Fisher, 2007) participants had to remember to phone the experimenter at a pre-arranged time after a 7-day interval, and to record the details of occasions when they thought about this intention during that interval. Obviously the intention could not be held in consciousness throughout the interval because participants had to pay attention to their everyday life. However the reports indicated that thoughts about the task emerged on occasions, mostly spontaneously, during the entire interval, with some intensification toward its end. This suggests that the intention persisted at a subconscious level and was brought into consciousness both occasionally and crucially nearer the critical time. Hence time-based prospective memory probably requires the operation of some kind of internal subconscious counter or clock, perhaps similar or identical to the internal clock which is supposed to allow people to wake up spontaneously at an unusual self-chosen time.

4. Time And The Body

Time is an abstract concept and according to some theorists (e.g. Lakoff and Johnson, 2003) abstract concepts are formulated and expressed using metaphors grounded in bodily interactions with the physical world. In current parlance, the past and the future appear to have opposite bodily correlates: the past is behind us, while the future is ahead of us. It is true, as Corballis (2014) mentions, that mental travel into either past or future is naturally imagined as a forward movement in both cases, but there is empirical evidence that at least under certain conditions, bodily motor dispositions or actual movements associated with mental travel into past and future differ from one another. In an experiment by Miles et al. (2010), blindfolded standing observers were to recall during a period of 15 seconds the circumstances and events of a typical day 4 years in the past, or imagine their everyday life circumstances of a typical day 4 years in the future. A sensor attached to one leg showed a small significant tendency, gradually increasing during the 15 second period, to sway backward when looking into the past and forward when looking into the future. In another experiment (Christian et al., 2013) observers "mind wandered" at will while viewing a display that elicited an illusion of backward or forward self-motion or vection. Thinking about the past predominated during backward vection, while thoughts oriented toward the future prevailed during forward vection. In yet another experiment (Hartmann and Mast, 2012), observers categorized verbal stimuli in relation to the concept of future or past while their whole body was actually displaced forward and backward. Future related verbal stimuli were categorized faster during forward as compared to backward motion, while there was no effect of direction of passive motion on past related word. Taken together, the results of these experiments support theories of embodied cognition, but it should not be inferred from them that mental time travel is impossible without feedbacks from the body. For example, patients with near-complete immobilization in the locked-in syndrome can communicate utilizing blinking, or up and down ocular movements, or, in a few cases, specific patterns of brain activity generated on command and visualized by neuroimaging. They retain a normal consciousness and memory of the past, are generally satisfied with their interactions with family and caretakers, and despite being fully aware of their huge behavioural limitations, they can have level-headed goals in their lives which can last for many years (Cappa et al.,

1985; Schnakers et al., 2008). Some of them have written sophisticated books or poetry on their condition as seen from the inside (e.g. Dudzinski, 2001; Sledz et al., 2007), leaving no doubt that they are fully capable of a realistic mental time travel.

5. Temporal Cognition And Death.

In concluding his paper Corballis (2014) remarks that human understanding of time leads unavoidably to awareness and fear of death. Human beings know that their predictions about the future can be fallible, except for the certainty of their own mortality. If some capacity for mental time travel exists in non-human animals, a possibility not excluded by Corballis (2014), it seems nonetheless quite unlikely that even the great apes can think about their death. Humphreys has argued that chimpanzees cannot conceive of their possible death because they lack a general notion of death as the end of life. He mentions the observations of the primatologist Tetsuro Matsuzawa regarding the treatment by a group of chimpanzees of the corpse of one of their members, suggesting that the dead animal was either thought to be asleep or regarded and used as an object. According to Humphreys (2011), if chimpanzees fail to appreciate that one of their kin has ceased to live, they can hardly imagine (and fear) that a similar fate can befall them. He maintains that fear of death is a uniquely human trait: it is the toll to be paid for possessing a consciousness which by encompassing the lifetime's remembered past and foreseeable future contributes to the biological fitness of humans more than to that of any other animal. But even in absence of the capacity for mental time travel, non-human animals can learn from their own dangerous experiences and by observing what happens to other animals in perilous or fatal situations. Strange as it may appear, there is some evidence that such learning can also occur in dreams during the stage of sleep associated with rapid eye movements (REM sleep), and the knowledge so acquired has nothing to do with extrasensory perception or other paranormal phenomena. The functional significance of sleep and dreams is still largely a mystery, but one function of REM sleep may be that of prospecting to consciousness potential predicaments which may never have been experienced as such, but which might happen in future real life. Convinced that such a function of sleep is real and has an evolutionary adaptive value, Revonsuo (2000; Valli and Revonsuo, 2009) believes that during dreams animals and humans alike undergo threatening experiences which prompt them to rehearse strategies for survival when facing actual challenges of the same type. Suggestive evidence for this belief is furnished by the predominant association of human dreams with negative emotions and by the types of behavior exhibited by animals during REM sleep. Normally REM sleep is mostly characterized by inaction because motoneurons are actively inhibited, but if inhibition is experimentally interfered with, the observed behaviour is indicative of reactions typical of fight or flight situations.

For example, during REM sleep cats so treated may assume predatory, hunting and stalking postures or motions, or exhibit signs of fear and rage as they occur in defensive situations (Morrison, 1983). Similar aggressive or defensive forms of behaviour observed during sleep or narcolepsia in humans suggest that the virtual reality of dreams may indeed supplement the mental time travels in the future which occur during conscious wakefulness (Siclari et al., 2010).

6. Brain Activity and Mental Time Travel.

Neuropsychological studies of brain damaged patients continue to afford evidence relevant to the understanding of temporal cognition (Rosenbaum et al., 2014), but present-day technological advances in neuroimaging make it possible also to study the intact human brain during remembering and imagining. Further, additional evidence comes from recordings of neuronal activities in experimental animals in structures believed to be critical for mental time travel in humans, such as the hippocampus and related structures. As mentioned by Corballis (2014), specific neuronal activities in the hippocampus of rats may signal perceived, imagined or even dreamed-of spatial positions or trajectories, and its is difficult to resist the temptation to consider them as brain correlates of mental time travel as well. That the hippocampus and related structures are pivotal structures in episodic memory and mental time travel in humans is shown by the causal role of lesions of these structures in producing amnesia, but neuroimaging findings implicate many other brain structures in these functions (Mullally and Maguire, 2013; Schacter et al., 2014). Frontal, temporal and parietal regions are no doubt involved and the most attractive finding is that a common network of these regions appears to be activated during both remembering the past and imagining the future. The emphasis on this commonality of brain regions presumably subserving memory and imagination alike is no doubt justified, but there must be a neural correlate of our chronesthesic capacity to distinguish the reactivation of an old memory from the simulation of a future event. As far as I know the available scanty evidence for this correlate paradoxically suggests that the hippocampal region is not involved in chronesthesia and subjective time travel (Nyberg et al., 2011).

REFERENCES

Cappa S.F., Pirovano, C. & Vignolo, L.A. (1985). Chronic 'locked-in' syndrome: psychological study of a case. European Neurology, 24, 107-111.

Christian, B.M,, Miles, L.K., Parkinson, C. & Macrae, C.N. (2013). Visual perspective and the characteristics of mind wandering. Frontiers in Psychology Oct 9;4:699.

Corballis, M. C,. (2014). Mental Time Travel: How The Mind Escapes From The Present, Cosmology 18. 139-145.

Corkin, S. (2013). Permanent Present Tense: The Man With No Memory, And What He Taught The World. Allen Lane Press, London.

Craver, C.F., Kwan, D., Steindam, C. & Rosenbaum, R.S. (2014). Individuals with episodic amnesia are not stuck in time. Neuropsychologia, 57, 191-195.

Dudzinski, D. (2001). The diving bell meets the butterfly: identity lost and remembered. Theoretical Medicine and Bioethics, 22, 33-46.

Hartmann, M. & Mast, F.W. (2012). Moving along the mental time line influences the processing of future related words. Consciousness and Cognition, 21, 1558-1562.

Humphreys, N. (2011). Soul Dust. The Magic of Consciousness. Quercus, London.

Jackson S.R., Parkinson, A., Kim, S.Y., Schüermann, M. & Eickhoff, S.B. (2011). On the functional anatomy of the urge-for-action. Cognitive Neuroscience, 2, 227-243.

James, W. (1890). The Principles of Psychology. MacMillan, London.

Klüver, H. & Bucy, P.C. (1939). Preliminary analysis of functions of the temporal lobes in monkeys. Archives of Neurology and Psychiatry, 42, 979-1000.

Kwan, D., Craver, C.F., Green, L., Myerson, J., Boyer, P. & Rosenbaum, R.S. (2012). Future decision-making without episodic mental time travel. Hippocampus. 22, 1215-1219.

Kwan, D., Craver, C.F., Green, L., Myerson, J. & Rosenbaum, R.S. (2013) Dissociations in future thinking following hippocampal damage: evidence from discounting and time perspective in episodic amnesia. Journal of Experimental Psychology General, 142, 1355-1369.

Kvavilashvili, L & Fisher, L. (2007) Is time-based prospective remembering mediated by self-initiated rehearsals? Role of incidental cues, ongoing activity, age, and motivation. Journal of Experimental Psychology General, 136, 112-132.

Lakoff, G. & Johnson, M. (2003). Metaphors we Live By. The University of Chicago Press, Chicago.

Miles, L.K., Nind, L.K. & Macrae, C.N. (2010). Moving through time. Psychological Science, 21, 222-223.

Morrison, A.R.(1983). A window on the sleeping brain. Scientific American, 248(4), 94-102.

Mullally, S.L. & Maguire, E.A. (2013). Memory, imagination, and predicting the future: A common brain mechanism? Neuroscientist, 20, 220-234.

Nyberg, L, Kim, A.S., Habib, R., Levine, B. & Tulving, E. (2010). Consciousness of subjective time in the brain. Proceedings of the National Academy of Sciences of the U S A, 107, 22356-22359.

Revonsuo, A. (2000). The reinterpretation of dreams: an evolutionary hypothesis of the function of dreaming. Behavioral Brain Sciences, 23, 877-901; discussion 904-1121.

Rosenbaum, R.S., Gilboa, A. & Moscovitch, M. (2014). Case studies continue to illuminate the cognitive neuroscience of memory. Annals of the New York Academy of Sciences, 1316, 105-133.

Schacter, D.L., Addis, D.R., Hassabis, D., Martin, V.C., Spreng, R.N. & Szpunar, K.K. (2012). The future of memory: remembering, imagining, and the brain. Neuron, 21, 677-694.

Schnakers, C., Majerus, S., Goldman, S., Boly, M., Van Eeckhout, P., Gay, S., Pellas, F., Bartsch, V., Peigneux, P., Moonen, G. & Laureys S. (2008). Cognitive function in the locked-in syndrome. Journal of Neurology, 255, 323–330.

Siclari, F., Khatami, R., Urbaniok, F., Nobili, L., Mahowald, M.W., Schenck, C.H., Cramer Bornemann M.A. & Bassetti, C.L. (2010). Violence in sleep, Brain, 133, 3494-3509.

Sledz, M., Oddy, M, & Beaumont, J.G. (2007). Psychological adjustment to locked-in syndrome. Journal of Neurology, Neurosurgery and Psychiatry, 78, 1407–1408.

Suddendorf, T., Addis, D.R. & Corballis, M.C. (2009). Mental time travel and the shaping of the human mind. Philosophical Transactions of the Royal Society of London, Series B Biological Sciences, 364, 1317-1324.

Suddendorf, T. & Corballis, M.C. (1997). Mental time travel and the evolution of the human mind. Genetic, social, and general psychology monographs, 123, 133-167.

Suddendorf, T. & Corballis, M.C. (2007). The evolution of foresight: What is mental time travel, and is it unique to humans? Behavioral and Brain Sciences, 30, 299-313; discussion 313-351.

Terzian, H. & Dalle Ore, G.D. (1955). Syndrome of Klüver and Bucy; reproduced in man by bilateral removal of the temporal lobes. Neurology, 5, 373-380.

Tulving, E. (2002a). Episodic memory: from mind to brain. Annual Review of Psychology, 53, 1-25.

Tulving. E (2002b). Chronesthesia: Conscious awareness of subjective time. In D.T. Stuss & R.T.Knight (eds.), Principles of Frontal Lobe Function. Oxford University Press., Oxford, pp. 311-325.

Valli K, Revonsuo A. (2009). The threat simulation theory in light of recent empirical evidence: a review. American Journal of Psychology, 122, 17-38.

16. Mental Time Travel And The Self-Concept
Commentary On Michael Corballis's Mental Time Travel

Liliann Manning

Strasbourg University and Laboratory of Cognitive Neuropsychology and Pathophysiology of Schizophrenia (INSERM, U 1114), Strasbourg, France

Abstract

The present commentary focuses on (i) theoretical models supporting the notion that autobiographical recollections are associated with episodic but also semanticized memories, and (ii) recent evidence showing that the human tendency to communicate thoughts and feelings to others engages neural and cognitive mechanisms associated with reward. On those bases, it is suggested that mental time travel, besides being the goal of human communication, as it is stated in the target article, could also be seen as the privileged means "used" to convey the messages to best suit our self-concept.

Keywords: Mental Time Travel, Self-Concept, Language, Reward, Episodic Memory,

1. Introduction

There are a variety of views as to the nature and purpose of "mental time time travel" the most prominent is that of Corballis (2014) Tulving (e.g. 1985; 2001; 2005) and Suddendorf and Corballis (1997; 2007), i.e. mental time travel enables episodes from memory to be re-experienced, and that language may have evolved to make possible the sharing of communications about the non-present, e.g., what has happened, what may happen. Mental time travel allows for communication about memories of "me" "the self" in time.

The present commentary suggests "moving the spotlight" from mental time travel per se to the traveller self in connection with the self-concept, aiming at understanding what and why we communicate when sharing our mental world with others. To that end, two issues from Corballis's (2014) excellent article will be highlighted in this commentary. (i) The exclusive association that is drawn between the capacity of mentally traveling in time and the episodic memory system, and (ii) the suggestion that language may have evolved precisely to share our mental time travels.

The pioneer proposals and investigations of episodic memory and mental time travel by Tulving (e.g. 1985; 2001; 2005) and Suddendorf and Corballis (1997; 2007) have been extremely influential in the past decades, stimulating a rich field of research. Their fundamental assumption was based on Tulving's (1985) laboratory and clinical attempts to relate different memory systems (procedural, semantic and episodic) to their corresponding levels of consciousness (anoetic, noetic and autonoetic, respectively). Consistent with this excellent and very under-utilized theoretical connection, one of the central premises in Corballis's (2014) article is that episodic memory is the sole memory system infused with a sense of one's self (see Klein and Lax, 2010 for a different point of view). In Corballis's (2014) words, episodic memory "allows one to remember episodes experienced by the self, so that these episodes can be re-experienced" (p. 10). This statement is not easily contradicted. However, some related issues could be considered. One of them regards the link between episodic memory and mental time travel. An instance of this kind of reflection was expressed by Klein and Gangi's (2010, p. 11) suggestion that "the ability to mentally travel back and forth in time is not wedded to a particular form of memorial experience; rather, there appear to be qualitatively different types of temporal experience associated with different forms of memory."

A further related issue is that the multifaceted and polysemous nature of the self is increasingly a subject of consideration in autobiographical memory studies (as well as the multifaceted sense of subjective time travel, see Klein, 2013a). These reflections have resulted, specifically, in different models that explicitly integrate the relationships between memory and the self (Conway and Pleydell-Pearce, 2000; Conway et al., 2004a; Conway, 2005, Klein, 2014a, Prebble et al., 2013). Although Klein's (2014a) model of the two selves is likely the most challenging, the ontological reality of the self, which is at the centre of Klein's theoretical framework, is outside the scope of the present commentary. Two other models will be briefly commented on, the Self Memory System (SMS) model, in which the role of the semantically-based notion of autobiographical knowledge is essential to form autobiographical memories (Conway and Pleydell-Pearce, 2000; Conway et al., 2004a; Conway, 2005), and the more recent Prebble et al.'s (2013) model that puts forward four distinct sense of self components, each of which relates differently to autobiographical memory. With regard to the latter, only two out of the four self components will be tackled in this commentary, the subjective sense of self and the self-concept (see below). The reason to comment on some sections of these two models is that they provide the theoretical arguments to suggest a complementary view of the relationships outlined by Corballis (2014) between episodic memory and the subjective sense of self.

Bearing in mind that Corballis (2014) considers language as the phylogenetically acquired means of transmitting thoughts from one individual to another, a further key message in his target article is that "language may well have developed precisely to enable the communication of the non-present" (p. 8). Consistent with Corballis's (2014) proposal, not surprisingly, the contents of our informal speech mostly consist of experiences of mental time travels told to one another (Suddendorf and Corballis, 1997; 2010). Again, this claim is not easily challenged. However, taking into account the two mentioned views (Conway's SMS and Prebble et al.'s self component models), as well as recent research on our daily informal conversations (see below), communicating the non-present appears to be modulated by different psychological factors. In the first place, the higher accessibility to those personal memories that underwrite our self-esteem (Conway and Pleydell-Pearce, 2000; Conway et al., 2004a; Conway, 2005), and, secondly, the constant desire to broadcast our self (Tamir and Mitchel, 2012). Perhaps besides seeing mental time travel as the ultimate goal of our social communication (Corballis 2014), it might also be seen as the privileged means by which humans unrelentingly diffuse messages about their self. In other (ordinary) words, the aim of this commentary is to "move" the focus of mental time travel per se, to its ever-present target, the conceptual self (Conway, 2005).

2. Relations Between Memory And The Self

The relationships between autobiographical memory and the self, largely based on common sense and intuition, have been pointed out at least since the 4th century (Manning et al., 2013): it is the rememberer, who remembers, the traveller self, who travels in time. Complementarily, the observations that, for instance, people use their implicit theories of self to construct their personal histories (Ross, 1989), indicate that human memory is a major component of the self. Lampinen et al.'s (2004, p. 255) summary is as follows: "Who we conceive ourselves to be depends crucially on how we remember our lives. How we remember our lives depends crucially on how we mentally represent who we are. And our mental (and perhaps physical) well-being depends crucially on both memory and self- concept". The bidirectional effects outlined in this quotation have been operationalized in the SMS model (Conway, 2005), in which autobiographical knowledge constrains what the self is, has been, and can be, whereas the self modulates access to long-term knowledge.

As to the part regarding the self, in the relations between memory and self, investigations on the self are known for being particularly problematic. A central difficulty is associated with the definition of the self, since the types of selves and the corresponding definitions are numerous (e.g. Klein, 2012; Lampinen et al., 2004). Furthermore, there is no consensual concept of the self despite important attempts to categorize its different aspects (e.g., Gallagher, 2000; Klein and Gangi, 2010; 2012; Klein and Gangi, 2010). Consistent with the scope of the present commentary, however, far from discussing those elusive issues, a conceptually agreed on notion and taxonomy of the self is commented on in the next paragraph.

The most fruitful and enduring definition and taxonomy of the self traces back to the 19th century and was proposed by William James (1890). In James's view, the self can be seen as 'I' (i.e., the self as subject, knower, experiencer), or as 'Me' (i.e., the self as object, known or experienced) (James, 1890; Klein, 2012;

2014a). Corballis's (2014) implicit notion of the self would be consistent with James's I-self notion since he points out that the capacity to escape the constraints of physical time depends on autonoetic consciousness.

Turning back to memory, Suddendorf and Corballis (1997) coined the expression "mental time travel" and developed the notion that episodic memory is not so much a memory system but rather part of a more general mechanism to travel backwards and forwards in time.

On these bases, and tackling now the relations between memory and self in the target article, episodic memory is unquestionably unique precisely because it plays host, so to speak, to the self. Tulving's claims are unambiguous in this respect: "Autonoetic consciousness is a necessary correlate of episodic memory. It allows an individual to become aware of his or her own identity and existence..." (Tulving, 1985, p. 388). Moreover, in Corballis's theoretical framework, episodic memory is also distinguished from other memory systems, based largely on its functional commitments. Tulving (2002) claimed that the storage of episodic memories must depend on the noetic semantic memory system, however, he argued, memories related to the self in subjectively sensed time are stored separately from the semantic system (Tulving, 1985; 2002; 2005; Wheeler et al., 1997).

3. Episodic And Semantic Memories May Contribute To Mental Time Travel

Tulving's and Corballis's theoretical framework focus on the subjective experience and autonoetic consciousness associated exclusively with episodic memory. Notwithstanding the importance of this framework, it would be interesting to consider the suggestion that in naturalistic autobiographical memory, not only episodic memories, but also generic, semantic personal information operate in tandem (Levine, 2004) and are difficult to tease apart in the healthy individual (Levine et al., 2002).

Conway's proposal of the SMS targets the contribution of the two memory systems to personal recollections and mental time travel. It emerged from the notion of memory as being essentially a motivated cognitive function, i.e., driven by goals (Conway and Pleydell-Pearce, 2000; Conway et al., 2004a; Conway, 2005). The SMS is thought to be hierarchically higher and phylogenetically more recent than the episodic memory system. It is composed by the "working self" and the "autobiographical memory knowledge base". The working self is a dynamic, goal-based component that modulates and influences encoding, storage and retrieval memory processes. The autobiographical memory knowledge base is a long-term memory system that hierarchically organises autobiographical information at different levels of abstraction. It is formed by two distinct types of representation: autobiographical knowledge and episodic memories.

In contrast to Tulving's (1985; 2005) and Corballis's (2014) concept of episodic memory, in Conway's theoretical model the autobiographical system evolved to support adaptive short-term goal processing and is cue driven. Episodic memories correspond to discrete moments and are transitory per se. It is only through the goal-based influence of the working self mechanisms that episodic memories will be conceptually organised within the SMS. The episodic memories thus transformed form the autobiographical memory system, which, in its turn, influences the self-concept. The self-concept is defined as a non-temporally specified set of abstracted knowledge self structures, such as possible selves (Markus and Nurius, 1986), attitudes, values or beliefs. The self-concept knowledge is independent of, but connected with, episodic memories, which activating specific instances, contextualize and integrate their concepts, and autobiographical knowledge.

Very briefly stated, an individual's goals in interaction with the self-concept knowledge guide and regulate current cognition, affect and behaviour. The central principle of the SMS is that memory and crucial aspects of the self constitute a system characterised by its coherence. More precisely, self-concept knowledge is supported by memories of experiences that are coherent with the self and as such are more accessible than dissonant memories, which may jeopardise the beliefs about the self. Set against the pervasive need for coherence between memories and the self, is the demand of correspondence between memory and the experience. Conway et al. (2004b) suggest that in the case of dissonant memories between the two contradictory requirements, i., e., coherence with the self and correspondence with the experience, a compromise can be achieved (the concept was called "adaptive coherence"). One way to achieve this compromise is through the conceptual representation of experiences, i.e., the semanticized memories, which implies retention of the meaning (the gist) of experiences without access to associated episodic memories (Conway, 2005).

The relationships between Corballis's Tulvinian theory and Conway's SMS model have been analyzed and developed by Prebble et al. (2013). The authors propose a framework that defines sense of self along two dimensions, (i) a personal dimension: subjective (phenomenological experience of selfhood) vs. objective (knowledge about ourselves), and (ii) a temporal dimension: present vs. temporally extended. Four components results by combining the two dimensions: two of them are categorised as pertaining to the Jamesian category 'I-self' (the present subjective sense of self, and the temporally extended

phenomenological continuity). The other two components correspond to James's 'Me-self' classification (the present self-concept, and the temporally extended semantic continuity). Comments limited to the "present moment", highlight (i) the subjective sense of self (Tulving, 1985; 2005), which allows introspection, reflection, and evaluation of subjective experience, and (ii) the self-concept (Conway and Pleydell-Pearce, 2000; Conway et al., 2004a; Conway, 2005), viewed as the way individuals internally conceptualize who they are. These two self components are not only at the same hierarchical level in Prebble et al.'s (2013) model, but the subjective sense of self is connected with, and actively constructs, the self-concept representation.

The selected sections of the two models of memory and self, briefly commented above, suggest that episodic and semantic memory systems influence one another and work together in the healthy individual to form autobiographical memories. The role of the subjective sense of self is thought to be crucial for the construction of abstracted, semantic autobiographical memories (Prebble et al., 2013). In the same vein, the connection from the subjective sense of self to the self-concept may imply that the two forms of memory to which these two self components are related are in turn connected. On the other hand, one of the salient contributions of the SMS model is to view personal memories as the information bank of the working self, which is conceptualised as a set of active goals and associated self-images constrained by the principle of coherence (Conway and Pleydell-Pearce, 2000; Conway et al., 2004a; Conway, 2005).

Tulving (2005, p. 15) states that "...there can be no travel without a traveller. If it is not 'self' that "does" the traveling, then who, or what?" No one else than the subjective sense of self does the traveling, which is enabled by the mechanisms of the episodic autonoetic recollection in healthy individuals. However, along the lines of the metaphor used by Tulving, it could be suggested that whenever the I-self does the traveling, the places it tends to visit seem to be influenced by the semantically-based self-concept before the travel begins. Following the SMS model (Conway, 2005), this knowledge-based system provides an organizing context for episodic memory. Importantly, it "provides an access route that locates memories and set of memories in meaningful ways for the self" (p. 622). The interactions between the working self and the autobiographical memory knowledge base (see above) would influence where our mental time travels have greater chances to lead us and, therefore, which recollections are more likely to be shared with others. The reason for the "selection" of memories, as already noted, would be the pursuit of coherence between memories and self-concept. The knowledge-based system, argues Conway (2005), is one in which coherence is the dominant force. Coherence would support long-term goals and keep our self-esteem relatively unobstructed. Bearing in mind that self-esteem refers to the degree to which an individual values and accepts him/herself, a coherent self will have a high self-esteem (Conway et al., 2004a).

4. A Few Clinical Comments

From a clinical standpoint, by exclusively ascribing autonoetic consciousness to episodic memory, Tulving's and Corballis's theory may have contributed to the assumption that without episodic memories of our past we would be left with no sense of our identity. However, research undertaken to investigate how the self is affected by the onset and progression of dementia suggest that Alzheimer's disease patients retain their sense of identity, despite massive damage to the episodic memory system (e.g. Caddle and Clare, 2011a; 2011b). The same conclusion was reached in single case studies that aimed at exploring how the self is perceived in amnesic brain damage patients, by testing semanticized memories and semantic personality trait summaries (e.g. Klein, 2014b; Klein and Gangi, 2010; Rathbone et al., 2009). In relation to semantic summary representations of one's personality traits, conclusions reached by different authors point to the same direction, that is, a functional independence between semantic trait summaries and episodic memory, however, different authors situate this theoretical independence at different levels. Addis and Tippett (2004) probed personality trait (conceptual self) knowledge by means of the Twenty Statements Test, in Alzheimer's disease patients. Based on their results they suggest that this type of conceptual self requires the participation of episodic memory. Indeed, they showed a reduction in the overall number of "I am..." statements together with a greater proportion of "abstract" self statements, which were related to episodic memory loss for the childhood period. In contrast, Conway (2005) suggested that the conceptual self is independent of, but connected with, episodic memories. Illustrating this connection, the author referred to results on the "reminiscence bump" phenomenon (that is, the widespread observation that a greater number of memories are recalled about events that took place in late adolescence and early adulthood). Failure in the formation of the reminiscence bump, he argued, could account for some key abnormalities observed in schizophrenia (see below). Finally, Klein and colleagues (Klein and Lax, 2010; Klein and Gangi, 2010) found that both the formation and maintenance of abstract trait self-knowledge are functionally viable without the participation of episodic memory. This unexpected and robust resilience of personality trait self-knowledge in patients presenting with different aetiologies (brain trauma, neural disease or autism), led Klein and Gangi (2010) to propose that (i) learning personality traits does not require access to episodic memories, (ii) in cases in which there is an inability to update personality self-knowledge, this does not interfere with the ability to retrieve information from an intact, preexisting semantic store of trait summaries, and (iii) results shown in different patients suggest that semantic trait summaries are formed and maintained by functionally isolable acquisition, storage and retrieval processes.

Prebble et al.'s (2013, p. 827) interpretation is "that an abstract framework of conceptual self- knowledge may exist independently of the episodic memory system, but that episodic memories (particularly those from early life) may provide the detail needed for a rich and nuanced self-understanding" (see Prebble et al., 2013 and Rathbone et al., 2008, for the distinction between abstract and concrete self-knowledge). Although future research is necessary to confirm and expand on these results, it is generally accepted that episodic memory impairment even if it is massive is not tantamount to losing one's self (e.g. Klein and Gnagi, 2010; Rathbone et al., 2009) or being unable to share about oneself with others (e.g. Caddle and Clare, 2011b).

In contrast to the above mentioned cases of episodic memory acquired impairment due to cerebral lesions or neurodegenerative disease, there are cases in which the subjective sense of self shows developmental abnormalities. In these latter cases, difficulties in phenomenological continuity will be observed, and, to be highlighted in the present commentary, the formation of the self-concept would likely be compromised. Damage to this self component (Prebble et al., 2013) has been regarded as a possible explanation for mental illness, most particularly, schizophrenia. The presence of delusions, auditory hallucinations, and thought insertion, has been linked with breakdown in sense of agency and ownership of experience (Gallagher, 2000; Klein et al., 2004; Sass and Parnas, 2003) and as such, it was thought to illustrate the I-self impairment (Prebble et al., 2013). The deficits in autobiographical recollection observed in schizophrenia (e.g. Danion et al., 2007) would therefore reflect problems with agency and ownership rather than loss of content in memory (Klein et al., 2004; Prebble et al., 2013). For his part, Conway (2005) has also argued that one of the likely features of the abnormal SMS found in cases of developmental failure is the weakening or impossibility of the grounding of conceptual autobiographical knowledge in episodic memories of formative experiences. This possibility is precisely what Prebble et al.'s (2013) hypothesized connection between the subjective sense of self and the self-concept would predict.

5. What Do We Talk About?

In healthy individuals, stated in a simple formulation, the episodic memories shared with others could be the result of those personal experiences that serve the well-known human tendency towards self-inflation, i.e., to represent the self in a positive light. This issue is addressed in the present section.

Corballis (2009) has suggested that language became more complex when the humans left the present as unique way of experiencing events. To talk about the past and future, he argued, places an important burden on the communication channel: "In order to represent or refer to episodic elements that are not available in the present, we need very large vocabularies of concepts, as well as of words to represent them" (Corballis, 2009, p. 556). In the context of the present commentary, it could be added that it is also in order to both refer to the self I was/or will be, and to communicate that past/future self at its best in terms of coherence with it's past, present or future goals (Gärdenfors, 2004). This addition might be corroborated by the observation that some languages have no tenses, and indicate time by prepositions (after, before) or adverbs (tomorrow, yesterday). In contrast, pronouns and most particularly the first person singular exists even in languages in which the system of pronouns is the simplest yet recorded in primitive languages (Corballis, 2009).

What pieces of information do we exchange in current everyday informal settings? Investigations analyzing the content of freely forming conversations indicate that people have a universal tendency to exchange evaluative opinions about an absent third party, i.e., what we usually label "gossip." Dunbar (2004) reports that approximately two thirds of conversation time is devoted to gossip and, considering gossip in a phylogenetic perspective, points out its important role as an evolutionary tool in the strengthening and maintenance of social groups. More to the point, bearing in mind that gossip is a social behaviour across culture, language and demographic characteristics such as age, sex, IQ or socioeconomic level (Mc Andrew et al., 2007), it has been suggested that gossip is used as a means of social comparisons to increase the users status (Mc Andrew et al., 2007; Yao et al., 2014). Diffusing negative news about rivals would help to control others' reputations and boost one's own self- esteem. From a social psychology approach, gossip can be considered as a status enhancing mechanism. Important as it is for the self, talking about others does not include the unique specificity that characterizes talking about one's own self. Indeed, it has been shown that introspecting about ourselves and sharing that information with others constitute independent sources of reward (Tamir and Mitchel, 2012). The time people devote to communicating their personal experiences, and personal relationships, has been estimated around 30-40%, in informal everyday speech. This percentage approaches 80% or more when people communicate using Internet social media sites.

Independently of the oral or written modality of communication, sharing stories about an absent third party or about one's self relies typically on personal memories and personal future plans. In other words, to control others and boost our own esteem or to indulge in the rewarding behaviour to broadcast our self, we "use" self coherent episodic memory-based mental time travel and also autobiographical semantic contents such as what we know about ourselves, our values, beliefs. The issue people are so focused on communicating concerns what they think about themselves rather than the experience dealing with the mental travel itself.

As a very cautious addition to the excellent article by Corballis (2014), I would suggest that communicating his/her own present feelings (e.g. "I am scared") and his/her present feelings about others (e.g. "he is dangerous") conveyed the social usefulness necessary to become associated with cerebral mechanisms of reward. Tamir and Mitchel (2012) comprehensive study demonstrates that self-disclosure acts per se as an intrinsic reward, clearly detected in the brain reward system, and that the tendency to broadcast one's thoughts and beliefs may have adaptive advantage consequences. Language as "an 'instinct for inventiveness' that may go beyond language per se" (Corballis, 2009, p. 557) may have invented how to multiply the benefits offered by such rewarding social behaviour, by sharing mental time travels. As quoted in the Introduction, "language may well have developed precisely to enable the communication of the non-present" (Corballis, p2014, p. 8; italics added).

Corballis (2009, p. 557) comments that "By telling you what happened to me, I can effectively create an imagined episode in your mind, and this added information might help you adapt more effectively to future conditions. And by telling you what I am about to do, you may form an image in your own mind, and work out a plan to thwart me." In light of Conway's (2005), and Tamir and Mitchel's (2012) works, it might be added that by telling another what happened to me and what I am about to do, I also am looking for a source of reward by conveying how clever/ worried/ honest/ funny/ sad... I was or will be.

As a clinical counterpart, it has been established that the brain reward system is altered in some conditions such as major depression (Naranjo et al., 2001), and schizophrenia (Nielsen et al., 2012). Moreover, deficits in social/communication skills, independently of cognitive performance have been demonstrated in schizophrenic patients (Dickinson et al., 2007), and also in clinically or mildly depressive patients (Segrin, 2010). Not surprisingly, it has also been demonstrated an altered response to autobiographical memory tests, not due to loss of memory, in schizophrenia (e.g. Klein et al., 2004) and clinical depression (Williams, 1996).

6. Conclusion

In conclusion, the present commentary aimed at complementing Corballis's (2014) view on episodic memory and mental time travel, by integrating explicitly different self-concepts in autobiographical memory. In the healthy person, the subjective sense of self (I-self, autonoetic consciousness) does the travel in time. However, where it goes, i.e., what personal memories are made more accessible than other seems to be "the job" of the Me-self. Moreover, some clinical observations strongly suggest that loss of episodic memory is not equivalent to loss of personal identity. Perhaps the Me-self could be thought as the decision maker in the healthy individual and the resilient function (in the shape of semanticized memories and summaries of personality self-traits), in the neurological patient. On these bases, the Me-self's role would be both influenced by the need of coherence (Conway, 2005) and influence on what will in fine be communicated. These proposals, together with the notion of cognitive and neural reward in sharing about us (Tamir and Mitchel, 2012), converge toward a caution suggestion to highlight the role of the Me-self in future research on mental time travel in healthy people, and also in clinical research. The reason for the latter is linked with Medved & Brockmaier's (2008, in Rathbone et al. 2009, p. 414) conclusion: "In the absence of their own episodic memory (the patients) would appropriate the memories of others, or simply speculate in their own inability to recall things, by way of generating a sense of self." The Me-self as interacting with, and complementing, Corballis's I-self would go in the direction pointed out by the definition of memory as a meaning-making system that we owe to Bartlett (1932).

REFERENCES

Addis, D. R., & Tippett, L. J. (2004). Memory of myself: Autobiographical memory and identity in Alzheimer's disease. Memory, 12: 56–74.

Caddle, L. and Clare, L. (2011a) Interventions supporting self and identity in people with dementia: A systematic review, Aging & Mental Health, 15: 797-810.

Caddle, L. and Clare, L. (2011b) I'm still the same person: The impact of early-stage dementia on identity Dementia, 10: 379–398.

Conway, M. A. (2005). Memory and the self. Journal of Memory and Language, 53: 594–628.

Conway, M., Meares, K., & Standart, S. (2004b). Images & goals. Memory, 12: 525–531.

Conway, M., & Pleydell-Pearce, C. (2000). The construction of autobiographical memories in the self-memory system. Psychological Review, 107: 261–288.

Conway, M., Singer, J., & Tagini, A. (2004a). The self and autobiographical memory: Correspondence and coherence. Social Cognition, 22: 491–529.

Corballis, M. (2009). Mental time travel and the shaping of language Exp Brain Res, 192: 553–560.

Corballis, M. (2014). Mental Time Travel: How The Mind Escapes From The Present, Cosmology, Vol 18, 139-145.

Danion, J.-M., Huron, C., Vidailhet, P., & Berna, F. (2007). Functional mechanisms of episodic memory impairment in schizophrenia. Canadian Journal of Psychology, 52: 693–701.

Dickinson, D., Bellack, A., Gold, J. (2007). Social/communication skills, cognition, and vocational functioning in Schizophrenia. Schizophrenia Bulletin, 33: 1213–1220.

Dunbar, R., Duncan, N. & Marriott, A. Human conversational behavior. Human Nature, 8: 231-246.

Gallagher, S. (2000). Philosophical conceptions of the self: Implications for cognitive science. Trends in Cognitive Sciences, 4: 14–21.

Gärdenfors P. (2004). Cooperation and the evolution of symbolic communication. (pp 237–256). In: Oller DK, Griebel U (eds), Evolution of communication systems, MIT Press, Cambridge.

James, W. (1890). The principles of psychology (Vol. 1). New York, NY: Holt.

Klein, S. (2012). The self and its brain. Social Cognition, 30: 474-516.

Klein, S. (2013a). The complex act of projecting oneself into the future. WIREs Cognitive Science, 4: 66-79.

Klein, S. (2013b). The Sense of Diachronic Personal Identity. Phenomonology and the Cognitive Sciences, 12:791-811.

Klein, S. (2014a). The two selves. New York: Oxford University Press.

Klein, S. (2014b). Sameness and the self: philosophical and psychological considerations Front. Psychol., doi: 10.3389/fpsyg.2014.00029.

Klein, S., & Gangi, C. E. (2010). The multiplicity of self: Neuropsychological evidence and its implications for the self as a construct in psychological research. Annals of the New York Academy of Sciences, 1191: 1–15.

Klein, S., German, T., Cosmides, L. & Gabriel, R. (2004). A theory of autobiographical memory: necessary components and disorders resulting from their loss. Social Cognition, 22: 460-490.

Klein, S., & Lax, M. L. (2010). The unanticipated resilience of trait self-knowledge in the face of neural damage. Memory, 18: 918–948.

Lampinen, J., Beike, D. & Behrend, D. (2004). The self and memory: It's about time (pp. 255-262). In D. Beike, J. Lampinen & D. Behrend (eds), The self and memory. Hove: Psychology Press.

Levine, B. (2004). Autobiographical memory and the self in time: Brain lesion effects, functional neuroanatomy, and lifespan development. Brain and Cognition, 55: 54–68.

Levine, B., Svoboda, E., Hay, J., Winocur, G. & Moscovitch, M (2002). Aging and autobiographical memory: Dissociating episodic from semantic retrieval. Psychology and Aging, 17: 677-689.

Manning, L., Cassel, D., Cassel, JC. (2013) St. Augustine's Reflections on Memory and Time and the Current Concept of Subjective Time in Mental Time Travel Behav. Sci.,3: 232–243.

Markus, H., & Nurius, P. (1986). Possible selves. American Psychologist, 41, 954–969.

McAndrew, F., Bell, E. & Contitta, M. (2007). Who do we tell and whom do we tell on? Gossip as a strategy for status enhancement. Journal of Applied Social Psychology, 37: 1562– 1577.

Naranjo, C., Tremblay, L. & Busto, U. (2001). The role of the brain reward system in depression Prog Neuropsychopharmacol Biol Psychiatry., 4: 781-823.

Nielsen, M., Rostrup, E., Wulff, S., Bak, N., Lublin, H., Kapur, S. & Glenthøj, B. (2012). Alterations of the brain reward system in antipsychotic naïve schizophrenia patients. Biol Psychiatry., 71: 898-905.

Prebble, S., Addis, R. & Tippett, L. (2013). Autobiographical memory and sense of self. Psychological Bulletin, 139: 815-840.

Rathbone, C. J., Moulin, C. J. A., & Conway, M. A. (2008). Self-centered memories: The reminiscence bump and the self. Memory & Cognition, 36: 1403–1414.

Rathbone, C. J., Moulin, C. J. A., & Conway, M. A. (2009). Autobiographical memory and amnesia: Using conceptual knowledge to ground the self. Neurocase, 15: 405–418.

Ross, M. (1989). The relation of implicit theories to the construction of personal histories. Psychological Review, 96: 341–357.

Sass, L. & Parnas, J. (2003). Schizophrenia, consciousness and the self.

Schizophrenia Bulletin, 29: 427-444.

Segrin, C. (2010). Depressive Disorders and Interpersonal Processes (pp. 425-448), in L. Horowitz and S. Strack (Eds), Handbook of Interpersonal Psychology: Theory, Research, Assessment, and Therapeutic Interventions. West Sussex: John Wiley & Sons.

Suddendorf, T. & Corballis, M. (1997). Mental time travel and the evolution of the human mind. Genet. Soc. Gen. Psych., 123, 133–167.

Suddendorf, T. & Corballis, M. (2007). The evolution of foresight: What is mental time travel, and is it unique to humans? Behav. Brain Sci., 30, 299–351.

Suddendorf, T. & Corballis, M. (2010). Behavioural evidence for mental time travel in nonhuman animals. Behav Brain Res., 215:292-8.

Tamir, D. & Mitchel, J. (2012) Disclosing information about the self is intrinsically rewarding. PNAS, 109: 8038-8043.

Tulving, E. (1985). Memory and consciousness. Canadian Psychology/Psychologie canadienne, Vol 26: 1-12.

Tulving, E. (2001) Episodic memory and common sense: how far apart? Phil Trans. of the Royal Soc. London. B. 356: 1505-1515.

Tulving, E. (2005). Episodic memory and autonoesis: Uniquely human? (pp. 3–56). In H. S. Terrace & J. Metcalfe (Eds.), The missing link in cognition: Origins of self-reflective consciousness. New York, NY: Oxford University Press.

Wheeler, M., Stuss, D., & Tulving, E. (1997). Toward a theory of episodic memory: The frontal lobes and autonoetic consciousness. Psychological Bulletin, 121: 331–354.

Williams, J. (1996). Depression and the specificity of autobiographical memory. In D. Rubin (Ed.), Remembering our past: Studies in autobiographical memory (pp. 244–267). Cambridge, England: Cambridge University Press.

Yao, B., Scott, G., O'Donnell, P. & Sereno, S. (2014). Familiarity with interest breeds gossip: contributions of emotion, expectation, and reputation. PLoS ONE DOI: 10.1371/journal.pone.0104916

17. Continuity In Hippocampal Function As A Constraint On The Convergent Evolution Of Episodic-Like Cognition

James M. Thom[1] and Nicola S. Clayton[2]

[1]Department of Philosophy, King's College London, Strand, London, UK
[2]Department of Psychology, University of Cambridge, Cambridge, UK

Abstract

Continuity of hippocampal function in mammals has been cited as indicative of continuity in some of the processes underpinning mental time travel. The main evidence presented for this continuity is the activity of rat hippocampal place cells outside of the spatial context to which they correspond. Research into food-caching in corvids has revealed memory and planning abilities that resemble some aspects of human mental time travel. While these abilities appear to have evolved independently of mental time travel in humans, dependence on the hippocampus may have constrained the evolution of both. Central to this claim is the substantial evidence that the hippocampus supports similar spatial cognition in both mammals and birds, and the importance of the hippocampus in food-caching.

Key words: Hippocampus, corvid, episodic, episodic-like, cognition

Corballis (2014) argues for the evolutionary continuity of Mental Time Travel, and suggests that it is the extent of input from language, rather than a qualitative difference in consciousness, that separates humans from the rest of the Animal Kingdom. As an example, he cites: "[our] increasing understanding of what happened at the origin of the universe itself, as well as what will happen some 7.5 billion years from now when the sun will grow so big that it will subsume the earth."

Certainly such things seem, literally, to be beyond the imagination of other animals. In imagining the origin of the universe, we must depend on language to represent scales of time and distance beyond what any human can experience. We agree that human mental time travel seems inextricably bound up with language. So when comparing mental time travel in humans and other animals, species differences in language ought to be an important consideration. This position runs counter to arguments for mental time travel as uniquely human, which propose a fundamental difference between human consciousness and that of other animals

(Suddendorf & Corballis, 1997). The emphasis of these arguments has been on autonoetic consciousness – the ability to experience simulations of our past and future, and chronesthaesia – the awareness of our own past and future.

Because we cannot ask a non-human animal about their experiences, phenomenological questions of conscious seem unanswerable. We are therefore restricted to behavioural and physiological measures. For this reason, the focus of comparative research has been on 'episodic-like' cognition (Clayton & Dickinson, 1998), which specifies behavioural criteria for an analogue of mental time travel, without making assumptions about consciousness. In our own research on corvids, the behaviour most examined for episodic-like cognition is food-caching. For example, Western scrub-jays remember what, where, and when they have cached (Clayton & Dickinson, 1998), and preferentially cache food items they will want to eat in the future, independently of their current motivational state (Correia et al., 2007).

Corballis's (2014) case for the continuity of Mental Time Travel rests on the apparent continuity of hippocampal function across the mammalian clade. In particular, he highlights firing of hippocampal place cells in rats, whilst outside of the environment and locations to which those cells correspond. We agree that this activity seems to be, at very least, an important building block for the evolution of hippocampal scene construction in human mental time travel (Hassabis & Maguire, 2007).

Continuity of hippocampal function may extend beyond just mammals. Colombo & Broadbent (2000) comprehensively review the case for the avian hippocampal complex as a functional homologue of the mammalian hippocampus. They report the strikingly similar deleterious effects of hippocampal lesions on spatial cognition tasks in mammals (rats and monkeys) and birds (pigeons). In contrast, performance on non-spatial tasks in either clade appears relatively unaffected by such lesions. In addition, the authors describe several similarities in development and connections to other basal ganglia, despite some anatomical differences. Crucially, hippocampal synapses in mammals and birds both show plasticity in the form of long-term potentiation. In other words, we have good reason to believe that the avian hippocampus performs a similar function in supporting spatial cognition, and may employ similar processes to do so. Colombo & Broadbent conclude: "...despite 300 million years of independent evolution, there are no degrees of freedom in the evolution of hippocampal function."

If hippocampal function is conserved across the mammalian and avian clades, we might expect avian episodic-like cognition to depend upon the hippocampus, just as human episodic cognition does. While this relationship has not been directly explored, we do know that food-caching – which has been the focus of much of the research into episodic-like cognition – does depend critically upon spatial processing in the hippocampus: Caching specialisation predicts spatial memory performance (Bednekoff et al., 1997; Kamil et al., 1994) and hippocampal volume (Basil et al., 1996; Lucas et al., 2004) between species. Furthermore, experimentally manipulated differences in experience of caching and retrieval predict variation in individual hippocampal volume (Clayton & Krebs, 1994). Lesion studies directly test reliance on the hippocampus for cache-retrieval: hippocampal lesions reduce retrieval accuracy to chance levels (e.g. Sherry & Vaccarino, 1989), and hippocampal tissue transplants reverse these spatial memory deficits (Patel et al., 1997).

In our view, the shared reliance on relatively conserved spatial processing in the hippocampus, in both mammals and birds, is a plausible constraint on the convergent evolution of episodic and episodic-like cognition. Given the evidence for the importance of the hippocampus in imagination generally (Hassabis et al., 2007a; Hassabis et al., 2007b), this argument might also be applicable to the evolution of other cognitive faculties. Indeed, it has been argued that imagination is one critical component of a domain-general cognitive 'toolkit' that evolved separately in corvids and the Great Apes, in order to solve similar problems (Emery & Clayton, 2004).

Speculation: If hippocampal function has indeed constrained the convergent evolution of episodic and episodic-like cognition, then the psychological processes underpinning these systems may be more similar than has previously been assumed. The hippocampus appears particularly to perform a crucial function supporting spatial cognition in both humans and other mammals, and also in birds. Those same processes necessary for spatial cognition also appear important in scene construction (Hassabis & Maguire, 2007) – a key component of human mental time travel and imagination. It may be that much of human mental time travel is inextricably tied to this spatial element. For example, visualizations of the past or future from a third-person perspective are associated with reduced salience of internal thoughts (Pronin & Ross, 2006), which in of themselves have no obvious spatial component. We might therefore expect avian recall of past events, and imagination of the future, to be as intrinsically spatial as these abilities are in our own species, perhaps even to the extent of being tied to a first-person perspective.

One way to test whether avian episodic-like cognition is inherently spatial is to look for spatial phenomena associated with human mental time travel. For example, people often remember seeing more of a scene than they did, extrapolating the borders of an observed stimulus, and leaving a representation of a scene appearing more 'zoomed out' (Intraub & Richardson, 1989). These boundary extension errors are associated with hippocampal activation (Chadwick et al., 2013), and are less severe in patients with hippocampal amnesia (Mullally et al., 2012). Evidence of boundary extension errors in corvids would implicate hippocampal scene construction as a central component process of both avian episodic-like cognition and human mental time travel.

References

Basil, Jennifer A, Kamil, Alan C, Russell, Balda, & Fite, Katherine V. (1996). Differences in hippocampal volume among food storing corvids. Brain, Behavior and Evolution, 47(3), 156- 164.

Bednekoff, P. A., Balda, R. P., Kamil, A. C., & Hile, A. G. (1997). Long-term spatial memory in four seed-caching corvid species. Animal Behaviour, 53, 335-341. doi: Doi 10.1006/Anbe.1996.0395

Chadwick, Martin J, Mullally, Sinéad L, & Maguire, Eleanor A. (2013). The hippocampus extrapolates beyond the view in scenes: An fMRI study of boundary extension. Cortex, 49(8), 2067-2079.

Clayton, N. S., & Dickinson, A. (1998). Episodic-like memory during cache recovery by scrub jays. Nature, 395(6699), 272-274.

Clayton, N. S., & Krebs, J. R. (1994). Hippocampal growth and attrition in birds affected by experience. Proceedings of the National Academy of Sciences, 91(16), 7410-7414.

Colombo, M, & Broadbent, N. (2000). Is the avian hippocampus a functional homologue of the mammalian hippocampus? Neuroscience and Biobehavioral Reviews, 24(4), 465-484.

Corballis, M. (2014). Mental Time Travel: How The Mind Escapes From The Present, Cosmology, Vol 18, 139-145.

Correia, S. P., Dickinson, A., & Clayton, N. S. (2007). Western scrub-jays anticipate future needs independently of their current motivational state. Current Biology, 17(10), 856-861. doi: Doi 10.1016/J.Cub.2007.03.063

Emery, N. J., & Clayton, N. S. (2004). The mentality of crows: convergent evolution of intelligence in corvids and apes. Science (New York, N Y), 306(5703), 1903-1907.

Hassabis, D., Kumaran, D., & Maguire, E. A. (2007a). Using imagination to understand the neural basis of episodic memory. The Journal of neuroscience : the official journal of the Society for Neuroscience, 27(52), 14365-14374.

Hassabis, D., Kumaran, D., Vann, S. D., & Maguire, E. A. (2007b). Patients with hippocampal amnesia cannot imagine new experiences. Proceedings of the National Academy of Sciences of the United States of America, 104(5), 1726-1731.

Hassabis, D., & Maguire, E. A. (2007). Deconstructing episodic memory with construction. Trends in cognitive sciences, 11(7), 299-306.

Intraub, Helene, & Richardson, Michael. (1989). Wide-angle memories of close-up scenes. Journal of Experimental Psychology: Learning, Memory, and Cognition, 15(2), 179.

Kamil, Alan C, Balda, Russell P, & Olson, Deborah J. (1994). Performance of four seed-caching corvid species in the radial-arm maze analog. Journal of Comparative Psychology, 108(4), 385-393.

Lucas, Jeffrey R, Brodin, Anders, de Kort, Selvino R, & Clayton, Nicola S. (2004). Does hippocampal size correlate with the degree of caching specialization? Proceedings of the Royal Society of London. Series B: Biological Sciences, 271(1556), 2423-2429.

Mullally, Sinéad L, Intraub, Helene, & Maguire, Eleanor A. (2012). Attenuated boundary extension produces a paradoxical memory advantage in amnesic patients. Current Biology, 22(4), 261-268.

Patel, Sanjay N, Clayton, Nicky S, & Krebs, John R. (1997). Hippocampal tissue transplants reverse lesion-induced spatial memory deficits in zebra finches (Taeniopygia guttata). The Journal of neuroscience, 17(10), 3861-3869.

Pronin, Emily, & Ross, Lee. (2006). Temporal differences in trait self-ascription: when the self is seen as an other. Journal of Personality and Social Psychology, 90(2), 197.

Sherry, David F, & Vaccarino, Anthony L. (1989). Hippocampus and memory for food caches in black-capped chickadees. Behavioral Neuroscience, 103(2), 308.

Suddendorf, T., & Corballis, M. C. (1997). Mental time travel and the evolution of the human mind. Genetic, Social, and General Psychology Monographs, 123(2), 133-167.

18. The Theory of MindTime
John T. Furey[1] and Vincent J. Fortunato[2]

[1]The MindTime Project; [2]Walden University and The MindTime Project

Abstract

According to modern cosmologists, the evolution of consciousness corresponded with the evolution of matter into increasingly complex, elaborate, and interactive systems, with the human brain providing the highest level of complexity known. Psychological research shows that just about all of human experience is dependent upon and influenced by how individuals perceive time, localize themselves consciously within space and time, process their temporally-based perceptions and experiences, and utilize their episodic and semantic memory structures to engage in mental time travel. We propose that over the course of evolution, sensitivities toward perceiving potentially pleasurable/ appetitive and aversive/harmful environmental stimuli and the motivation to approach and/or avoid such stimuli moved beyond reflexive, innate, and learned associative neural networks and became increasingly influenced by, and in turn influenced, the cognitive structures associated with organisms' ability to perceive and conceptualize time. In this paper, we present a theory of consciousness and psychology in which we propose that three general yet distinct cognitive patterns, or thinking perspectives, exist, which we refer to as Past, Present, and Future thinking, and that these three patterns are universal conditions of consciousness and form the foundation and framework for understanding, in particular, all of human thought and interaction, from the individual to the collective, and from the formation of an idea to the creation of cultures and artifacts based on those ideas.

Keywords: Thinking perspective, Past thinking, Present thinking, Future thinking, mental time travel, time, temporal perspective, perception, consciousness, self-awareness, MindTime

1. Introduction

The ability to consciously localize experience temporally and engage in what Tulving (1985a) referred to as mental time travel is considered to be one of the most important evolutionary advancements of consciousness (e.g., Corballis, 2013; Jaynes, 1976; Liljenström, 2011; Suddendorf & Corballis, 1997, 2007). Although mental time travel—the ability to mentally project one's mind forward and backward in time—has been demonstrated in many animals (e.g., Corballis, 2013), in human beings, the ability to engage in mental time travel is especially advanced. Human beings, in particular, have the ability to "internalize" worlds through the formation of spatial and temporal patterns and form an understanding of the relationship between cause and effect (Liljenström, 2011).

Mental time travel appears to have coincided with the evolution of complex memory systems and self-awareness (e.g., Liljenström, 2011; Suddendorf & Corballis, 1997, 2007; Tulving, 1985a, 1985b) and, according to Suddendorf and colleagues (Suddendorf, 1999; Suddendorf & Corballis, 1997, 2007), provided Homo sapiens with a cognitive flexibility to respond to the cyclical nature of temporally-based experiences, draw upon past experiences and stored knowledge, creatively imagine an infinite set of hypothetical future possibilities and engage in "what if" theorizing, set short-term and long-term goals, develop strategies and action plans, and organize the resources needed to obtain long-term results; all with the goal of maximizing survival. Mental time travel also provides human beings with the boundless flights of imagination that typify the human species (e.g., Hejazi, 2012; Lombardo, 2006).

Mental time travel has become an increasingly popular construct of interest among developmental psychologists (e.g., Atance & O'Neill, 2005; Atance & Jackson, 2009), cognitive psychologists (e.g., Trope & Liberman, 2003, 2010), social psychologists (e.g., Gilbert & Wilson, 2007), and neuropsychologists (e.g., Addis, Wong, & Schacter, 2007; Hassabis, Kumaran, & Maguire, 2007) and there is increasing evidence that just about all of human experience is dependent upon and influenced by how individuals perceive time, process their temporally-based perceptions and experiences, and utilize their episodic and semantic memory structures. For example, the ability of human beings to project their minds forward and backward in time has been shown to influence the emotions people experience (D'Argembeau & Van der Linden, 2007), their goals and motivational intentions (Oettingen & Mayer, 2002), their attributions of others (Warner, VanDeursen, & Pope, 2012), their perceptions of physical, social, and psychological distance (Trope & Liberman, 2003, 2010), their performance on creative and analytical tasks (Förster & Becker, 2012; Förster, Friedman, & Liberman, 2004), how they perceive and process cognitive information (Liberman & Trope, 1998), and how they perceive themselves and imagine their future selves (Wilson, Buehler, Lawford, Schmidt, & Yong, 2012).

Moreover, the ability to localize experience temporally and engage in mental time travel is implicated in the dynamic social constructive process in which language evolved as a means of social interaction and collaboration (Corballis, 2011; Perrot-Clermont & Lamboltz, 2005), the development of personal and collective narratives (i.e., life scripts: Berntsen & Bohn, 2010; Bronckart, 2005), the development of social schemas, norms, and expectations (e.g., Levine, 2006), and the development and evolution of culture and cultural artifacts (Vale, Flynn, & Kendal, 2012; Trompenaars & Hampden-Turner, 1997). Indeed, Tattersall (2011) suggested that human beings are, arguably, the only creatures on earth capable of mentally disassembling the world into a vocabulary of abstract symbols and then recombining them to create alternative versions of reality and that the fact that each human being recreates the world differently is at the root of the complexities of human experience and of human society.

Relatedly, Gilead, Liberman, and Maril (2012) proposed that mental time travel in human beings is a special case of the ability to traverse psychological distance and that many higher order cognitive functions such as the ability to imagine hypothetical realities, deceive the mind of others, and engage in prospective theorizing and counterfactual thinking are all based on a common underlying mechanism that involves the human capacity for abstract–linguistic and disembodied representations. Indeed, cognitive psychology research (Liberman & Trope, 1998; Trope & Liberman, 2003, 2010) has shown that when individuals are asked to engage in mental time travel, the temporal direction (i.e., past or future) and distance (e.g., near-term or distal future) with which they imagine themselves profoundly affects the degree of abstraction with which objects of consciousness are mentally represented and how individuals perform on a variety of cognitive tasks. The greater the temporal distance an object or event is imagined to exist in time, either into the future or into the past, the more likely it is that individuals will mentally represent that object or event abstractly.

However, despite the inextricable relationship between time, consciousness, and human experience, with perhaps the exception of Lewin (1942, 1951), who proposed that individuals' develop their own personal psychology based on the totality of their views of their own psychological past and psychological future, little theoretical development exists incorporating human perception of time and awareness and the ability to engage in mental time travel into a general model of human psychology. In this manuscript, we present a framework—the theory of MindTime—that we believe provides a foundation for understanding human consciousness, personality, cognition and information processes, preferences, and behavior.

2. The Theory of MindTime

King (2011) stated that "the evolutionary key to consciousness may lie in the survival advantage it could provide in anticipating threats and strategic opportunities." King goes on to describe how these two characteristics of life—the motivation to avoid environmental threats and the motivation to approach opportunities—exist at all levels of life, from the eukaryote cell to human beings. Each form of life contains sense organs that are sensitive to feedback about its environment. However, over time, life evolved more intricate and complex methods of survival, including complex nervous systems, such as those associated with episodic and semantic memory systems, that allowed for increased capacity of information storage and processing (Liljenström, 2011), which in turn provided life forms the ability to, in varying degrees, (a) perceive the passage of time, (b) localize their awareness in time, and (c) engage in mental time travel (e.g., Corballis, 2013).

As Lombardo (2011) noted, consciousness is temporally based in that it is always opening into the future or looking backwards into the past. According to Lombardo, "conscious beings are aware of duration, relative stability, and patterns of change; of becoming and passing away; and of an experiential direction to time. The conscious now ... may be anchored at the level of perception, and contextualized within consciousness of the past (memories) and conscious anticipation of the future, all three phenomenologically blurring together at the edges."

We propose that over the course of evolution, sensitivities toward perceiving potentially pleasurable/appetitive and aversive/harmful environmental stimuli and the motivation to approach and/or avoid such stimuli moved beyond reflexive, innate, and learned associative neural networks and became increasingly influenced by, and in turn influenced, the cognitive structures associated with organisms' ability to perceive and conceptualize time.

Specifically, we propose that at the foundation of consciousness are three temporally- based, cognitive–perceptual, abstract–linguistic patterns of thinking, which we refer to as Past, Present, and Future thinking (see Fig. 1). These three thinking perspectives correspond with (a) innate representations of the past, present, and future as temporal realities; (b) the conceptual representations and concepts that emerge from those representations; and (c) how organisms, particularly human beings, localize themselves in time. Past, Present, and Future thinking form the foundation for understanding all of human perception, thought, and interaction, from the individual to the collective, and from the formation of an idea to the creation of cultures and artifacts based on those ideas. Past, Present, and Future thinking influence the personal (and collective) narratives that individuals (and groups) develop and how individuals (and collectives): (a) perceive the world around them; (b) process and then encode information in semantic and episodic memory; (c) mentally represent objects of consciousness; (d) formulate goals and intentions; (e) develop their preferences; and (f) communicate and interact with others.

At the level of the individual, we propose that the foundation of a personal identity (i.e., a me) and just about all stable trait-based individual differences, such as the personalities that individuals manifest, as well as individuals' perceptions, intentions, values, beliefs, motivations, and behaviors can be understood as the interaction of individuals' Past, Present, and Future thinking. We also propose that the theory of MindTime can be used to describe and explain the behaviors of increasingly aggregated higher-order and complex collections of individuals such as groups and work teams, organizations, and nations as well as the temporal origins of the cultural artifacts created by these aggregates. Thus, Past, Present, and Future thinking operate as collective cognitive filters that form the basis for the development of a group culture and the cultural norms and expectations that follow. In this way we believe these three patterns of thought to be universal in scope and fractal in nature.

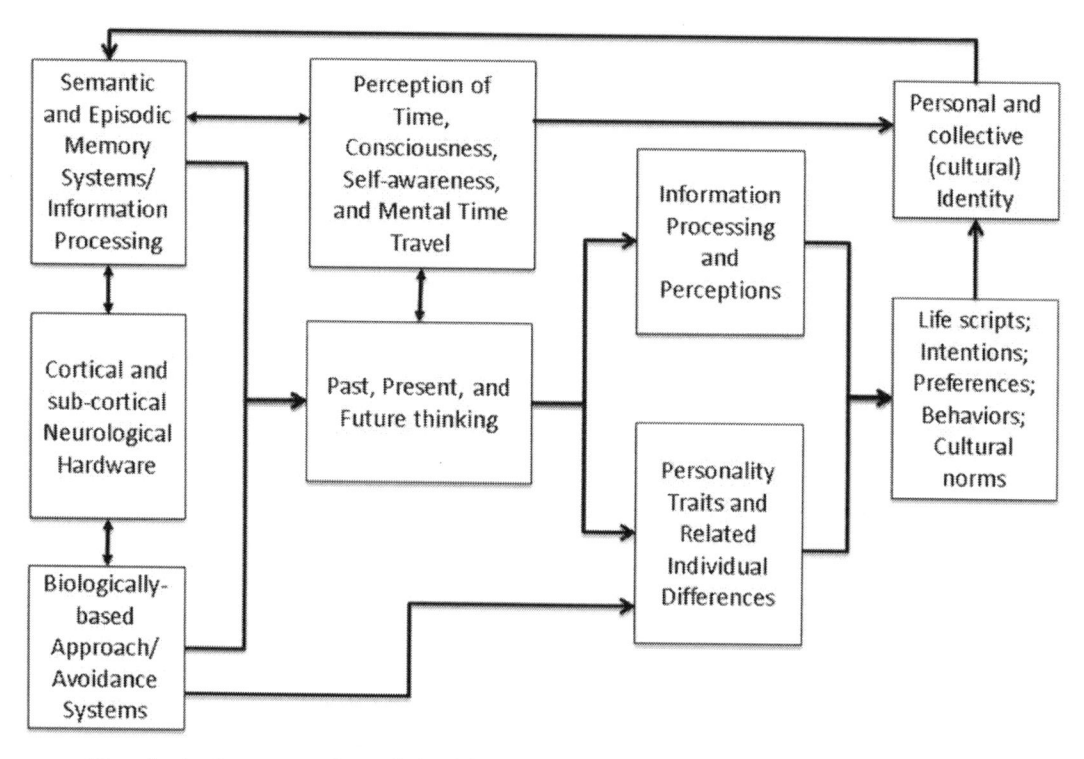

Fig. 1. An integrated model of human psychology and individual differences.

Finally, we propose that the degree to which any individual or collection of individuals resonates with any other individual or collection of individuals or with any artifact or product of human endeavor (e.g., culture, political and economic system[s], institution[s], technology, language, symbol[s], message[s], manufactured object[s], architecture, design[s]) will depend on the degree of similarity that exists between the thinking perspective characteristics of the subject and that of the object. That is, the theory of MindTime provides a basis for understanding the quality and nature of the interaction between any two individuals or collectives of individuals as well as the quality and nature of the interactions among the members of any collective and between any individual or collective and the products of human endeavor.

In the sections that follow, we briefly describe the three thinking perspectives, review some empirical evidence that support our descriptions and propositions involving Past, Present, and Future thinking, discuss a few theoretical and practical implications, and, in our closing remarks, seek to place our theory into a cosmological context.\

Table 1

Consciousness, Neuroscience, Time Travel

Table 1. Key characteristics of Past, Present, and Future thinking

Past Thinking	Present Thinking	Future Thinking
Abstract and concrete mental representations of objects	Concrete mental representations of objects	Abstract representations of objects
Qualitative aspect of abstraction oriented toward understanding "why?"	Qualitative aspect of thinking oriented toward understanding "how?"	Qualitative aspect of abstraction oriented toward understanding "what if?"
Recollection, reconstruction, analysis, and critical evaluation of past experiences and knowledge stored in episodic and semantic memory	Organization of data, information, people, objects, and events into conceptual schemas stored in memory	Imagination of alternative realities and the envisioning of novel and innovative solutions to intractable problems
Encodes in memory knowledge and truth determined via analytical thinking	Encodes in memory spatial and temporal organizational frameworks	Encodes in memory the interconnections observed via systems-level perception
Reflective thinking oriented toward sense-making (meaning)	Functional thinking oriented toward controlling the environment (form)	Forward thinking oriented toward opportunity (movement)
Analytical thinking oriented toward knowledge, truth, and relevance	Methodical thinking oriented toward creating predictable outcomes	Systems level thinking oriented toward understanding interconnections among objects of perception
Oriented toward evaluating validity of personal, social, and cultural schemas	Oriented toward maintaining personal, social, and cultural schemas (the status quo)	Oriented toward challenging and possibly disrupting personal, social, and cultural schemas
Oriented toward risk avoidance	Oriented toward control and maintaining equilibrium and continuity of stable, secure, and predictable environment	Oriented toward possibility
Perceptually sensitive to environmental threats and to errors and gaps in knowledge	Perceptually sensitive to actual or anticipated discrepancies between expectations and current reality and deviations from plans	Perceptually sensitive to the big-picture and to opportunities
Motivational impulse occurs when threats to personal and collective safety are perceived	Motivational impulse occurs when discrepancies are perceived between expectations and observations	Motivational impulse occurs when an opportunity is perceived
Drives understanding	Drives conformity	Drives change
Action: Information gathering, sense-making	Action: Reduce discrepancies	Action: Create forward movement
Tools: reasoned discourse, logic, Socratic debate, inductive and deductive reasoning, propaganda, historical storytelling	Tools: Normalized frames of reference, processes, plans, rules, laws, schedules, procedures, agendas	Tools: Ideation (brainstorming), persuasion, articulation of abstract ideas, metaphors, imagery, visualization, storytelling
Creates: Scientific, academic, and knowledge-based institutions	Creates: Logistical institutions; bureaucracies	Creates: Innovative advancements; visionary ventures; new forms of artistic expression

2.1 The Three Thinking Perspectives. Although it is customary to refer to the past, present, and future in that order (as we do throughout this manuscript), we discuss Past and Future thinking before we discuss Present thinking for reasons we hope will be apparent to the reader. Table 1 lists the key characteristics of the three thinking perspectives and Fig. 2 visually some key higher-order concepts that emerge from the interaction of Past, Present, and Future thinking.

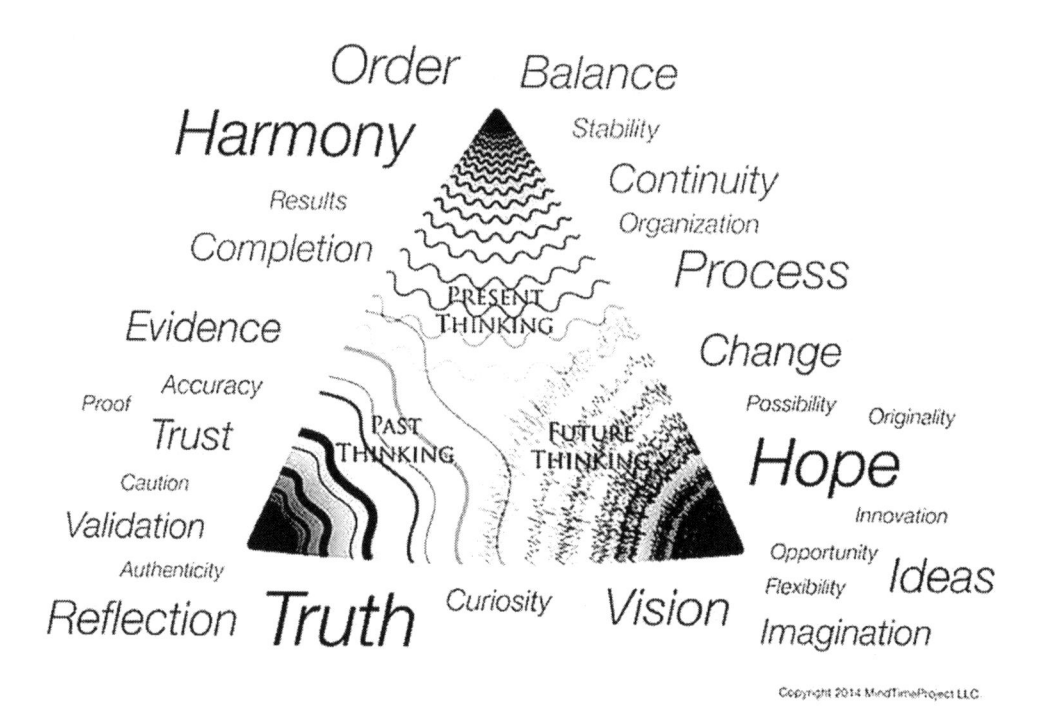

Fig. 2. A visual depiction of the key higher-order concepts that emerge from the interaction of Past, Present, and Future thinking. Figure used with permission from the MindTime Project, LLC. Illustrated by Alden Bevington.

2.1.1 Past thinking. We use the term Past thinking to refer to the cognitive patterning that exists when memory processes are used by organisms to differentiate and dichotomize experienced reality. In human beings, the products of that dichotomization are the conceptual and social schemas, measurements, perceived phenomena, and qualities that facilitate our understanding of the world, the universe, and ourselves within it. Past thinking involves both abstract and concrete mental representations of objects and is used when individuals reflect, recollect, reconstruct, analyze, and critically evaluate experienced events and knowledge stored in memory. Past thinking is used to avoid threats, minimize risks, and maximize survival, by determining the relevance, validity, and truth of information and knowledge as well as those of prevailing personal, social, and cultural schemas, expectations, and norms. Past thinking is also used when individuals integrate ideas, observations, and information into conceptual schemas through assimilation and accommodation processes.

Our construct of Past thinking is similar to Dewey's (1910/1933) construct of reflective thinking. Dewey believed that reflective thinking had two critical components: sense-making and continuity. The former refers to the process in which individuals develop a deep understanding of the relationships and connections to experiences and ideas. The latter refers to how individuals make sense of each new experience based on the meaning derived from past experiences as well as prior knowledge of the world, such as what we have heard about and read of others' experiences and ideas. Although Dewey conceptualized reflective thinking as a process to be cultivated within the educational system in order to facilitate the moral, intellectual, and emotional growth of the individual, we conceptualize Past thinking as a stable individual difference variable, which, along with Present and Future thinking, influences how individuals perceive, process, and utilize information.

At the individual level, we hypothesize that Past thinking manifests as: (a) a sensitivity to the presence of potentially negative environmental stimuli and to gaps in knowledge; (b) a propensity to avoid risks; (b) slow and deliberate thinking; (c) principled, judicious, reflective, and thoughtful decision making; (d) a propensity to second guess decisions once made; (e) a cautious, skeptical outlook on life; (f) fair-mindedness; and (g) a propensity to experience psychological distress, such as depression and anxiety. Past thinking also manifests as the Big Five personality traits (e.g., Costa & McCrae, 1992a, 1992b) of introversion and neuroticism, a preference for quiet and studious environments, opportunities for contemplation, the need for few but deeply trusting friendships, a somewhat ideological leadership style, and careers that allow for independence of thought, opportunities for applying scientific and analytical inquiry, and working with information and knowledge.

We also propose that it is Past thinking at work when collections of individuals engage in the recollection, reconstruction, analysis, and critical evaluation of collectively experienced and/or communicated events and when such information is used to establish and maintain a sense of collective continuity as well as a sense of collective efficacy. Through the tools of science, reasoned discourse, logic, Socratic debate, inductive and deductive reasoning, propaganda, and historical storytelling, societies establish and maintain a sense of cultural identity and continuity as well as a sense of collective well-being and social efficacy. For example, the scientific method is the formal use of Past thinking to understand the physical universe and theology is the formal use of Past thinking to understand the spiritual universe. Past thinking is that aspect of consciousness that human beings utilize to understand reality and derive meaning regarding our experience of reality.

2.1.2 Future thinking. We use the term Future thinking to refer to the cognitive patterning that exists when organisms utilize their memory structures to perceive and imagine novel and innovative arrangements or solutions to personal and environmental challenges and opportunities. In human beings, Future thinking is systems-level thinking in which the higher-order properties of objects of perception, both physical and abstract, the relationships and interconnections among those objects of perception and mental activity are observed. Future thinking involves the abstract mental representation of objects and events and is used to speculatively generate alternate realities and possible future ("what if") scenarios.

Future thinking occurs when novel and innovative solutions to immediate problems are envisioned, solutions that are not limited by preexisting conceptual and social schemas or by the contents of memory. Future thinking also manifests as abstract, creative, open-ended, big-picture thinking oriented toward perceiving and pursuing opportunities, forward motion, the generation and exploration of future possibilities, ingenuity, the constructive disruption of current personal, social, and collective schemas, and the adaptation and reinvention of such schemas to accelerate the pace of personal, social, technological, cultural, and spiritual evolution. Thus, boundless uses of imagination, such as intuitive leaps in scientific thinking, the creation of novel art, music, dance, literature, and poetry, positive psychology, the visionary evolution of human ventures, inventive and innovative technological advancements, and the reimagining of human physical and mental possibilities are all products of Future thinking.

Our construct of Future thinking encompasses many related and overlapping constructs, such as episodic future thinking (e.g., Atance & O'Neill, 2005), episodic foresight(Martin- Ordas, Atance, & Louw, 2012; Suddendorf & Moore, 2011), episodic simulation (Schacter & Addis, 2007), semantic future thinking (Atance & O'Neill, 2005), semantic prospection (e.g., Gilbert & Wilson, 2007), and semantic fantasy (Merker, 2007). Our construct of Future thinking also encompasses the constructs of forward thinking, which is often used colloquially to describe new and innovative ideas, conceptual schemas, and/or methods as they apply to specific fields of inquiry or areas of application, and futures thinking, which involves thinking about the future and future trends from a broad contextual or societal perspective (a systems level of understanding) and the examination of the evolutionary processes associated with specific fields of endeavor or social trends that allow for the speculative imagination of a wide variety of future possibilities (Hejazi, 2012; Lombardo, 2006). Similarly, the term future thinker is often used to describe individuals who are considered pioneers in their respective fields and far ahead of their times (e.g., Lutzo, 2004), and who offer innovative approaches that transform thinking, research, and/or application in their respective fields (Honan, 1997).

At the individual level, we hypothesize that Future thinking manifests as (a) visionary, speculative thinking; (b) creativity and creative problem solving; (c) a perception of new environmental opportunities; (d) idealistic expectations; (e) flexibility and adaptability; (f) quick and often impulsive decision making; (g) a hopeful, optimistic, and resilient outlook on life; and (h) a tendency to engage in impulsive, spontaneous, and hedonistic behavior. We also hypothesize that Future thinking manifests as the Big Five personality traits of openness and extraversion as well as the tendency to be adventurous, charismatic and charming, dynamic, energetic, inventive and innovative, inspirational, and persuasive. Future thinking also manifests as a preference for flexible, social, and unrestricted environments with few rules and regulations and as a need to form relationships with a broad network of people with diverse ideas and interests.

At the collective level, we propose that Future thinking uses the mental spaciousness created by the concept of future for the generation and exploration of new ideas, approaches, and relationships as well as the generation of alternative conceptual, social, and cultural schemas, which in turn, may capture the imagination of the collective and stimulate cultural, technological, social, and political upheaval, innovation, and change. To paraphrase Hajazi (2012), Future thinking expands the universe of ideas, vistas, and realities, entertains the unthinkable and makes it thinkable, and opens the human mind to considering alternative visions of reality and existence (p. 4).

2.1.3 Present thinking. We use the term Present thinking to refer to the cognitive pattern that exists when organisms utilize their memory systems to impose control mechanisms on the environment in order to maximize biological survival. Present thinking is functional thinking in which the lower- order properties of objects and relationships among objects are observed. In human beings, Present thinking occurs when individuals impose complex cognitive and behavioral control mechanisms, such as plans, structures, processes, rules, and schemas on the physical environment and when cultural norms and expectations and individual and cultural life scripts based on those norms also are developed. Moreover, Present thinking occurs when individuals form concrete, contextualized, goal-oriented mental representations of objects and when individuals develop action plans and organize the resources needed to execute those plans.

Similarly, Present thinking occurs when individuals attempt to restore and/or impose order when shifts in the status quo, cultural norms, and goals are perceived or anticipated.

Present thinking is also the cognitive patterning involved when the abstract products of Past and Future thinking are woven and manifested into reality. For example, Present thinking involves the organization of data, information, people, objects, and events into the conceptual schemas, standards, and best practices deemed valid by Past thinking. Present thinking is also involved when action and contingency plans are created that weave and manifest into reality existing ideas, solutions, and technologies that are logically derived from prevailing schemas and consistent with the status quo as well as novel and innovative ideas, solutions, and inventions (i.e., the creative products of Future thinking) that alter the status quo.

We propose that all planning activities are the domains of Present thinking, including those oriented toward (a) current circumstances or future goals; (b) maintaining harmony with the status quo and existing prevailing schemas; (c) applying the innovative ideas and solutions of Future thinking to existing processes and control systems; (d) making manifest the conceptual schemas and knowledge of Past thinking or the creative ideas and solutions generated by Future thinking; or (e) achieving personal or collective goals. As such, we argue (as did Merker, 2007; and Tulving & Kim, 2007) that activities such as planning for the future, which have been considered to belong to the construct domain of episodic future thinking and strategic planning, and have often been confused with strategic thinking (Mintzberg, 1981, 1994), are not activities that belong to the domain of Future thinking, but rather to the domain of Present thinking.

At the individual level, we hypothesize that Present thinking manifests as a tendency to: (a) classify objects and events concretely; (b) perform well on detailed-oriented tasks; (c) organize, plan, and structure one's environment and activities; (d) adopt and maintain predefined cultural, social, and personal schemas and life scripts; (e) make pragmatic decisions; (d) create stability, harmony, and good relations with others; and (f) approach life with pragmatic resilience and positive well-being. Present thinking also manifests as (a) the Big Five personality traits of conscientiousness and agreeableness; (b) the tendency to be highly practical, pragmatic, predictable, compliant, dependable, dogmatic, efficient, inflexible, methodical, task-oriented, and resourceful; and (c) the need to force conformity, both within oneself and among others, with prevailing cultural norms and expectations that have been adopted as well as to one's own plans and goals. Present thinking also manifests as a natural inclination to be drawn to social activities that involve community or social groups and memberships in clubs and organizations.

At the collective level, we propose that Present thinking is involved in the implementation and enactment of and conformity to cultural expectations, laws, and norms and occurs when communities, organizations, societies, and nations attempt to successfully navigate change and execute collective goals (e.g., achieving the goal of linking the Atlantic and Pacific oceans through Panama). Similarly, we propose that Present thinking involves attempts to create social equilibrium by synchronizing specific culturally shared expectations, life scripts, and agendas into a cohesive community (clubs, organizations, institutions, social networks), in order to facilitate cooperative, collaborative, and productive interactions among members of those communities. Present thinking at the collective level manifests as bureaucracies, logistical institutions (e.g., world governing bodies, national governments, education, civil defense, physical and technological infrastructure, public utilities), and planning and control systems (e.g., law enforcement, armed defense, auto and air traffic control systems, navigational systems, computer systems, bodies that manage and regulate financial markets, legal agreements, and national and international standards) as well as systems and institutions such as the United Nations, the World Bank, The World Trade Organization, and the International Organization for Standardization, which exist to maintain stability and order both within and between populations.

2.2 Empirical Support. Research findings have generally supported our propositions involving Past, Present, and Future thinking (Fortunato and Furey, 2009, 2010, 2011, 2012). For example, we found that scores on a measure of Future thinking correlated positively with scores on measures of extraversion and openness, hedonism and risk-taking, optimism and resilience, and negatively with neuroticism, and anxiety and depression. Conversely, scores on a measure of Present thinking correlated positively with scores on measures of conscientiousness and agreeableness, resiliency and optimism, and negatively with scores on measures of openness, cynicism, anxiety, and depression. Finally, scores on measures of Past thinking correlated positively with scores on measures of neuroticism, cynicism, anxiety and depression, and negatively with extraversion and agreeableness,, and optimism and resiliency.

Indirect support for some of our propositions comes from research that has shown that scores on measures of extraversion and openness—personality traits that we hypothesize are mediated by Future thinking—correlated positively with scores on measures of creative and divergent thinking (e.g., Furnham &Bachtiar, 2008; Gelade, 2002; George & Zhou, 2001), sexual risk taking (Turchik, Garske, Probst, & Irvin, 2010), financial risk taking (Hunter & Kemp, 2004), and risky decision-making (Lauriola & Levin, 2001; Nicholson, Soane, Fenton-O'Creevy, & Willman, 2005). Conversely scores on measures of introversion and neuroticism— personality traits that we hypothesize are mediated by Past thinking—correlated negatively with scores on measures of optimism and creativity (Fink & Neubauer, 2008; Gelade, 2002), sensation seeking and risk-taking behavior (Zuckerman & Kuhlman, 2000), and positively with analytical thinking (Gelade, 2002), and anxiety and depression (Costa & McCrae, 1992b; Williams, 1992). Moreover, individuals who score high on extraversion show greater cortical activation during creative tasks than individuals who score high on introversion (Fink & Neubauer, 2008). Conversely, individuals who score high on introversion show greater cortical activation during mental reasoning and memory tasks (Fink & Neubauer, 2008: Fink, Grabner, Neuper, & Neubauer, 2005). Finally, scores on measures of conscientiousness—a personality trait that we hypothesize is mediated by Present thinking—correlated positively with conformity and rule-following (Gelade, 2002), whereas low scores on openness, a characteristic of Present thinking, tends to be observed in individuals who prefer the familiar, the routine, and that status quo (McCrae, 1996) and who tend to be rigid, inflexible to change, and have difficulty adapting to dissimilar points of view (e.g., McCrae, 1987).

Indirect support for our hypotheses also comes from an extensive series of studies by Liberman, Trope, and associates (see Liberman & Trope, 1998; Trope & Liberman, 2003; 2010 for reviews of this literature). For example, when individuals are asked to adopt a distant future temporal perspective, compared to individuals who are asked to adopt a near-term temporal perspective, they tend to (a) classify objects and events into broad, rather than specific categories; (b) form super-ordinate, idealistic goals; (c) perform better on creative tasks than on analytical tasks; (d) display high levels of optimism and overconfidence when making decisions; (e) engage in risk-taking behavior; (f) minimize any delay of gratification of outcomes; and (g) be persuaded by messages that speak to possible opportunities inherent in the situation, context, or product.

Similarly, research has shown that nostalgic recollections of past events are characterized by both abstract and concrete mental representations (Stephan, Sedikides, & Wildschut, 2012).

According to Stephan et al. (2012), on the one hand, because past events are often idealized versions of events that endure over time and are characterized as having a distal psychological perspective, nostalgic recollections of the past tend to consist of high-level construals. On the other hand, because nostalgic recollections are often activated by current stimuli that engage comparative processing of the relevance of the event to present circumstances, nostalgic recollections of events involve low-level construals that are rich in details.

Finally, brain imaging studies have also provided indirect support for our theory. For example, Addis et al. (2007) found opposite patterns of activation when individuals engaged in mental time travel into the past versus mental time travel into the future. Specially, they found that although reconstructions of past events and episodic future thinking both engage the left hippocampus, episodic future thinking uniquely engages the right hippocampus, the right frontopolar cortex and left ventrolateral prefrontal cortex. Similarly, Weiler, Suchan, & Daum (2010, 2011a, 2011b) observed differences in both left hemispheric temporo-parietal and right hemispheric parieto-occipatal activation depending on whether individuals were engaged in reconstructing past events or imagining future ones. Weiler et al. (2011) also noted that the thalamus appeared to play a more important role in the executive aspects of memory required for the imagination of novel scenarios above and beyond those brain areas involved in episodic future thinking.

In summary, there is ample theoretical and empirical evidence to support the general propositions of our theory. However, further research is needed examining the direct relationships between Past, Present, and Future thinking and individual differences on personality traits, cognitive and perceptual tasks, and behaviors as well as corresponding brain activity differences using enhanced imaging techniques.

2.3 Integration. Although we propose that Past, Present, and Future thinking are three distinct patterns of temporally-based consciousness, we also propose that each is inextricably intertwined with the others. Past, Present, and Future thinking form three essential codependent and interdependent cognitive patterns understood to be human thinking. Thus, we propose that it is only when Past, Present, and Future thinking are consciously utilized in combination that they are truly effective in maximizing current and future survival at both the individual and collective levels. For example, when the products of Future thinking (ideas, solutions, and novel approaches) are made available to Past thinking for evaluation, understanding, and analysis, and the synthesized products of Past and Future thinking are made available to Present thinking for manifestation into reality, the highest level of adaptive and survival value of human thinking is realized. Human beings, perhaps not uniquely among all species, have the ability to become consciously aware of both conscious and previously unconscious thought processes and direct them as desired. Thus, human beings are capable of choice and of consciously applying, adapting, and utilizing their Past, Present, and Future thinking when needed in response to environmental, personal, and social contingencies. As Merker (2007) stated, the ability to speculatively imagine a variety of future scenarios is useful only to the extent to which the utility, probability, or certainty of each scenario can be determined.

We further propose that Past, Present, and Future thinking are the foundation for one's own personal identity that is located in space and time. Tulving (1985a, 1985b, 2002) made a similar point when he argued that episodic memory, in particular, and semantic memory, in general, provide the basis for the development of a personal identify that links all of the temporally-based experiences that form the foundation for one's own personal history. We propose that Past, Present, and Future thinking are the elements of and inseparable from one's own self or ego.

It is our view that the physical, cognitive, social, and cultural achievements of human beings could not have occurred without the combined use of all three thinking perspectives. Indeed, it is the interaction and resulting entanglement of these three discreet patterns of thought that has led to an ever-increasing level of abstraction in human beings' understanding of the world and the language used to symbolize and communicate that understanding with others, which in turn, has resulted in ever-increasing technological and social advancements and other artifacts of culture. It is only when Past, Present, and Future thinking are used in combination that human thought is brought into unity, for each of these thought patterns is but one aspect of a trilogy that, together, forms the entirety of human conscious awareness.

It is also our view that the ability to look backwards into the past, focus attention on the present, and imagine the future allowed human beings to direct their attention toward increasingly abstract and higher-order goals (e.g., Maslow, 1943) (see Fig. 3). Human beings, arguably, unique among all animals, seek to understand the nature of consciousness and existence and have learned to transcend their own physical and information processing limitations.

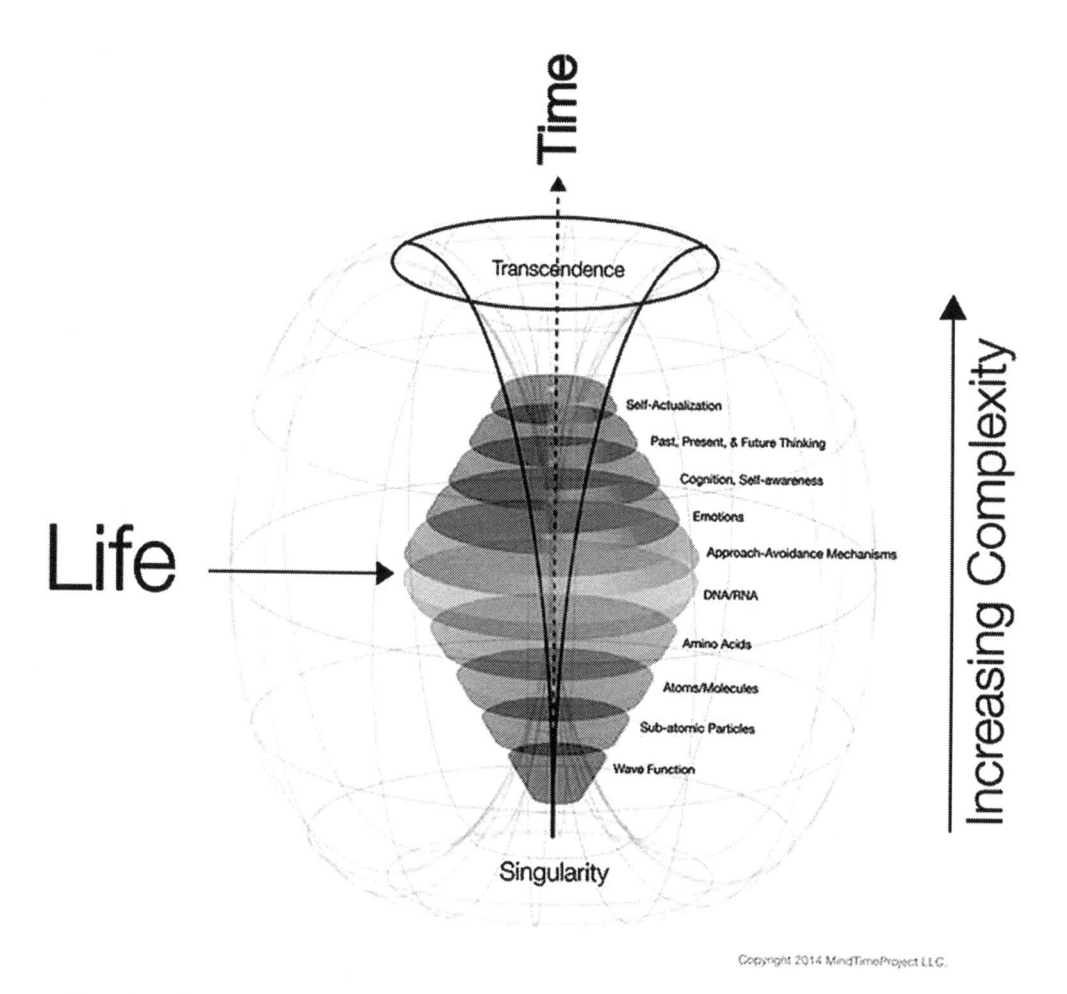

Fig. 3. Physical and temporal expansion of space–time and the evolution of differentiated consciousness from the singularity of the Big Bang to transcendence. Figure used with permission from the MindTime Project, LLC. Illustrated by Alden Bevington.

3. Theoretical and Practical Implications

The theory of MindTime can be applied across a broad spectrum of inquiry and that there are many theoretical and practical applications of the theory. For example, we suggest that the theory of MindTime may function as a meta-theory by providing a foundation for developing a holistic understanding of human psychology at both the individual and collective level and for connecting seemingly disparate fields of inquiry, such as psychology, education, sociology, social economics, cultural anthropology, political theory, governance, service science, organizational theory, and communication, even though each utilize different epistemologies, vocabulary, taxonomies, and methodologies.

At the practical level, we argue that the theory of MindTime has utility in understanding the interactions between people and the environmental domains in which they function. For example, according to Walsh, Craik, and Price (2000), the concept of person-environment fit is perhaps the most dominant conceptual force for understanding human behavior, and research in a variety of disciplines shows that the extent to which the characteristics of an individual are aligned, congruent, match, fit, or resonate with the characteristics of their surrounding physical, social, learning, and working environments influences how they react to that environment. Thus, by measuring individuals' Past, Present, and Future thinking as well as by identifying the Past, Present, and Future characteristics of environments and cultures, effective interactions between and among individuals and their environments can be facilitated. For example, in organizations, the theory can be used for hiring, placement and promotion of employees, leadership development, developing internal and external communications strategies, developing appropriate reward mechanisms, and facilitating work team effectiveness. In educational environments, the theory can be used to facilitate student learning by matching students with instructors based on their thinking perspective and by developing personalized instructional materials, curricula, and strategies based on how student think. In clinical and counseling contexts, the theory can be used to facilitate an enhanced understanding of clients' perceptual constructs and world views and to develop new therapeutic tools. In global politics, the theory of MindTime can be used to facilitate an understanding of the fundamental world views and cultures of different political parties, ideologies, peoples, and nations. This understanding can be used for both enhancing diplomatic relations and for resolving conflict.

Our theory also has implications for the digital age. Today, people are a part of an exponentially growing and continuous flow of digital data, in what has been called big data. In a recent book by Frank, Roehrig, and Pring (2014), the authors address the growing phenomenon whereby people have vast amounts of data attached to their identity and pertaining to their behavior, habits, preferences, educational, occupational, and health histories, interests, purchases, demographics, psychographics, and every other data point that can be captured digitally. The authors referred to these data as code halos. What has been missing from big data is a unifying theory that facilitates an understanding of the underlying patterns that exist within that data, both at the individual and collective levels. We propose that by examining individual and collective level data through the lens of the theory of MindTime, patterns will emerge that are consistent with Past, Present, and Future thinking, from which a whole new level of understanding of human behavior can be achieved. Preliminary cross-cultural data from North America, Europe, and Asia suggests that the theory of MindTime is generalizable across situations, domains, and cultures (MindTime Techologies, Inc., 2012, 2013, 2014).

4. Closing Comments

In this paper, we presented a theory of consciousness and psychology in which perception of time is the fundamental process by which human beings perceive and process information and interact with the world and others. Specifically, we proposed that variation exists in how individuals utilize episodic and semantic memory structures to process information, form mental representations of temporally located objects and events, and engage in mental time travel. We proposed that Past, Present, and Future thinking are abstract–linguistic, perceptual–cognitive patterns that mediate the relationships between individuals' neurological responses to environmental and social stimuli and their perceptions, personality, intentions, values, beliefs, motivations, and behaviors. We also proposed that Past, Present, and Future thinking represent complex and abstract methods by which human beings are able to maximize survival by (a) avoiding negative stimuli, (b) approaching positive stimuli, and (c) controlling their environment.

Although we have focused in this paper on human beings, we speculate that Past, Present, and Future thinking also exist in other animals in various degrees. That is, Past, Present, and Future thinking are the mechanisms of consciousness, with human consciousness representing, arguably, the most complex form of consciousness known at this time (Liljenström, 2011). We surmise that perception of time, whether conscious or unconscious, narrow or infinite in extension, and the cognitive patterns associated with time perception, which we referred to as Past, Present, and Future thinking, no matter how evolutionarily advanced, are the fundamental driving mechanisms of consciousness and behavior as well as intra-species behavioral variation.

In a recent volume of this journal, several authors (e.g., Kafatos, Tanzi, & Chopra, 2011; Liljenström, 2011; Lombardo, 2011; Mitchell & Staretz, 2011; M. Cabanac, R. Cabanc, & Hammel, 2011) argued that consciousness and matter are two complementary aspects of one reality. On the one hand, for example, Kafatos et al (2011) proposed that consciousness is a field phenomenon, which they referred to as primordial consciousness that exists outside of space and time and which is characterized by generalized quantum physics principles, such as complementarity, non-locality, scale-invariance, and undivided wholeness. Similarly, Mitchell and Staretz (2011) wrote that "all matter seems interconnected with all other matter and this interconnection even transcends [italics are ours] space and time...and that at the lowest level resides the most basic aspects of undifferentiated awareness [italics are ours] built upon the quantum principles of entanglement, non-locality, and coherent emission/absorption of photons."

On the other hand, consciousness also exists within space and time. Kafatos et al (2011), King (2011), Liljenström, (2011), Mitchell and Staretz (2011) also all similarly proposed that the evolution of consciousness involved the evolution of matter into increasingly complex, elaborate, and interactive systems, with the human brain providing the highest level of complexity known. For example, Mitchell and Staretz wrote "We postulate that ...the most fundamental aspect of consciousness which we describe as undifferentiated awareness [forms the] mechanism of basic perception [and] extends up the entire evolutionary chain of increasing complexity of living organisms." According to Mitchell and Staretz, consciousness itself becomes increasingly complex as matter becomes increasingly complex; and that all organisms from the simplest to the most complex are interconnected at a very fundamental level using information obtained by nonlocal quantum coherence.

From the point of view of the human mind, we note a confounding and recursive paradox. According to quantum theory (see Penrose & Hammeroff, 2011, for example), the physical universe does not exist except as wave functions that collapse only when an observation is made by an observer or an instrument as an extension of an observer. Thus, any perception of space–time reality and, similarly, any attempt by any organism within this reality (e.g., human beings) to understand its own consciousness and the nature of the universe, are products of the very differentiation made possible by the perceptual apparatus that perceives space and time, apparatus which itself can only exist when observed. Thus, it is differentiating consciousness that is itself responsible for the apparatus that is doing the observing. Differentiating consciousness (for example, the human mind) creates itself and the very world it perceives.

Kafatos et al. (2011), based on a speech given by Werner Heisenburg in 1932, wrote that "the universe presents the face that the observer is looking for and when she looks for a different face, the universe changes its mask." Previously, we proposed that Past, Present, and Future thinking operate as cognitive patterns that form the foundation for individuals' perceptions of the world. We now take this one step further and suggest that Past, Present, and Future thinking, because they are integral in the encoding of our memories and the development of personal identities and belief systems, are the very mechanisms by which organisms collapse quantum wave functions and create the realities they observe and experience.

REFERENCES

Addis, D. R., Wong, A. T., & Schacter, D. L. (2007). Remembering the past and imagining the future: Common and distinct neural substrates during event construction and elaboration. Neuropsychologia, 45(7), 1363–1377. doi:10.1016/j.neuropsychologia.2006.10.016

Atance, C. M., & Jackson, L. K. (2009). The development and coherence of future-oriented behaviors during the preschool years. Journal of Experimental Child Psychology, 102(4), 379–391.

Atance, C. M., & O'Neill (2005). The emergence of episodic future thinking in humans. Learning and Motivation, 36(2), 126–144.

Berntsen, D., & Bohn, A. (2010). Remembering and forecasting: The relation between autobiographical memory and episodic future thinking. Memory & Cognition, 38(3), 265–278. doi:10.3758/MC.38.3.265

Bronckart, J. (2005). The temporality of discourses: A contribution to the reshaping of human actions. In A. Perret-Clermont (Ed.) Thinking time: A multidisciplinary approach. Cambridge, MA: Hogrefe & Huber.

Buckner, R. L., & Carroll, D. C. (2007). Self-projection and the brain. Trends in Cognitive Sciences, 11(2), 49–57. doi:10.1016/j.tics.2006.11.004

Cabanac, M., Cabanac, R., & Hammel, H. T. (2011). Consciousness: The fifth influence. Journal of Cosmology, 14. Retrieved from http://journalofcosmology.com/Consciousness127.html

Carver, C. S., Sutton, S. K., & Scheier, M. F. (2000). Action, emotion, and personality: Emerging conceptual integration. Personality and Social Psychology Bulletin, 26(6), 741–751. doi:10.1177/0146167200268008

Corballis, M. C. (2011). The recursive mind: The origins of human language, thought, and civilization. Princeton, NJ: Princeton University Press.

Corballis, M. C. (2013). Mental time travel: A case for evolutionary continuity. Trends in Cognitive Sciences, 17(1), 5–6. doi:10.1016/j.tics.2012.10.009

Costa, P. T., & McCrae, R. R. (1992a). Four ways five factors are basic. Personality and Individual Differences, 13(6), 653–665. doi:10.1016/0191-8869(92)90236-I

Costa, P. T., & McCrae, R. R. (1992b). Normal personality assessment in clinical practice: The NEO Personality Inventory. Psychological Assessment, 4(1), 5–13. doi:10.1037/1040-3590.4.1.5

D'Argembeau, A., & Van der Linden, M. (2007). Emotional aspects of mental time travel. Behavioral and Brain Sciences, 30(3), 320–321. doi:10.1017/S0140525X07002051

Dewey, J. (1933). How we think. Buffalo, NY: Prometheus Books. (Original work published 1910)

Fink, A., & Neubauer, A. C. (2008). Eysenck meets Martindale: The relationship between extraversion and originality from the neuroscientific perspective. Personality and Individual Differences, 44(1), 299–310. doi:10.1016/j.paid.2007.08.010

Fink, A., Grabner, R. h., Neuper, C., & Neubauer, A. c. (2005). Extraversion and cortical activation during memory performance. International Journal of Psychophysiology, 56(2), 129–141.

Förster, J., & Becker, D. (2012). When curiosity kills no cat—but mediates the relation between distant future thoughts and global processing across sensory modalities. European Journal of Social Psychology, 42(3), 334–341. doi:10.1002/ejsp.1856

Förster, J., Friedman, R. S., & Liberman, N. (2004). Temporal construal effects on abstract and concrete thinking: Consequences for insight and creative cognition. Journal of Personality and Social Psychology, 87(2), 177–189. doi:10.1037/0022-3514.87.2.177

Fortunato, V. J., & Furey, J. T. (2009). The theory of MindTime and the relationships between thinking perspective and the Big Five personality traits. Personality and Individual Differences, 47(4), 241–246.

Fortunato, V. J., & Furey, J. T. (2010). The theory of MindTime: The relationships between thinking perspective and time perspective. Personality and Individual Differences, 48(4), 436–441. doi:10.1016/j.paid.2009.11.015

Fortunato, V. J., & Furey, J. T. (2011). The theory of MindTime: The relationships between Future, Past, and Present thinking and psychological well-being and distress. Personality and Individual Differences, 50(1), 20–24.

Fortunato, V. J., & Furey, J. T. (2012). An examination of thinking style patterns as a function of thinking perspective profile. Personality and Individual Differences, 53(7), 849–856. doi:10.1016/j.paid.2012.06.017

Frank, M., Roehrig, P., & Pring, B. (2014). Code Halos: How the digital lives of people, things, and organizations are changing the rules of business. Hoboken, NJ: John Wiley & Sons.

Furnham, A., & Bachtiar, V. (2008). Personality and intelligence as predictors of creativity. Personality and Individual Differences, 45(7), 613–617. doi:10.1016/j.paid.2008.06.023

Gelade, G. A. (2002). Creative style, personality and artistic endeavor. Genetic, Social, and General Psychology Monographs, 128(3), 213–234.

George, J. M., & Zhou, J. (2001). When openness to experience and conscientiousness are related to creative behavior: An interactional approach. Journal of Applied Psychology, 86(3), 513–524. doi:10.1037/0021-9010.86.3.513

Gilbert, D. T., & Wilson, T. D. (2007). Prospection: Experiencing the future. Science, 317(5843), 1351–1354. doi:10.1126/science.1144161

Gilead, M., Liberman, N., & Maril, A. (2012). Construing counterfactual worlds: The role of abstraction. European Journal of Social Psychology, 42(3), 391–397. doi:10.1002/ejsp.1862 Hassabis, D., Kumaran, D., & Maguire, E. A. (2007). Using imagination to understand the neural basis of episodic memory. Journal of Neuroscience, 27(52), 14365–14374. doi:10.1523/JNEUROSCI.4549-07.2007

Hejazi, A. (2012). Futures metacognition: A progressive understanding of futures thinking. World Future Review, 4(2), 18-27.

Honan, W. H. (1997, November 2). Looking back at forward thinkers. The New York Times. Retrieved from www.nytimes.com.

Hunter, K., & Kemp, S. (2004). The personality of e-commerce investors. Journal of Economic Psychology, 25(4), 529–537. doi:10.1016/S0167-4870(03)00050-3

Jaynes, J. (1976). The origin of consciousness and the breakdown of the bicameral mind. Houghton Mifflin: Boston, MA.

Kafatos, M., Tanzi, R. E., & Chopra, D. (2011). How consciousness becomes the physical universe. Journal of Cosmology, 14. Retrieved from http://journalofcosmology.com/Consciousness140.html

King, C. (2011). Cosmological foundations of consciousness. Journal of Cosmology, 14, 3706– 3725. Retrieved from http://journalofcosmology.com/Consciousness103.html

Lauriola, M., & Levin, I. P. (2001). Personality traits and risky decision-making in a controlled experimental task: An exploratory study. Personality and Individual Differences, 31(2), 215–226. doi:10.1016/S0191-8869(00)00130-6

Levine, R. (2006). A geography of time. Oxford, England: Oneworld Publications.

Lewin, K. (1942). Time perspective and morale. In G. Watson (Ed.), Civilian morale. Second yearbook of the S.P.S.S.L. Boston, MA: Houghton Mifflin.

Lewin, K. (1951). Field theory in social science. New York, NY: Harper and Row.

Liljenström, H. (2011). Intention and attention in consciousness dynamics and evolution. Journal of Cosmology, 14, 4839–4847. Retrieved from http://journalofcosmology.com/Consciousness138.html

Liberman, N., & Trope, Y. (1998). The role of feasibility and desirability considerations in near and distant future decisions: A test of temporal construal theory. Journal of Personality and Social Psychology, 75(1), 5–18. doi:10.1037/0022-3514.75.1.5

. Lombardo, T. (2006). The evolution of future consciousness, Bloomington, IN: AuthorHouse. Lombardo, T. (2011). The ecological cosmology of consciousness. Journal of Cosmology, 14, 4859–4868. Retrieved from http://journalofcosmology.com/Consciousness141.html

Lutzo, E. (2004). Are you a forward thinker? Weatherhead Coaches Corner, 1(9).

Martin-Ordas, G., Atance, C. M., & Louw, A. (2012). The role of episodic and semantic memory in episodic foresight. Learning and Motivation, 43(4), 209–219. doi:10.1016/j.lmot.2012.05.011

Maslow, A. H. (1943). A theory of human motivation. Psychological Review, 50(4), 370–396. Retrieved from http://psychclassics.yorku.ca/Maslow/motivation.htm

McCrae, R. R. (1987). Creativity, divergent thinking, and openness to experience. Journal of Personality and Social Psychology, 52(6), 1258–1265. doi:10.1037/0022-3514.52.6.1258

McCrae, R. R. (1996). Social consequences of experiential openness. Psychological Bulletin, 120(3), 323–337. doi:10.1037/0033-2909.120.3.323

Merker, B. (2007). Memory, imagination, and the asymmetry between past future. Behavioral and Brain Sciences, 30(3), 325–326. doi:10.1017/S0140525X07002117 MindTime Techologies, Inc. (2012, 2013, 2014). Unpublished technical reports.

Mintzberg, H. (1981). What is planning anyway? Strategic Management Journal, 2(3), 319–324.

Mintzberg, H. (1994). Planning and Strategy. In H. Mintzberg (Ed.), Rise & fall of strategic planning (pp. 5–34). New York, NY: The Free Press.

Mitchell, E. D., & Staretz, R. (2011). The quantum hologram and the nature of consciousness. Journal of Cosmology, 14, Retrieved from http://journalofcosmology.com/Consciousness149.html

Nicholson, N., Soane, E., Fenton-O'Creevy, M., & Willman, P. (2005). Personality and domain- specific risk taking. Journal of Risk Research, 8(2), 157–176. doi:10.1080/1366987032000123856

Oettingen, G., & Mayer, D. (2002). The motivating function of thinking about the future: Expectations versus fantasies. Journal of Personality and Social Psychology, 83(5), 1198– 1212. doi:10.1037/0022-3514.83.5.1198

Penrose, R., & Hameroff, S. (2011). Consciousness in the universe: Neuroscience, quantum space-time geometry and Orch OR theory. Journal of Cosmology, 14. Retrieved from http://journalofcosmology.com/Consciousness160.html

Schacter, D. L., & Addis, D. R. (2007). On the constructive episodic simulation of past and future events. Behavioral and Brain Sciences, 30(3), 331–332. doi:10.1017/S0140525X07002178

Stephan, E., Sedikides, C., & Wildschut, T. (2012). Mental travel into the past: Differentiating recollections of nostalgic, ordinary, and positive events. European Journal of Social Psychology, 42(3), 290–298. doi:10.1002/ejsp.1865

Suddendorf, T. (1999). The rise of the metamind. In M. C. Corballis & S. E. G. Lea (Eds.), The descent of mind: Psychological perspectives on hominid evolution (pp. 218–260). Oxford University Press: New York, NY.

Suddendorf, T., & Corballis, M. C. (1997). Mental time travel and the evolution of the human mind. Genetic, Social, and General Psychology Monographs, 123(2), 133–167.

Suddendorf, T., & Corballis, M. C. (2007). Mental time travel across the disciplines: The future looks bright. Behavioral and Brain Sciences, 30(3), 335–351. doi:10.1017/S0140525X0700221X

Suddendorf, T., & Moore, C. (2011). Introduction to the special issue of episodic foresight. Cognitive Development, 26(4), 295–298. doi:10.1016/j.cogdev.2011.09.001

Tattersall, I. (2011). Evolution of modern human consciousness. Journal of Cosmology, 14, 4831–4838. Retrieved from http://journalofcosmology.com/Consciousness151.html

Trompenaars, F., & Hampden-Turner, C. (1997). Riding the waves of culture: Understanding cultural diversity in business. Finland: Werner Söderström Oy.

Trope, Y., & Liberman, N. (2003). Temporal construal. Psychological Review, 110(3), 403–421. doi:10.1037/0033-295X.110.3.403

Trope, Y., & Liberman, N. (2010). Construal-level theory of psychological distance. Psychological Review, 117(2), 440–463. doi:10.1037/a0018963

Tulving, E. (1985a). How many memory systems are there? American Psychologist, 40(4), 385– 398. doi:10.1037/0003-066X.40.4.385

Tulving, E. (1985b). Memory and consciousness. Canadian Psychology/Psychologie canadienne, 26(1), 1–12. doi:10.1037/h0080017

Tulving, E. (2002). Episodic memory: From mind to brain. Annual Review of Psychology, 53(1), 1–25. doi:10.1146/annurev.psych.53.100901.135114

Tulving, E., & Kim, A. (2007). The medium and the message of mental time travel. Behavioral and Brain Sciences, 30(3), 334–335. doi:10.1017/S0140525X07002208

Turchik, J. A., Garske, J. P., Probst, D. R., & Irvin, C. R. (2010). Personality, sexuality, and substance use as predictors of sexual risk taking in college students. Journal of Sex Research, 47, 411-419. doi: 10.1080/00224490903161621.

Vale, G. l., Flynn, E. g., & Kendal, R. l. (2012). Cumulative culture and future thinking: Is mental time travel a prerequisite to cumulative cultural evolution? Learning and Motivation, 43(4), 220–230. doi:10.1016/j.lmot.2012.05.010

Walsh, W. B., Craik, K. H., & Price, R. H. (Eds.). (2000). Person–environment psychology: New directions and perspectives (2nd ed.). Mahwah, NJ: Lawrence Erlbaum.

Warner, R. H., VanDeursen, M. J., & Pope, A. R. D. (2012). Temporal distance as a determinant of just world strategy. European Journal of Social Psychology, 42(3), 276–284. doi:10.1002/ejsp.1855

Weiler, J. A., Suchan, B., & Daum, I. (2010a). Foreseeing the future: Occurrence probability of imagined future events modulates hippocampal activation. Hippocampus, 20(6), 685– 690. doi:10.1002/hipo.20695

Weiler, J. A., Suchan, B., & Daum, I. (2010b). When the future becomes the past: Differences in brain activation patterns for episodic memory and episodic future thinking. Behavioural Brain Research, 212(2), 196–203. doi:10.1016/j.bbr.2010.04.013

Weiler, J. A., Suchan, B., & Daum, I. (2011). What comes first? Electrophysiological differences in the temporal course of memory and future thinking. European Journal of Neuroscience, 33(9), 1742–1750. doi:10.1111/j.1460-9568.2011.07630.x

Williams, D. G. (1992). Dispositional optimism, neuroticism, and extraversion. Personality and Individual Differences, 13, 475-477.

Wilson, A. E., Buehler, R., Lawford, H., Schmidt, C., & Yong, A. G. (2012). Basking in projected glory: The role of subjective temporal distance in future self-appraisal. European Journal of Social Psychology, 42(3), 342–353. doi:10.1002/ejsp.1863

Zuckerman, M., & Kuhlman, D. M. (2000). Personality and risk-taking: Common biosocial factors. Journal of Personality, 68, 999-1029.

Made in the USA
Columbia, SC
24 July 2017